GASTROINTESTINAL HORMONES AND PATHOLOGY OF THE DIGESTIVE SYSTEM

ADVANCES IN EXPERIMENTAL MEDICINE AND BIOLOGY

Recent Volumes in this Series

GASTROINTESTINAL HORMONES AND PATHOLOGY OF THE DIGESTIVE SYSTEM

Edited by

Morton Grossman

Veterans Administration
Wadsworth Hospital Center
Los Angeles, California

and

V. Speranza, N. Basso, and E. Lezoche

Institute of 3rd Surgical Pathology
University of Rome

SPRINGER SCIENCE+BUSINESS MEDIA, LLC

Library of Congress Cataloging in Publication Data

Symposium on Gastrointestinal Hormones and Pathology of the Digestive System, Rome, 1977.
    Gastrointestinal hormones and pathology of the digestive system.

    Proceedings of the conference held July 13–15.
    Includes index.
    1. Gastrointestinal hormones–Congresses. 2. Digestive organs–Diseases–Congresses. I. Grossman, Morton I. II. Title. [DNLM: 1. Gastrointestinal hormones–Congresses. 2. Digestive system–Pathology–Congresses. WK170 S98lg 1977]
PQ572.G35S94 1977   616.3'3   78-17547

    ISBN 978-1-4684-7250-9   ISBN 978-1-4684-7248-6 (eBook)
    DOI 10.1007/978-1-4684-7248-6

Proceedings of the Symposium on Gastrointestinal Hormones
and Pathology of the Digestive System
held in Rome, Italy, July 13–15, 1977

© Springer Science+Business Media New York 1978
Originally published by Plenum Press New York in 1978
Softcover reprint of the hardcover 1st edition 1978

Preface

The discovery that the same or similar peptides are present in endocrine cells and in neurons is one of the most exciting and provocative recent developments in biology. Suddenly neurophysiologists and endocrinologists have found that they have a great deal to discuss with each other. Substances originally isolated as hypothalamic hormones turn out to be abundantly present in neurons of other parts of the brain and in endocrine cells and neurons of the gut and pancreas. Similarly, substances originally isolated as gut hormones are found not only in gut endocrine cells but also in gut neurons and in brain neurons. It turns out that the group of peptides that we are accustomed to call gastrointestinal hormones are not all confined to the gastrointestinal tract and are not all solely hormones. We are learning that the chemical transmitters of the neurocrine, endocrine, and paracrine systems form a single group of related substances. This volume contains the latest installments in this fascinating story. It tells how these peptides were isolated and their amino acid sequences determined, how the heterogeneity of most, perhaps all, of these peptides is being revealed as variant forms of them are discovered, how antibodies to these peptides are used as powerful tools to measure their concentrations in body fluids and to localize the cells in which they are synthesized and stored, and, finally, how the role of these substances in normal physiology and in pathological states is being unraveled. This book contains contributions from most of the leading authorities in this exciting field of study.

Morton I. Grossman

Contents

CONTENTS

THE GASTROINTESTINAL HORMONES : AN OVERVIEW

R. A. Gregory

University of Liverpool, The Physiological Laboratory

Liverpool L69 3BX, England

In ancient times all roads led to Rome; and so it is on the present occasion when we are gathered here from many countries and continents to consider some of the hormonal activities in health and disease of what is now recognised to be the largest and most complex endocrine organ in the body - the digestive system. I am sure we are all conscious of the privilege of meeting in the capital city of that country in which originated the great revival of culture and learning in Europe after the long period of the 'Dark Ages'; and no doubt Dr. Grossman will remind us that this year is the 75th anniversary of the discovery by Bayliss and Starling of the "messenger function" of hormones as exemplified by secretin. This discovery brought to an end the Pavlovian era of the 19th century in which the gastrointestinal mechanisms were explained in terms of nervous reflexes; but although great advances soon followed in respect of other endocrine organs, 60 years were to elapse before the study of the gastrointestinal hormones could enter upon the astonishing expansion of knowledge of their nature and understanding of their functions which we are all now playing some part in furthering.

The truly remarkable developments of the past 15 years are clearly due to a combination of two circumstances. First of all there came the successful isolation of what are generally regarded as the major gastrointestinal hormones - and since then many peptides whose status has not yet been clarified - with the resultant provision of supplies of pure peptides, natural or synthetic, so that their physiological properties could be widely studied. Secondly, and deriving from those achievements, came the introduction of immunological methods of study which have made possible the measurement of the hormones in tissues and body fluids in health and

disease and the positive identification of their cells of origin in
the gastrointestinal tract and elsewhere.

The old idea of 'one hormone, one function', which was widely
assumed to follow from Bayliss and Starling's concept of the 'messenger' role of hormones, has long gone. The numerous actions on
the glands and muscle of the digestive system which are exerted by
most of the gastrointestinal hormones and the fact that these actions are often shared by more than one hormone means (1) that each
hormone can be recognised by more than one target cell, and (2) that
each target cell is capable of recognising more than one hormone.
In the natural circumstances of the digestive response to a meal,
all of the hormones are likely to be in circulation and exerting
to some degree their characteristic actions on their target sites;
and the problem is to decide, on the basis of experimental tests of
hormonal actions (alone or combined), and the measurements of their
circulating amounts by radioimmunoassay, what may be said to be the
physiological role of each hormone, acting as it does not alone but
in concert with the others.

This formidable problem is further complicated by the fact that
probably all of the hormones are to some degree heterogeneous in
their circulating forms. There was until fairly recently a general
assumption, again based on Bayliss and Starling's original concept,
that an endocrine cell released only a single version of its 'chemical messenger'. They at first believed that secretin was released
by acid from an inactive 'prosecretin' in the mucosa; but they
dropped the idea for lack of evidence, and there is nothing to show
that they believed that 'prosecretin' might be secreted along with
secretin. However, commencing with the classical discovery of
proinsulin by Steiner and the later demonstration by radioimmunoassay of its release along with insulin (and the C-peptide) into
the circulation, has come the recognition that probably all peptide
hormones are heterogeneous in their tissues of origin and in the
circulation because of the way in which they are synthesised in the
cell, a large molecule being sequentially cleaved by specific enzymes
to produce the final major active form. Besides the latter, the
precursor forms and discarded fragments are likely to be released
into the circulation, and their presence there may lead to errors in
the measurement of the hormone by radioimmunoassay and in turn to
errors in attempts to evaluate the physiological actions of a hormone by reproducing experimentally what is believed to be its normal postprandial level in the circulation. This situation, best
exemplified so far by the case of gastrin which circulates in two
major forms of very different activity, calls for increased sophisti-
cation of radioimmunoassays so that the different forms can be
measured separately and their contributions to the final physiological response thus evaluated more precisely. There is already evidence, of which we shall hear in this Symposium, that CCK will pre-

sent a problem similar to that of gastrin in respect of heterogeneity and radioimmunoassay of circulating forms; and I have little doubt that a similar problem will to some degree unfold in turn for the other peptides with which we have to deal.

The highly active area of immunocytology and ultrastructural studies of the endocrine cells themselves, a field in which Professor Solcia and his colleagues have played such an eminent role, provides the indispensible basis for our physiological ideas. On the basis of their APUD and ultrastructural characteristics, it has defined a large number of endocrine cell-types and in most cases has successfully assigned to them peptides of established or putative hormonal role already identified by chemical isolation. It has shown us the precise location and general distribution of these cell-types in the digestive system and has offered us new and challenging ideas as to their functional roles. The suggestion that an endocrine cell may influence its near neighbours by local humoral transmission rather than by the circuitous route of the general circulation - 'paracrine' influence - awaits the decisive experimental evidence which will establish the existence of such a mechanism in the gut.

A particularly exciting area of study which leans heavily upon immunocytology is that of the brain-gut relationship. It was discovered many years ago that the characteristic pharmacological activity on smooth muscle attributed to an unidentified principle ('substance P') in crude extracts of intestine (particularly muscle) could also be found in brain; but the possible general significance of this observation never dawned on us until after the observation that the hypothalamic peptide somatostatin could also be found in the gastrointestinal tract and pancreas. Now there are appearing in rapid succession examples of peptides already known in the one tissue being present also in the other, e.g., VIP, neurotensin, enkephalin and what is probably CCK-octapeptide. What these peptides - and no doubt others yet to be discovered - are doing in these two situations is a fascinating problem. Obviously, activity of 'neuroendocrine' character is at least in part involved, for several of them are present in nerve-cells of the gut; but what their role may be in terms of neurotransmission or possibly influence of trophic character, we can hardly guess until further evidence appears, probably from the neurophysiologist and perhaps also the embryologist. Endocrine secretion by neuronal-type cells is a very primitive activity being found in the simplest forms of multicellular organisms and becoming diversified, though never lost, as the development of the "typical" activity of the nervous system proceeds in higher forms; and insofar as ontogeny recapitulates phylogeny, the study of the embryological development of hormones in brain and gut may throw some light on their role in these tissues in postnatal life.

Finally, the field of comparative studies of the endocrine cells and their peptides undoubtedly has further rich rewards to offer. The brilliant success of Professor Erspamer and his school in the exploration of the peptides present in amphibian skin has shown us the great predictive value of comparative studies of this kind, for each of the families of peptides they have discovered has proved to have its counterpart in mammalian forms; and it is safe to anticipate that further exciting discoveries will come from him along this line.

In the foregoing brief survey I have touched upon only a few of the many topics we shall discuss in the symposium; the physiological activities of the established and putative hormones, the measurement of them in health and disease and the clinical significance of the results in relation to diagnosis and treatment, the problem of heterogeneity, the brain-gut relationships and the prospects of further knowledge and understanding still to come. It never ceases to surprise me that each Symposium held in this field can offer a wealth of new work and ideas, always interesting and often challenging, and the present occasion will prove no exception, judging by the program.

A SHORT HISTORY OF DIGESTIVE ENDOCRINOLOGY

M. I. Grossman

Veterans Administration Wadsworth Hospital Center

Los Angeles, California, USA

In digestive endocrinology, as in other branches of science, most of the contributions have stemmed from a few epoch-making discoveries:

1. *PHYSIOLOGICAL ORIGINS.* On January 16, 1902, Bayliss and Starling[1] performed what they quite appropriately perceived to be "the crucial experiment" showing that putting acid into the jejunum still stimulated pancreatic secretion after all nervous connections between the two organs had been cut. Correctly deducing what the nature of the non-nervous mechanism must be, they made an extract of jejunal mucosa and showed that it stimulated pancreatic secretion when given intravenously whereas an extract of ileal mucosa did not. Bayliss and Starling recognized that they had not only discovered a new substance, secretin, but had also introduced a new concept, the regulation of bodily activities by blood borne chemical messengers or hormones. And thus was born the science of endocrinology in general as well as digestive endocrinology in particular. In due course, almost every digestive function was studied to determine whether it might have an endocrine component in its regulation. In addition to secretin, other studies with crude extracts which eventually led to isolation of chemically characterized peptides are: Edkins[2] 1905, gastrin; Ivy[3] 1927, cholecystokinin; Euler and Gaddum[4] 1931, substance P; Harper[5] 1941, pancreozymin (later shown to be identical to CCK); Brown[6] 1967, motilin; once again Brown[7] 1970, gastric inhibitory peptide (GIP); and Said and Mutt[8] 1970, vasoactive intestinal peptide (VIP). Many additional mucosal extracts have been made and given names but we don't know whether they contain new peptides since most of their biological actions are shared with known peptides.

2. *THE BIOCHEMICAL ERA*. Completely new avenues were opened when gastrointestinal hormones became available as pure substances of known chemical structure. In 1964, Gregory and coworkers[9] announced the amino acid sequence of gastrin and since that time the sequences of secretin (Mutt[10] 1970), cholecystokinin (Mutt[11] 1971), substance P (Chang[12] 1973), GIP (Brown[13] 1971), motilin (Brown[14] 1973), and VIP (Mutt[15] 1974) have been reported. Proof of structure by synthesis of fully biologically active peptides has been accomplished for gastrin,[16] secretin,[17] motilin,[18] and VIP.[19] Homologies of amino acid sequences place gastrin and CCK in one chemical family and secretin, GIP, VIP, and pancreatic glucagon in another. The shared carboxyl-terminal fragment of gastrin and CCK has all the biological actions of both hormones. Gastrin and CCK display molecular heterogeneity in the form of molecules with different chain lengths; 3 forms of gastrin and 2 of CCK have been sequenced but all have the same carboxylterminal pentapeptide sequence. Once the hormones were available in pure state the remarkable range of their actions and interactions was revealed, including their trophic actions.[20] The availability of pure hormones makes it possible to make antibodies with which to measure their concentrations in blood and to identify the cells of origin. The pure hormones are now being applied to studies on the nature of the receptors on cell surfaces and the second messengers released by activation of these receptors.

3. *RADIOIMMUNOASSAY*. While studying the metabolism of insulin in 1956, Berson and Yalow[21] serendipitously discovered that insulin-treated patients had antibodies to insulin in their blood. They quickly recognized that such antibodies could be used to measure the amount of insulin in body fluids and thus was born the science of radioimmunoassay which has now been applied to every hormone, including those from the gut, as well as to many other substances. Reliable radioimmunoassay of gastrin is well established and assays for all of the other gut peptides are being perfected.

4. *CELLULAR LOCALIZATION*. Antibodies are also the tools used to determine which cells make and store the various peptides. Using well established principles of immunocytochemistry, McGuigan[22] identified the gastrin containing or G-cells in antral mucosa in 1968 and since that time Pearse, Polak, Solcia and others have demonstrated cells that react with antibodies to each of the peptides that have been extracted from the gut (secretin,[23] CCK,[24] GIP,[25] motilin,[26] VIP,[27] substance P[28]). There appear to be separate cells of origin for each peptide and the cells storing each peptide have a distinctive distribution along the gut, some being confined to small areas, others being found throughout the tract. Once an antibody to a peptide is available, the entire body can be surveyed to determine which organs contain cells that react with it. Such immunofluorescent surveys have given some surprising results.

Organs outside the gut have been shown to have cells reactive to antibodies to gut peptides and conversely the gut has cells that react with antibodies to peptides isolated from other organs. Thus cells reactive with antibodies to glucagon[29] from the pancreas, bombesin[30] from frog skin, and somatostatin[31] and neurotensin[32] from the hypothalamus have been found in intestinal mucosa. Similarly, substances reactive with antibodies to gastrin[33] and to VIP[34] have been found in the brain.

   5. *THE BRAIN-GUT AXIS*. When Euler and Gaddum[4] attempted in 1931 to study the distribution of acetylcholine in tissue extracts, they discovered an interfering substance in extracts of brain and intestine that was later to be identified as the peptide substance P. This was the first instance in which a peptide was shown to be present in both brain and gut. Now there are five additional examples of peptides found in both brain and gut: somatostatin,[31] gastrin,[33] VIP,[34] neurotensin,[32] and enkephalin.[35] The list is certain to grow. These findings that certain peptides occur in both brain and gut and in some instances in both endocrine and neural cells of the gut indicate that the chemical messenger cells of the body comprise a single system in which the same or similar peptides or amines may be utilized for neurocrine, paracrine, or endocrine transmission. Thus all of the major mechanisms involved in coordinating bodily activity can be viewed as belonging to a unified system.

REFERENCES

1. Bayliss WM, Starling EH: The mechanism of pancreatic secretion. J Physiol 28: 325-353, 1902

2. Edkins JS: On the chemical mechanism of gastric secretion. Proc R Soc Lond (B) 76: 376, 1905

3. Ivy AC, Oldberg E: Contraction and evacuation of gall-bladder caused by highly purified "secretin" preparation. Proc Soc Exp Biol Med 25: 113-115, 1927

4. Euler US, Gaddum JH: An unidentified depressor substance in certain tissue extracts. J Physiol 72: 74-87, 1931

5. Harper AA, Raper HS: Pancreozymin, a stimulant of the secretion of pancreatic enzymes in extracts of the small intestine, J Physiol 102: 115-125, 1943

6. Brown JC: Presence of a gastric motor-stimulating property in duodenal extracts. Gastroenterology 52: 225-229, 1967

7. Brown JC, Pederson RA: A multiparameter study on the action of
    preparations containing cholecystokinin-pancreozymin. Scand
    J Gastroenterol 5: 537-541, 1970

8. Said SI, Mutt V: Polypeptide with broad biological activity:
    isolation from small intestine. Science 169: 1217-1218,
    1970

9. Gregory H, Hardy PM, Jones DS, et al: Structure of gastrin.
    Nature 204: 931-933, 1964

10. Mutt V, Jorpes JE, Magnusson S: Structure of secretin. The
    amino acid sequence. Eur J Biochem 15: 513-519, 1970

11. Mutt V, Jorpes E: Hormonal polypeptides of the upper intestine.
    Biochem J 125: 57P-58P, 1971

12. Chang MM, Leeman SE, Niall HD: Amino-acid sequence of substance
    P. Nature (New Biol) 232: 86-87, 1971

13. Brown JC, Dryburgh JR: A gastric inhibitory polypeptide. II.
    The complete amino acid sequence. Can J Biochem 49: 867-
    872, 1971

14. Brown JC, Cook MA, Dryburgh JR: Motilin, a gastric motor stim-
    ulating polypeptide: The complete amino acid sequence. Can
    J Biochem 51: 533-537, 1973

15. Mutt V, Said SI: Structure of the porcine vasoactive intestinal
    octacosapeptide. The amino acid sequence. Use of kallikrein
    in its determination. Eur J Biochem 42: 581-589, 1974

16. Anderson JC, Barton MA, Gregory RA, et al: Synthesis of gastrin.
    Nature 204: 933-934, 1964

17. Bodanszky M, Ondetti MA, Levine SD, et al: Synthesis of secre-
    tin. II. The stepwise approach. J Am Chem Soc 89: 6753-
    6757, 1967

18. Wunsch E, Brown JC, Deimer KH, et al: Zur synthese von Norleucin
    13-Motilin. Z Naturforsch 28c: 235-240, 1973

19. Bodanszky M, Klausner YS, Said SI: Biological activities of
    synthetic peptides corresponding to fragments and to the
    entire sequence of the vasoactive intestinal peptide. Proc
    Natl Acad Sci USA 70: 382-384, 1973

20. Johnson LR: The trophic action of gastrointestinal hormones.
    Gastroenterology 70: 278-288, 1976

21. Berson SA, Yalow RS, Bauman A, et al: Insulin-I^{131} metabolism in human subjects: demonstration of insulin binding globulin in the circulation of insulin-treated subjects. J Clin Invest 35: 170-190, 1956

22. McGuigan JE: Gastric mucosal intracellular localization of gastrin by immunofluorescence. Gastroenterology 55: 315-327, 1968

23. Polak JM, Bloom S, Coulling I, et al: Immunofluorescent localization of secretin in canine duodenum. Gut 12: 605-610, 1971

24. Polak JM, Pearse AGE, Bloom SR, et al: Cellular localization of cholecystokinin. Lancet 2: 1016-1018, 1975

25. Polak JM, Bloom S, Kuzio M, et al: Cellular localization of gastric inhibitory polypeptide in the duodenum and jejunum. Gut 14: 284-288, 1973

26. Pearse AGE, Polak JM, Bloom SR, et al: Enterochromaffin cells of the mammalian small intestine as the source of motilin. Virchow's Archiv (Cell Pathol) 16: 111-120, 1974

27. Polak JM, Pearse AGE, Garaud JC, et al: Cellular localization of vasoactive intestinal peptide in the mammalian and avian gastrointestinal tract. Gut 15: 720-724, 1974

28. Heitz P, Polak JM, Timson CM, et al: Enterochromaffin cells as the endocrine source of gastrointestinal substance P. Histochemistry 49: 343-347, 1976

29. Polak JM, Bloom S, Coulling I, et al: Immunofluorescent localization of enteroglucagon cells in the gastrointestinal tract of the dog. Gut 12: 311-318, 1971

30. Polak JM, Bloom SR, Hobbs S, et al: Distribution of a bombesin-like peptide in human gastrointestinal tract. Lancet 1: 1109-1110, 1976

31. Polak JM, Pearse AGE, Grimelius L, et al: Growth-hormone release-inhibiting hormone in gastrointestinal and pancreatic D Cells. Lancet 1: 1220-1222, 1975

32. Orci L, Baetens O, Rufener C, et al: Evidence for immunoreactive neurotensin in dog intestinal mucosa. Life Sci 19: 559-562, 1976

10 M.I. GROSSMAN

33. Vanderhaeghen JJ, Signeau JC, Gepts W: New peptide in the
vertebrate CNS reacting with antigastrin antibodies. Nature 257: 604-605, 1975

34. Bryant MG, Bloom SR, Polak JM, et al: Possible dual role for
vasoactive intestinal peptide as gastrointestinal hormone
and neurotransmitter substance. Lancet 1: 991-993, 1976

35. Smith TW, Hughes J, Kosterlitz HW, et al: Enkephalins:
isolation, distribution and function. In Opiates and
Endogenous Opioid Peptides. S Archer (ed). Amsterdam,
Elsevier/North-Holland Biomedical Press, 1976, p 57-62

ENDOCRINE CELLS OF THE GASTROINTESTINAL TRACT: GENERAL ASPECTS, ULTRASTRUCTURE AND TUMOUR PATHOLOGY

E. Solcia, C. Capella, R. Buffa, L. Usellini,
P. Fontana, B. Frigerio

Centro di Diagnostica Istopatologica
Universita di Pavia a Varese

21100 Varese (Italy)

The endocrine-like cells scattered among epithelial cells of the gastrointestinal mucosa differ in several aspects from the cells concentrated in endocrine glands. In the latter, abundant sinusoidal capillaries are in close contact with endocrine cells, thus ensuring prompt and massive release of secretory products into blood; close contacts with nerve endings are also found, particularly in the islets of the dog and cat. In the gut, various amounts of connective tissue are interposed between the basal membrane of the glands or villi and blood capillaries or nerve endings. In particular, no special or preferential relationship is found between endocrine-type cells and underlying blood vessels, although some long-distance correspondance between pyloric G cells and fenestrated capillaries seems to occur. The differences of innervation and blood supply may help explain the failure of gastric A cells to respond to stimuli known to be active on pancreatic A cells.[1,2]

Most endocrine-like cells of the pyloric and intestinal mucosa directly contact the lumen in a narrow, specialized area likely acting as the receptor pole of the cell, which also shows supranuclear Golgi and basal secretory granules, i.e. a clear-cut morphologic polarity[2,3] (Fig. 1,2,5). The lumen may offer to the cell important information on the amount, nature and state of foods and digestive products. In the fundic mucosa endocrine cells lack luminal contacts. The fact that these fundic cells of "closed" type fail to respond to stimuli known to be active on pyloric endocrine cells[4] seems to confirm the importance of the luminal endings for the reception of luminal stimuli.

TABLE 1 – Varese 1977 classification of endocrine cells of the gut and related tissues

Respiratory tract	Pancreas	Stomach Oxyntic	Stomach Pyloric	Sm. intestine Upper	Sm. intestine Lower	Large Intestine	Urethra	Product proposed or determined
P	(P)	P	P	P		(P)	(P)	Bombesin? dopamine? subst. $P=EC_1$
(EC)	EC	EC	EC	$EC_{1,2}$	EC_1	EC_1	EC	5HT motilin $=EC_2$ others?
(Type 3)	D_1	D_1	D_1	D_1	D_1	D_1		VIP or VIP-like peptide?
	F	(F)	F	(F)	F	F		Pancreatic peptide
	D	D	D	D	(D)	(D)		Somatostatin
	B							Insulin
	A	A						Glucagon
(?)	(X)	X	(X)	(?)				Unknown
		ECL						HE or 5HT; peptide?
	(G)	G	G	(G)				Gastrin
				S	(S)			Secretin
				I	I			Cholecystokinin
				K	K			GIP
				VL	VL			Intestinal phase hormone?
				(N)	N	(N)		Neurotensin
				(L)	L	L		GLI

( ) = Only few cells or restricted to few species or to foetal life

Secretory granules are released at the basal surface of the cell or along the lower part of its lateral surface. Above the basal lamina, intervening cells form interstitial spaces and canaliculi which are closed by junctional complexes in the upper (juxtaluminal) part of the lateral surface of epithelial cells (Fig. 1,2). Local endocrine-exocrine as well as interendocrine correlations may occur through these intraepithelial intercellular spaces, besides by direct cell to cell contacts. The junctional complexes should prevent diffusion of secretory products to the lumen, at least in physiologic conditions. Granule release at the luminal surface has never been observed, although a few granules approaching the lumen have occasionally been seen in L cells. However, vesicles are often found just below the luminal surface of the cell and occasionally in the supranuclear cytoplasm between the Golgi complex and the luminal endings.[2] The functional meaning of these vesicles is presently unknown. Likely, they work as a transport system whose direction will only be established by tracer or labelled precursor experiments. They might originate from the luminal surface and allow the cell to engulf luminal contents including secretagogues, or they might originate from the reticulum (or even from the Golgi) and move to the luminal surface to release some secretory product. The latter possibility, if proven, may provide a morphologic basis for the luminal release of gastrin and other hormones recently observed by Vinik[5] and Uvnas-Wallensten.[4]

Below the basal lamina of the glands or villi, active peptides and amines released by endocrine-like cells may interact with some targets before entering the blood, including nerve endings, smooth muscle cells (with special reference to those of the villi and muscularis mucosa) and vessel walls. The reported action of somatostatin (from gut D cells) on neighbouring G, A, I, S and K cells or of 5HT (from EC cells) on intestinal nerve endings may be good examples of such local "paracrine" effects of gut endocrine-like cells.

Although some active substances might display both blood-mediated (truly endocrine) and local (paracrine) effects, the products of several cells (G, A, S, I and K cells) seem to display mainly endocrine effects, while those of other cells (D and D_1 cells) might have mainly paracrine functions. This could allow us to distinguish, among endocrine-like cells of the gut, paracrine cells, more closely fulfilling the concepts of Feyrter,[6] from truly endocrine cells. Unfortunately, information in this respect is quite incomplete. The nature of the active products of some cells (see Table 1) is still unknown or rather hypothetical and more substances than those so far identified might be produced by cells whose function seems now ascertained (Figs. 3,4,5).

The endocrine-like cells so far described in the lung, trachea and urethra display the same general morphological patterns as those of gut endocrine-like cells.[7,8] Both kind of cells belong to the so

called "diffuse or dispersed endocrine system (DES)". No truly endocrine functions have been identified so far in the lung or urethra;
the endocrine-like cells occurring in these tissues (P, D_1 or EC
cells) are likely to be interpreted as paracrine cells. From Table
1 it appears that - as noted before[8] - several cells of putative
paracrine function are widely distributed in various tissues (see
P, D_1, EC and D cells), while various cells to which a primarily
endocrine function has been assigned (B, A, G, S, I and K cells)
occur only in specific and restricted areas of the gut and/or pancreas. It might be that cells of more or less diffuse pattern like
F and L cells also belong to paracrine cells; the function of other
cells such as N, ECL or X cells remains uncertain. For the latter
cells even the nature of the product(s) remains unknown.

At present, cytological studies on endocrine-like tumours of
the gut or DES tumours are largely incomplete. Cases so far studied
show that the endocrine cell types occurring in such tumours can be
precisely identified when accurate histochemical and ultrastructural
researches are performed. Unfortunately, in most cases only paraffin blocks are available and no electron microscopy and only few
histochemical tests can be applied. Thus, for the time being, it
seems opportune to retain the time-honoured term carcinoid tumours
for the tumours of paracrine cells and endocrine-like cells of unknown function, and to use the term gut endocrine tumours for tumours of G, A, S, I and K cells (of which only G cell tumours have
been described so far). A "functional" endocrine syndrome seems
to appear more frequently in association with the latter tumours
than with carcinoids. Carcinoids can be divided into argentaffin
(EC cell) carcinoids - easily identified based on conventional
histologic patterns and a few popular histochemical tests - and
non-argentaffin carcinoids. Some of the former tumours, besides
5-hydroxytryptamine, produce substance P.[9] The latter should be
subdivided further on topographical ground (see Table 2); earlier[10]
as well as recent[11] evidence suggests that the nature of tumour cell
types, histologic patterns and prognosis are partly dependent on
the site from which the tumour arises. Of course, whenever possible
the exact cytological identification of tumour cell types should be
attempted.

Just as with normal endocrine cells of the pancreas, endocrine
tumours of this tissue could be divided into islet (cell) tumours
and tumours of the diffuse endocrine system (DES tumours). Tumours
related with A and B cells, the most typical endocrine cells of the
islets, belong to the first group; tumours related with the remaining cells, including both carcinoids and gut-related endocrine tumours (gastrinomas), fit in the second group. Major differences of
behaviour and prognosis - much more favourable for islet cell tumours than for DES tumours - justify the separation of these two
groups of tumours[8].

TABLE 2 - Tumours of the diffuse endocrine system

CARCINOIDS:

EC cell argentaffin carcinoids
1. Substance P (EC_1 cell) tumour
2. Other EC cell tumours.

Non-argentaffin carcinoids
1. Upper primitive gut
   (lung, thymus, oesophagus)
2. Stomach and duodenum
3. Pancreas, liver, biliary tree
4. Small bowel and appendix
5. Large bowel
6. Urogenital system

GUT ENDOCRINE TUMOURS:

1. G cell tumour (gastrinoma)
2. Others?

The interpretation of tumours producing "ectopic hormones" of the pancreas, gut, lung and thymus is still uncertain. Production of ACTH-like, ADH-like or CRF-like peptides by non-tumour tissues outside the pituitary-hypothalamic area has been suggested by some investigations[12,13,14] If proven, this may well represent the source of tumours producing "ectopic" peptides, some of which might prove in the future to be in fact "orthotopic".

Ectopic carcinoids, both argentaffin and non-argentaffin, may arise in teratomas (ovary, testicles) or in areas of intestinal metaplasia occurring in the stomach, gallbladder, ovary, cervix and urinary tract. An origin from intestinal (colonic type) metaplasia seems likely for the kidney "enteroglucagonoma" described by Gleeson et al.[15] Ultrastructurally this case showed cells closely resembling intestinal L(GLI) cells - with special reference to those of human large bowel - and obviously different from both A cells and pancreatic glucagonoma cells.

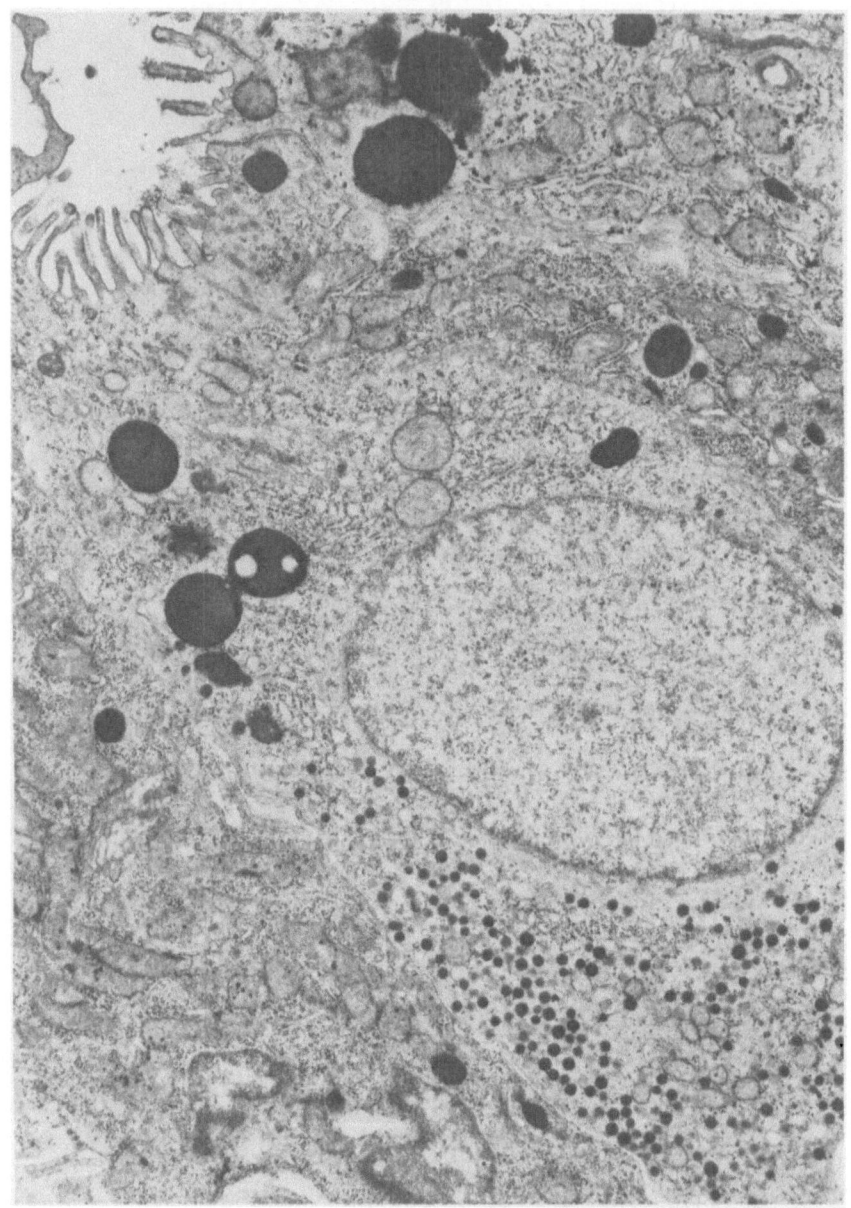

FIGURE 1. P cell of the human duodenal bulb showing small secretory
granules (mean diameter 125 nm) grouped at the base of the cell;
note the luminal ending covered with microvilli. The distribution
of this cell in the gut parallels the distribution of bombesin-
immunoreactive material. X 14,000

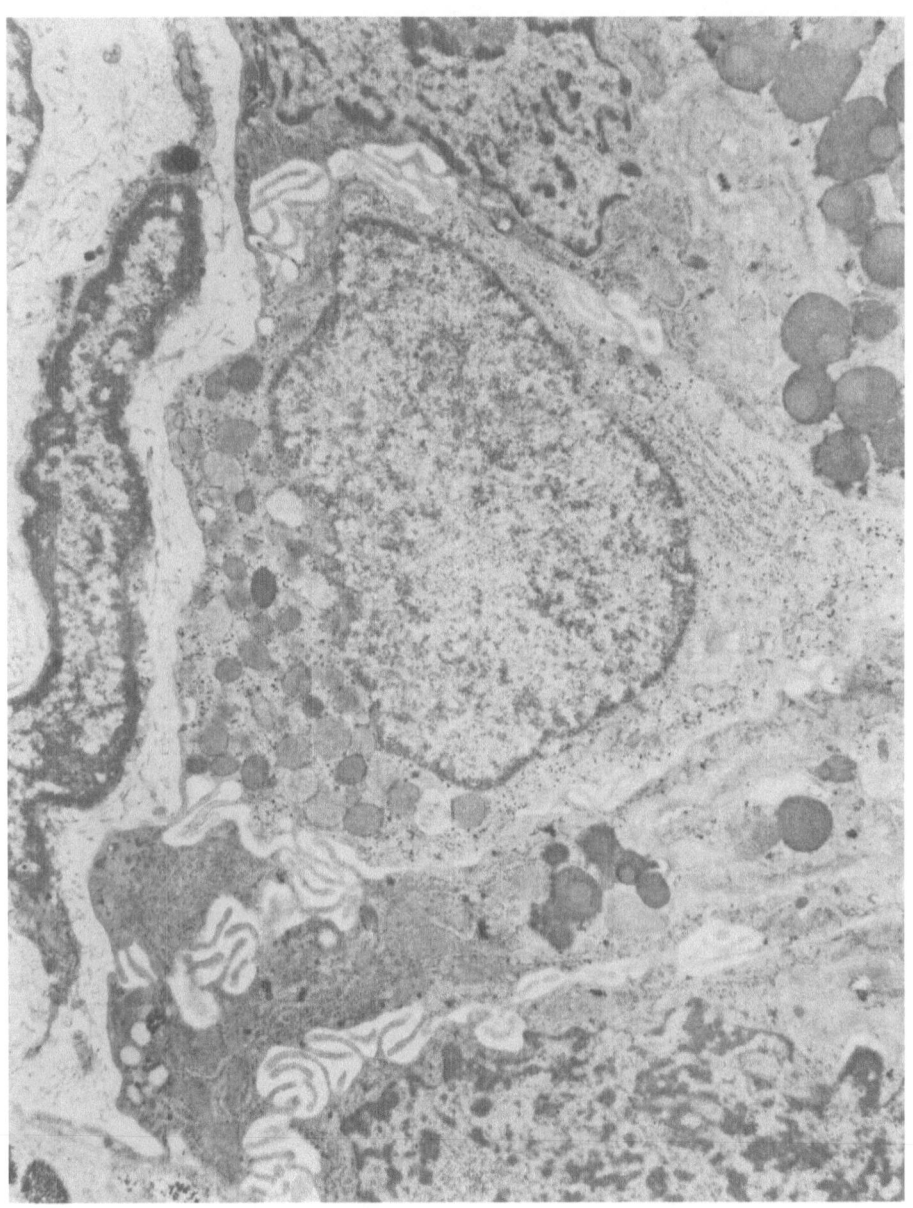

FIGURE 2. D cell in the human pyloric mucosa showing large, poorly osmiophilic, basal granules (m. d. 300 nm) storing somatostatin, and well developed intercellular spaces along its lateral surface. X 14,000

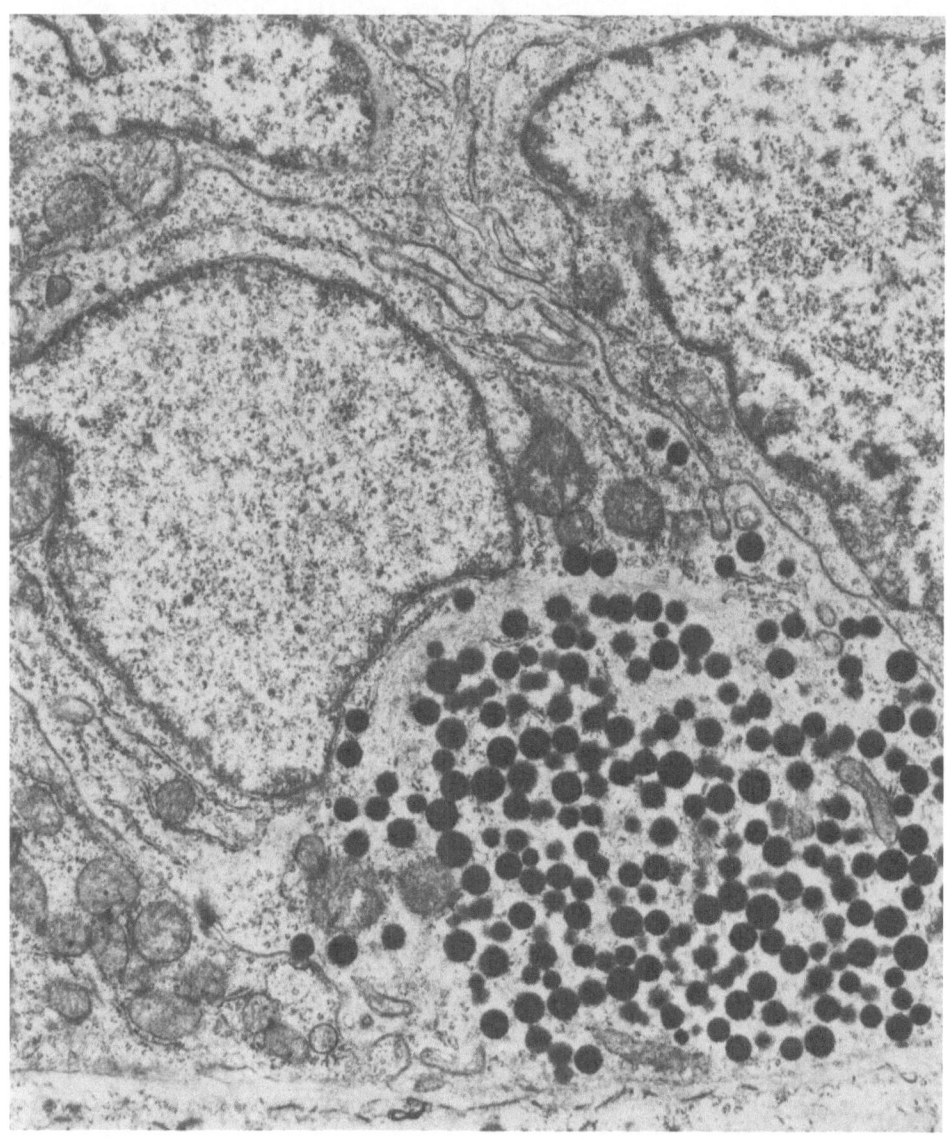

FIGURE 3. L cell of the human ileum showing moderately large
granules (m. d. 270 nm); these granules are smaller than those
of dog L (large granule) cells. The L cell is reputed to store
GLI. X 14,000

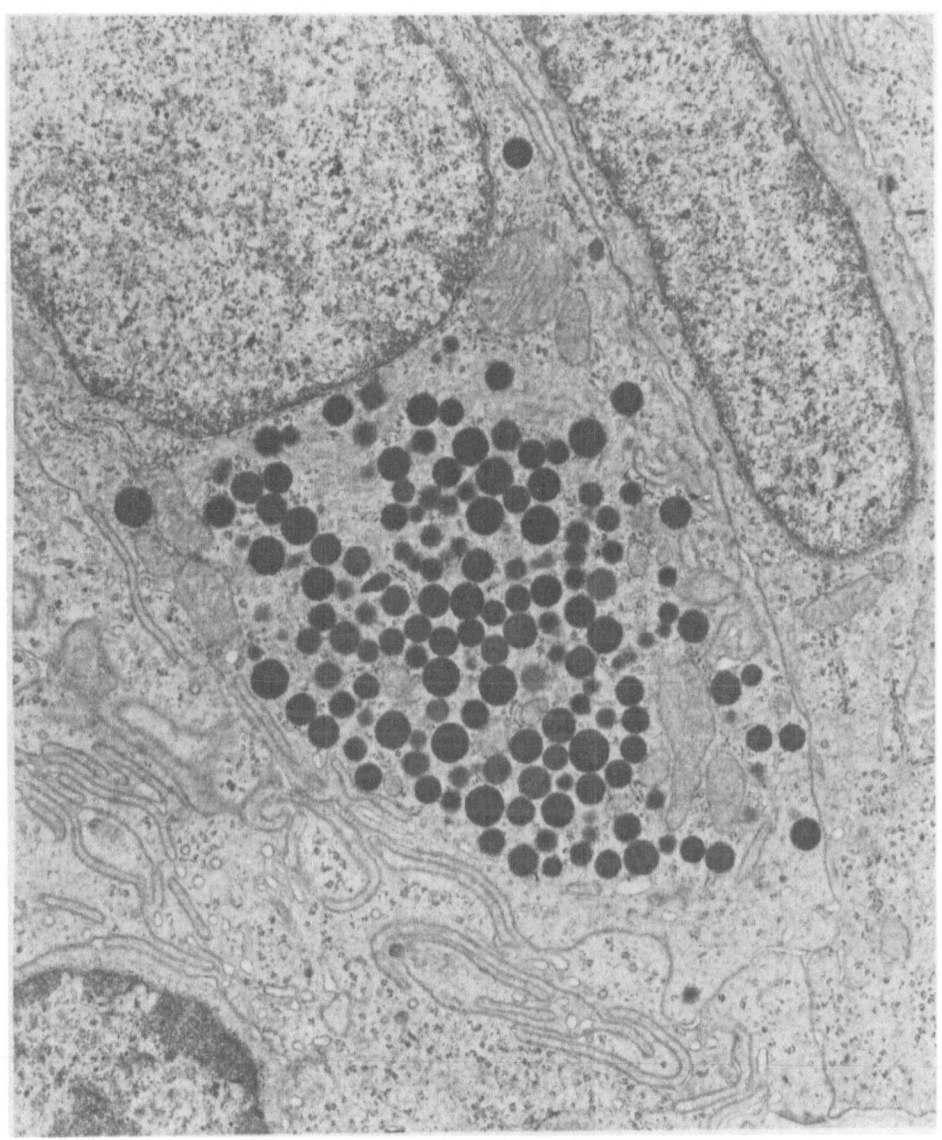

FIGURE 4. *N cell of the human ileum showing large granules (m. d. 300 nm), reputed to store neurotensin. X 14,000*

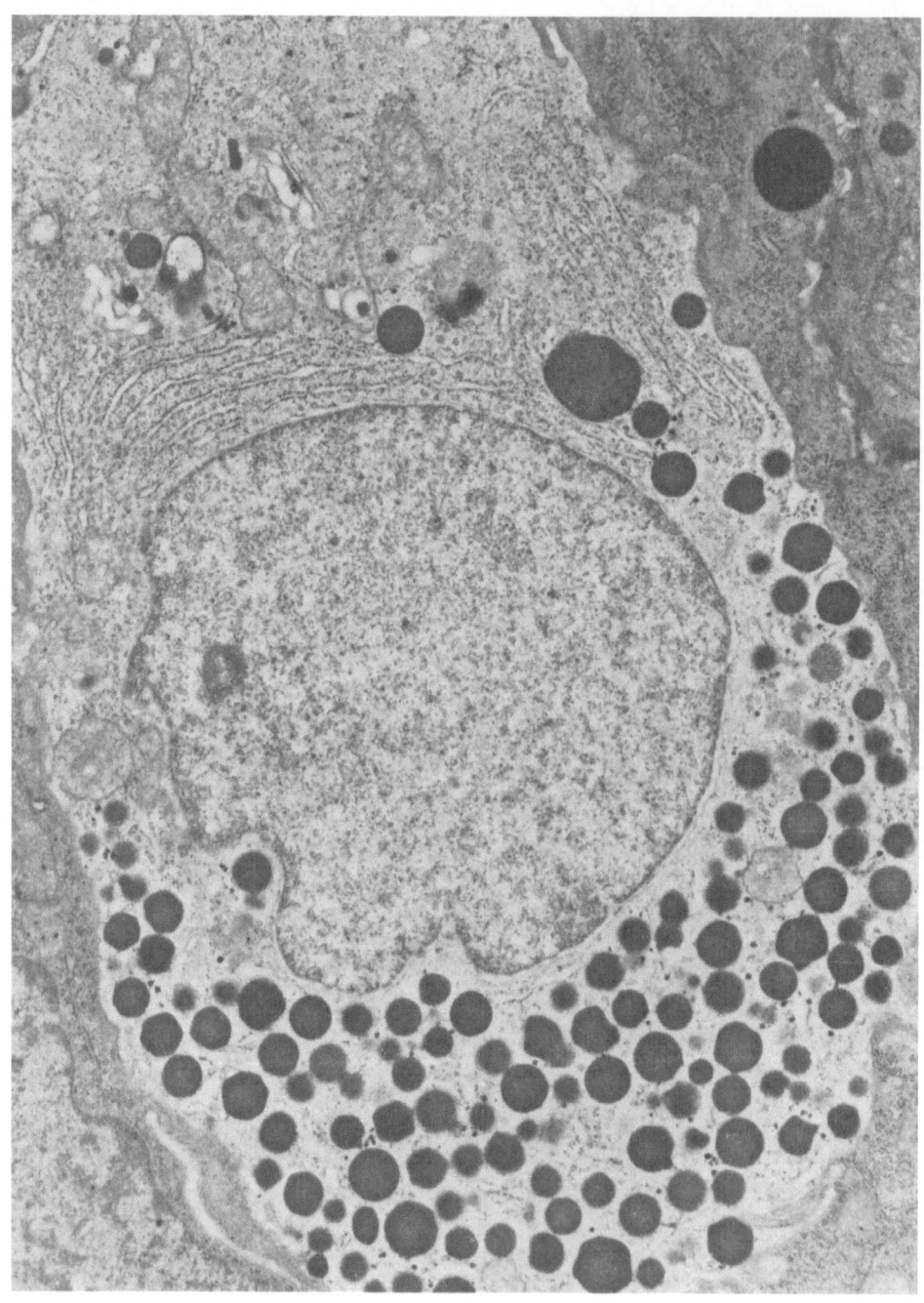

FIGURE 5.   VL cell of the human jejunum showing very large granules
(m. d. 450 nm).   The function of this cell is unknown.   X. 14,000

REFERENCES

1. Lefebvre PJ, Luyckx AS: Factors controlling glucagon release by the isolated perfused dog stomach. EASD 13th Annual Meeting: Geneva, Switzerland, September 28-30, 1977

2. Solcia E, Capella C, Vassallo G, Buffa R: Endocrine cells of the gastric mucosa. Internat. Rev. Cytol. 42: 223-286, 1975

3. Solcia E, Vassallo G, Sampietro R: Endocrine cells in the antro-pyloric mucosa of the stomach. Z Zellforsch. 81: 474-486, 1967

4. Uvnas-Wallensten K: Secretion of gastrin into the antral lumen of cats and its regulation by intra-antral pH. Paper delivered at the "International Symposium on Gastrointestinal Hormones and Pathology of the Digestive System". Rome, June 13-15, 1977

5. Vinik AI: Characterization of 'gastrin' released from mucosal surface of isolated human antra. Paper delivered at the "International Symposium on Gastrointestinal Hormones and Pathology of the Digestive System". Rome, June 13-15, 1977

6. Feyrter F: Uber die peripheren endokrinen (parakrinen) Drüsen des Menschen. 2. Aufl. Wien-Düsseldorf: W Maudrich, 1953

7. Capella C, Hage E, Solcia E, Usellini L: Ultrastructural similarity of endocrine-like cells of the human lung and some related cells of the gut. Cell Tiss. Res. (in press, 1977)

8. Capella C, Solcia E, Frigerio B, Buffa R, Usellini L, Fontana P: The endocrine cells of the pancreas and related tumours. Virchows Arch. A Path. Anat. Histol. 373: 327-352, 1977

9. Alumets J, Hakanson R, Ingemansson S, Sundler F: Substance P and 5HT in granules isolated from an intestinal argentaffin carcinoid. Histochemistry 52: 217-222, 1977

10. Williams ED, Sandler M: The classification of carcinoid tumours. Lancet 1: 238-239, 1963

11. Godwin JD: Carcinoid tumours. An analysis of 2837 cases. Cancer 36: 560-569, 1975

12. Vorherr H, Massry SG, Fallet R, et al: Antidiuretic principle in tuberculous lung tissue of a patient with pulmonary tuberculosis and hyponatremia. Ann. Intern. Med. 72: 383-387, 1970

13. Gewirtz G, Yalow RS: Ectopic-ACTH production in carcinoma of
     the lung. J. Clin. Invest. 53: 1022-1032, 1974

14. Yasuda N, Greer MA: Demonstration of corticotrophin-releasing
     activity in rat and human peripheral blood. Acta Endocr.
     84: 1-10, 1977

15. Gleeson MH, Bloom SR, Polak JM, Henry K, Dowling RH: Endocrine
     tumour in kidney affecting small bowel structure, motility,
     and absorptive function. Gut 12: 773-782, 1971

HOW DOES A CANDIDATE PEPTIDE BECOME A HORMONE?

M. I. Grossman

Veterans Administration Wadsworth Hospital Center

Los Angeles, California, USA

Peptides, like people, have identity crises. We ask for them, as we do for ourselves, what is my role? Once we have isolated and chemically identified a peptide, we inject it and observe its various biological actions. This tells us the various things it can do. Only when we are satisfied that we have determined which of these things it actually does do when it is endogenously released under physiological conditions do we confer the title of hormone on it. A hormone is a peptide with a physiological role.

The rigor of the evidence demanded to establish a given proposition varies with the state of the art. In an earlier day, before gastrointestinal hormones were chemically identified, three steps sufficed to establish that a hormonal mechanism was involved in any given physiological event during digestion. First was the demonstration that a stimulus applied to a part of the tract produced a response in a distant target. Second was the demonstration that the effect persisted after all nervous connections between the site of stimulation and the target had been severed. Third was the demonstration that the physiological event could be mimicked by injection of an extract of the part to which the stimulus had been applied but not by extracts from other tissues. Now that the chemical nature of the gastrointestinal hormones is known, and now that it is possible to measure the blood concentration of these hormones by radioimmunoassay, an additional kind of proof is demanded. We now ask whether the amount and kind of the peptide hormone that is released when the stimulus is applied can account for the observed response. In other words, if a similar increase in amount and kind of hormone were produced by infusion of exogenous hormone, would the response be reproduced?

The amino acid sequences of 6 peptides isolated from gastrointestinal mucosa are known: gastrin, secretin, cholecystokinin
(CCK), gastric inhibitory peptide (GIP), motilin, and vasoactive
intestinal peptide (VIP). The first 4 of these may be regarded as
"established" by the 3 nonquantitative criteria discussed above.
These criteria are fulfilled for acid secretion in the case of gastrin, pancreatic bicarbonate secretion in the case of secretin,
pancreatic enzyme secretion and gallbladder contraction in the case
of CCK, and release of insulin during hyperglycemia in the case of
GIP.

Only for gastrin and GIP are the radioimmunoassays sufficiently refined to allow application of the quantitative criterion.
Walsh and coworkers[1] have recently shown that much of the acid secreted in response to a protein meal can be accounted for by the
gastrin released. Copying the increase in serum gastrin by infusing exogenous gastrin reproduced the acid secretory response to a
meal. Increases in radioimmunoassayable GIP occur after feeding
and infusion of exogenous GIP to copy these increases stimulates
insulin release when hyperglycemia like that produced by a meal is
present.[2]

Radioimmunoassays for CCK are not sufficiently perfected to
be useful in establishing its contribution to responses to a meal.
Radioimmunoassayable VIP and motilin do not increase after feeding.
Whether this is because the assays are inadequate, or because factors other than meals control release, or because these peptides
serve a paracrine rather than an endocrine role, remains to be
determined. In the case of motilin, it appears that factors operating during fasting cause its phasic release in synchrony with the
interdigestive myoelectric complex.[3] VIP is present in both nerves
and endocrine cells so it may be as much under neurocrine as under
endocrine control.

The criterion that an effect should be reproduced by infusion
of exogenous hormone in amounts that copy the postprandial increase
could lead to false positive or false negative answers. For example,
infusion of exogenous gastrin to give blood concentrations like
those seen after meals produces an increase in the frequency of
pacesetter potential of the antrum of the stomach.[4] However, after
feeding, a decrease rather than an increase in frequency of antral
pacesetter potential is seen. Presumably, some of the many other
factors elicited by a meal oppose this action of gastrin and prevent
its expression. Infusion of the amount of secretin needed to give
the rate of pancreatic bicarbonate secretion seen after a meal produces blood secretin levels greater than are found after eating.[5]
The likely explanation for this is that cholecystokinin and vagal
reflexes attending a meal potentiate the response to secretin so
much that only a low concentration in blood, detectable by only the
most sensitive radioimmunoassays, is needed to produce the observed

rate of pancreatic bicarbonate secretion. These examples emphasize that the action of any one hormone may be greatly modified by the other hormonal and neural mechanisms that are set in motion by a meal. It may be helpful to study the effect of infusing the exogenous hormone after a meal rather than in the fasting state so that all of the additional factors that are operating in the fed state will be present.

The questions addressed here are as much philosophical as they are scientific, so there is a large element of subjectivity in what is said here. It is unlikely that a single set of universally applicable criteria can be devised to deal with the question of which actions of a peptide are "physiological." Some general guidelines may help, but we shall probably have to continue to deal with the problem on a case by case basis. In this, as in other scientific problems, it is the convergence of a wide variety of lines of evidence that convinces us.

REFERENCES

1. Feldman M, Richardson CT, Walsh JH: Mechanisms of acid secretory response to amino acid solutions in normal man. Clin Res 25: 467A, 1977

2. Andersen DK, Elahi D, Raizes G, et al: The role of gastric inhibitory polypeptide (GIP) in glucose tolerance and intolerance. Gastroenterology 72: 811, 1977

3. Itoh Z, Takeuchi S, Aizawa I, et al: Recent advances in motilin research: it physiological and clinical significance. In Abstracts of International Symposium on Gastrointestinal Hormones and Pathology of the Digestive System, Rome, June 13-15, 1977, p 125-128

4. Strunz UT, Schlegel JF, Grossman MI, Code CF: Action of gastrin on antral electrical activity. Clin Res 25: 112A, 1977

5. Fender HR, Curtis PJ, Rayford PL, Thompson JC: Simultaneous bioassay and radioimmunoassay of secretin in dogs. Gastroenterology 72: 1058, 1977

PEPTIDERGIC INNERVATION OF THE GASTROINTESTINAL TRACT

J. M. Polak and S. R. Bloom

Departments of Histochemistry and Medicine, The Royal
Postgraduate Medical School, Hammersmith Hospital

London W12 OHS, England

INTRODUCTION

The presence of peptides in gut endocrine cells has been
recognised for more than a decade but only in the last few years
has it become increasingly apparent that several of these peptides
are also found in the central and autonomic nervous system. In
fact, the recognition that peptides can occur in at least 2 types
of tissue is considerably older and dates from the discovery[1] that
the same peptide (substance P) could be extracted from both brain
and gut. The list of these peptides found in both brain and gut
is probably not yet complete but at present comprises VIP, CCK/
gastrin, neurotensin, somatostatin, bombesin, substance P, and
enkephalin. In the gut mucosa these peptides may be found either
in typical endocrine cells, where they may act as circulating or
local hormones, or in the autonomic nerves. Here they may function
as neurotransmitters or they may modify the responses of other
cells by stimulation or inhibition.

The term 'peptidergic neuron' was first used for those hypothalamic neurons which originated in the supra-optic and paraventri-
cular nucleii and synthesised oxytocin and vasopressin.[2] Recent
studies using highly specific radioimmunoassay as well as immunocytochemical methods have shown an increasing number of hormonally
active peptides in the brain, specifically localised to the neuronal
cell bodies or their peripheral elements. The brain thus contains
a widely distributed peptidergic system in addition to the established cholinergic and aminergic neurons.[3]

The nerve endings of the autonomic nervous system are classic-
ally divided into two types, adrenergic and cholinergic, each type
containing small secretory granules. The existence of other types
of neurons was first suggested by Dodgiel[4] and later fully supported
by the ultrastructural studies of Baumgarten et al[5] showing the
presence of a number of nerve terminals in the gut with much larger
and more electron dense secretory granules which resembled quite
closely those of the peptidergic vasopressin and oxytocin-producing
nerves of the neurohypophysis. In the autonomic nervous system of
the gut these nonadrenergic noncholinergic nerves have recently
been shown by morphology and immunocytochemistry to consist at least
in part of VIP containing nerves.[6] Recently Burnstock[7] has pointed
out that ATP is present in many autonomic nerve endings and has
suggested that these are therefore purinergic nerves which release
the purine nucleotide ATP as their transmitter. It seems equally
likely, however, that ATP is merely the energy source used for the
metabolism and release of the neuropeptides and thus its presence
in the secretory granules is quite nonspecific.[8] The physiological
control of these peptidergic nerves appears to be separate from
that of either the cholinergic or adrenergic nerves, which explains
the numerous observations of nerve-mediated physiological actions
which cannot be altered by cholinergic or adrenergic blockade. For
instance, vagal stimulation results in an immediate discharge of
the neuronal peptides VIP, somatostatin and enkephalin and this
discharge is unaffected by cholinergic blocking agents.[8]

THE APUD CONCEPT AND FEYRTER'S DIFFUSE ENDOCRINE SYSTEM

In 1966, Pearse[9] proposed a unifying hypothesis for the cells
responsible for the production of polypeptide hormones. It was
noted that almost all these cells possessed several similar cytochemical and functional characteristics. The term APUD is an
acronym taken from the initial letters of the most important of
these characteristics (Amine content or Precursor Uptake and Decarb-
oxylation). When the APUD concept was formulated it was observed
that the cells of the system were probably derived embryologically
from the neuroectoderm, largely on account of their amine storage
mechanism, and the presence of cholinesterases, which are also
found in nerve cells. Nowadays it is known that the cells of the
APUD series probably derive from a single region of the neuroendo-
crine-programmed ectoblast.[10] The APUD system as originally formulated was similar to that of the diffuse endocrine system proposed
independently by Feyrter.[11] The forerunner of both, however, was
the idea put forward by Masson in 1914 of an intestinal neuroendocrine gland closely related to the myenteric plexus. When the
APUD concept was postulated it referred almost exclusively to
classical polypeptide-producing endocrine cells. It was then not
known that many neural cells were also capable of producing such
polypeptides. Since 1975, the number of peptides recognised as

common to the brain and gut has greatly increased and thus the APUD
concept has had to be extended to include not only endocrine cells
but also neural and neuroendocrine elements.

With the knowledge that we have today, we might suggest that
the peptides produced by the neurons would have a neurotransmission
function and those produced by epithelial endocrine cells would
have a classical hormonal function. However, the situation is
probably more complicated than this and the functions of these
peptides in the various locations have not yet been fully worked
out. It is possible that they all have a spectrum of activity
ranging from that of neurotransmitter to that of a true hormone
and thus the APUD, diffuse, or neuroendocrine system could have
at least four different functions. The product of a peptidergic
neuron may influence another neuron (neurocrine) or may be secreted
directly into the bloodstream (neuroendocrine). The peptide-producing, non-neuron, endocrine cell may also secrete directly into the
bloodstream (endocrine) or into the inter-cellular fluid to affect
only the neighboring cells (paracrine). It does seem that the APUD
concept and that of a peripheral diffuse neuroendocrine system are
synonymous, and that the cells which constitute it have different
functions which vary with their location.

TECHNOLOGY

Histological methods have been used for many years for staining
the neuropeptides in both the central and peripheral tissues. In
the early days the aldehyde fuchsin Gomori technique was widely
used for staining vasopressin and oxytocin-producing cells in the
hypothalamus and a similar technique was employed to stain the
beta cells of the pancreas. A range of techniques has been used
for many years for the staining of the peptide-producing cells in
the gastrointestinal tract. However, the most useful and reliable
technique for the staining and differentiation of both the gastrointestinal endocrine cells and the peptidergic neurons in central
and peripheral tissues is immunocytochemistry at the light and
electron microscopical levels in combination with morphological
and ultrastructural methods.

Immunocytochemistry. Immunocytochemistry allows the intra-
cellular demonstration of the product of a cell, in this case the
peptide hormone, by the use of specific antibodies. A limitation
of the method which must be borne in mind is that the demonstration
of a hormone antigen by immunocytochemistry does not necessarily
imply that the hormone is in a biologically active state.

One of the first requirements for a successful immunocytochemical technique is the selection of a suitable fixative. This
must make the peptides insoluble without altering their tertiary

structure, which is important for the immunoreaction. The conventional aldehyde fixatives of histological laboratories are often only partially successful for immunocytochemical staining. The reason for this is that if a powerful cross-linking fixative such as glutaraldehyde is used at high concentration and for the long period necessary for satisfactory preservation of intra-cellular structures for ultrastructural morphology, the antigenic site of peptide hormones becomes totally unreactive to the antibody. Less powerful aldehyde fixatives such as formaldehyde fail to insolubilise most of the peptide hormones. Alternatives to these fixatives have been developed in recent years. Examples include bifunctional reagents, such as diethyl pyrocarbonate (DEPC) and p-benzoquinone.[12] The latter, which acts by cross-linking peptides to produce disubstituted quinones, has opened up new possibilities for the study of the gut neuroendocrine system. It has been extensively used in vapour form for the fixation of peptides in true endocrine (APUD) cells of the gastrointestinal tract. Unfortunately, peptidergic nerves have proved refractory to this treatment, perhaps because there is a different storage form for the peptides in the nerves which is affected by freeze-drying and wax embedding. A satisfactory technique has recently been developed[13] which combines the use of a solution of benzoquinone as a fixative with the classical method of cryostat sectioning which has long been routinely employed for the localisation of brain neuropeptides. This latter technique (benzoquinone in solution and cryostat sectioning) has shown nerves containing neuropeptides to be present in all areas of the gut, pancreas and gall bladder. This newly developed method has established convenient conditions for staining neuropeptides in both nerves and cells with a single fixation and sectioning technique.

Since the first work on immunostaining by Coons and co-workers, there have been important methodological advances which provide more sensitive and versatile modifications of the original techniques The use of enzyme-conjugated antibodies in place of the original fluorescein conjugates has some advantages and recently Sternberger[14] has modified the immunoperoxidase method by introducing a further stage, which uses a peroxidase anti-peroxidase complex. This is used as a purified gamma-globulin antigen which combines specifically only with the 2nd layer (unconjugated antibody). The high ratio of peroxidase to antibody makes the final histochemical enzyme reaction particularly strong, thus greatly increasing the staining sensitivity (Fig. 1).

An immunocytochemical reaction can be considered as specific only if it is prevented by prior absorption of the antiserum with a specific antigen and by no other antigen, and if the structure stained does not react with inappropriate antisera or non-immune serum.

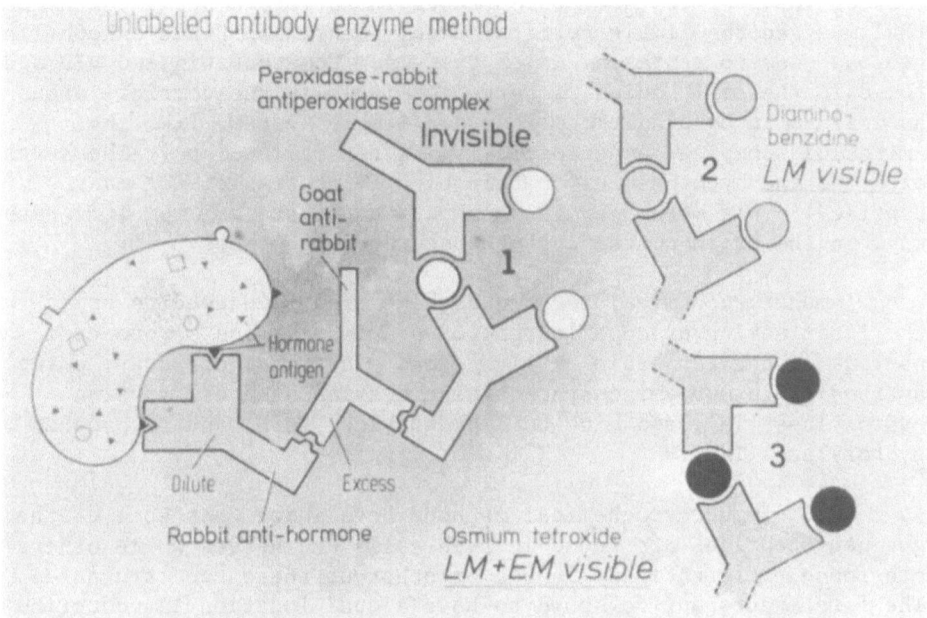

FIGURE 1. Schematic representation of the Sternberger's peroxidase antiperoxidase (PAP) immunocytochemical method.

Radioimmunoassay. Radioimmunoassay (RIA) has revolutionised the field of gut endocrinology by providing a precise measurement of the blood or tissue levels of a particular hormone. The method is based on the competition for a fixed amount of highly specific antibody between a known amount of radioactive antigen labelled with ^{125}I (usually attached to a tyrosine residue) and the unknown antigen in the sample to be investigated. The ratio of antibody-bound radioactive antigen to unbound radioactive antigen gives the measure of the amount of competing antigen in the sample. The techniques of immunocytochemistry together with radioimmunoassay have recently been used in the study of neuropeptides in the brain and gut and the combined approach has yielded much more information than could have been obtained by each technique separately.

MORPHOLOGY

Central Nervous System. Immunocytochemistry has revealed that each of peptidergic neurons has a characteristic and unique

distribution pattern although in many cases the patterns overlap.
Most of the peptides are concentrated in the brain stem, the hypothalamus and the limbic cortical area. Substance P and enkephalin
neurons seem to represent major systems. Somatostatin and VIP are
found in the cell bodies of nerve terminals in the cortical areas
as well as in hypothalamic areas and limbic system, like the
amigdaloid complex, and especially its central nucleus. The septal
area and the hypothalamus contain high concentrations of many
peptides. This method has also shown that there are specific path-
ways in the brain containing neurosecretory peptides.

Immunocytochemical techniques have also been used to study
the localisation of dopaminergic, monoaminergic and serotonergic
neurons in the rat brain by light- and electron microscopy, using
antibodies to monoamine-synthesising enzymes such as tyrosine
hydroxylase (TH), dopamine beta-hydroxylase (DbH) and tryptophan
hydroxylase (TRH).[3]

Gut. Immunocytochemical methods have shown that some of the
gut neuropeptides are found in both cells and nerves while others
are found preferentially in one or other of these two structures.
The first neuropeptide shown to have a dual location in endocrine
cells of the gut as well as in nerves was substance P in 1975.[17]
This was followed by the localisation of somatostatin which was
found mainly in endocrine cells of the gut and pancreas, characterised as D cells. A lesser amount of somatostatin can also be found
in scattered fine submucosal nerve fibres in the small and large
intestine, as well as around cell bodies in the myenteric plexus.[18]
No evidence has yet been provided for the localisation of neurotensir
in gut neurons, whereas its localisation in a well characterised
endocrine cell type (N cell) has been firmly established.[16] One of
the most interesting and impressive findings is the localisation of
VIP, not only in scattered endocrine cells but also as an important
component of the gut innervation. It has been shown to be an important constituent of the peptidergic division of the autonomic nervou:
system and to be involved in gastrointestinal diseases concerning
motility and secretory problems. Comparative studies of the relative
frequency of the various peptidergic nerves of the gut indicate that
the largest component is the VIPergic innervation, followed by that
of substance P, and then by somatostatin, enkephalin and bombesin
nerve fibres of the gut wall.

Immunocytochemical techniques applied at the ultrastructural
level seem to indicate that all neuropeptides are stored in small
membrane-bound cytoplasmic structures, the neurosecretory granules.
In the gut endocrine cells the various types of secretory granules
have an internationally accepted classification (Fig. 2).[16] However
the discovery of the many types of granules in the gut peptidergic
innervation is still so new that their description has not yet been
agreed upon. Brain neuropeptides seem to be produced by different

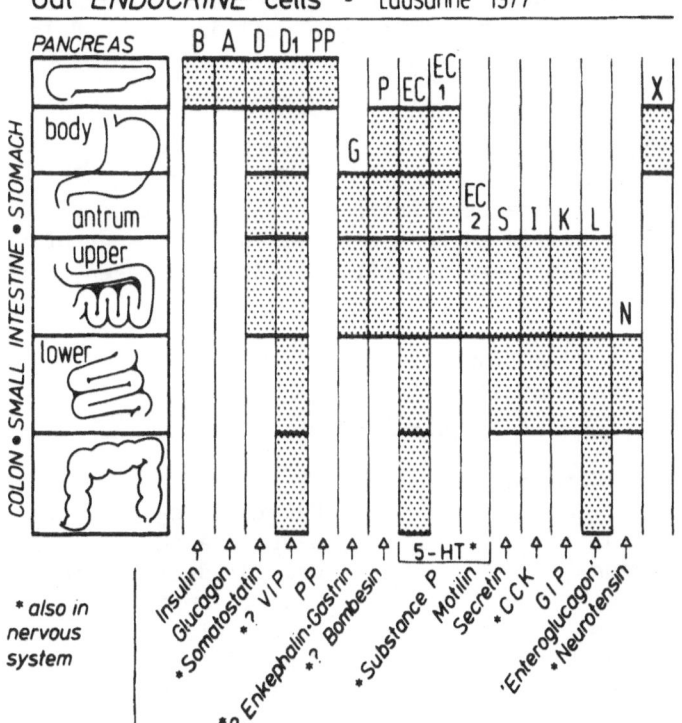

FIGURE 2.   *Lausanne international agreement (1977) on nomenclature and classification of gut endocrine cells.*

cell types. However, both the peripheral and central neural peptides are always associated with large and highly electron dense secretory granules, easily distinguishable from those of the cholinergic or adrenergic types.[5] It is not yet clear whether all or only some of the neuropeptides in the nerves of the gut are stored in what is now known as the Baumgarten p type of granules. We can only say that there are several divisions of the autonomic nervous system and further work will be necessary to unravel the complexity of the peptidergic innervation and the ultrastructural features of the autonomic innervation as a whole.

At present there is no good evidence for concomitant storage of two peptides in the same cell. It is clear that substance P is stored in a cell capable of storing an amine (serotonin) as well as the peptide.[17] Somatostatin has been reported to be present in the cell bodies of the sympathetic chain which also stores noradrenalin.[18] However, it is not yet known whether there is simultaneous

release of the two products, whether differential release results
from different stimuli, or whether one cell product modulates the
release of the other.

INDIVIDUAL HORMONES

    Somatostatin. Somatostatin is a cyclic 14 amino acid peptide
and is now widely available in synthetic form.[19] Most of the known
actions of somatostatin are inhibitory and can be seen in the
following table.

ACTIONS OF SOMATOSTATIN

Inhibition of

Growth Hormone	PP
TSH	Enteroglucagon
Insulin	Gastric Acid
Glucagon	Gall Bladder contraction
Gastrin	Gastric emptying
Secretin	Pancreatic bicarbonate
GIP	Pancreatic enzymes
Motilin	Coeliac axis blood flow
Neurotensin	

    Somatostatin distribution has been found to extend far beyond
the hypothalams from which it was originally isolated.[22] It was
subsequently found in other parts of the brain in even larger
amounts. Large amounts of somatostatin are found in the stomach
and smaller amounts in the duodenum. The cells which secrete
somatostatin are the D cells of the stomach, duodenum, jejunum and
pancreas (Fig. 3). The pancreatic D cell has been recognised since
1931 but its secretory product was not known until the discovery
of somatostatin.

    In the stomach somatostatin is a powerful inhibitor of gastrin
release and gastric acid secretion. This inhibitory action in the
stomach suggests that somatostatin may play a role in diseases when
there is abnormal gastric secretion. In the normal antrum somatostatin cells are located in the mid-zone of the antral mucosa,
scattered among epithileal non-endocrine and other endocrine cells.
In normal conditions there are eight times as many G cells as there
are inhibitory D cells in the antrum. In most subjects with duodena
ulceration (DU), the normal ratio (D:G) of 1/8 remains unchanged;
but in cases which have G cell hyperplasia this ratio is altered to
1 to 80. The number of somatostatin containing cells is decreased
(43%) in the antral mucosa in achlorhydria. In this condition, a
great increase in the number of G cells is also seen; thus, the
ratio of D to G cells in the antrum reaches 1/160. The relative

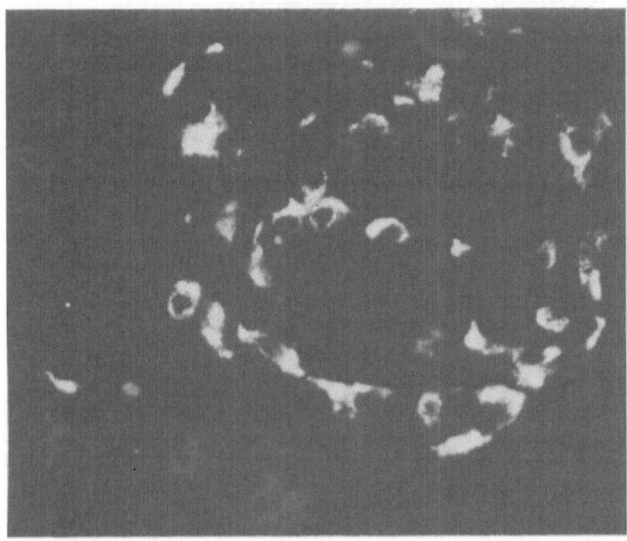

FIGURE 3.   Human pancreatic D cells stained with antibodies to
             somatostatin (x 952).

somatostatin deficiency in duodenal ulcer patients with G cell
hyperplasia could explain the over-reactivity of G cells to their
releasing stimuli in some of these patients. The G cell hyperplasia which is found in achlorhydric patients has previously been
viewed as being due to a failure of the acid feedback mechanism.
This is probably correct but the finding of somatostatin deficiency
suggests that there is in addition a failure of paracrine inhibition
of gastrin release. One may thus postulate that many disorders are
due to subtle failure of the normal balance of control mechanisms
rather than any single absolute deficiency.

    Two somatostatinomas have recently been reported.[20,21] Both
tumours originated in the pancreas and contained very high levels
of somatostatin when the tissues were analysed. The clinical
abnormalities included hypochlorhydria, steatorrhoea and a diabetic
glucose tolerance. Our group has found large quantities of somatostatin in 5 neural and 23 pancreatic tumours (Fig. 4). The presence
of somatostatin in some of the metastases of these tumours suggests
that this substance is an intrinsic tumour product. Chromatographic
analysis of the tissue extract indicates that the somatostatin from
the tumours is probably identical to ordinary somatostatin. The
effect of this peptide on the function of the tumour is unknown;
but it is of interest that some of the tumours with the highest
somatostatin content secreted very little of their main product,
suggesting a possible inhibitory influence.

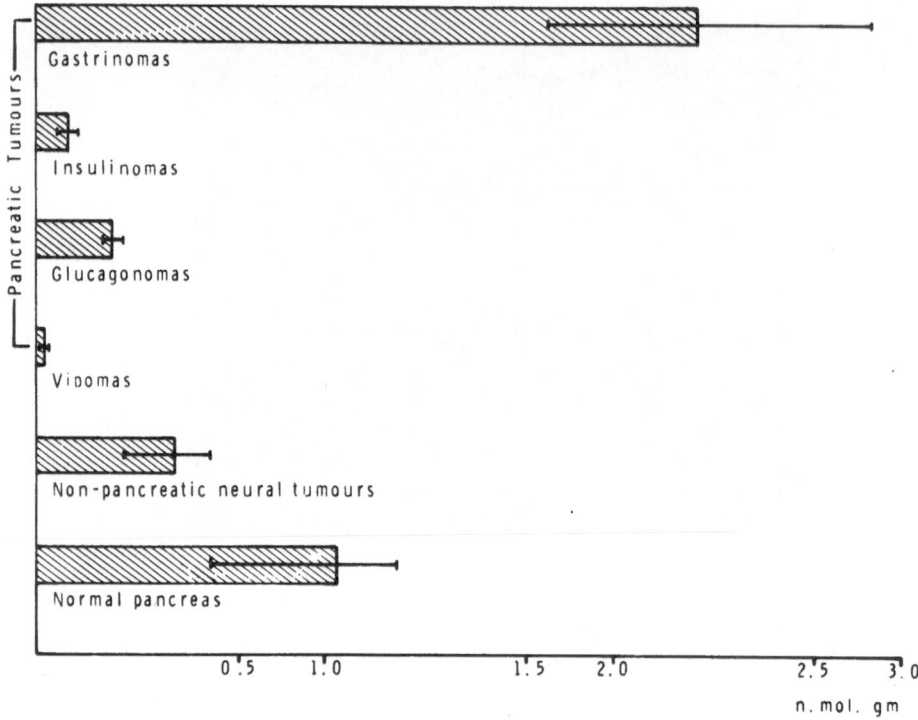

FIGURE 4. RIA somatostatin content of tissue extracts of various types of APUDomas, expressed in nmol/g.

Vasoactive Intestinal Peptide (VIP). VIP was first purified in 1972 by Said and Mutt. It was extracted from porcine gut and shown to have powerful vasodilatory properties, hence the name. VIP was found to have 28 amino acid residues and to be a member of the family of peptides which includes secretin, glucagon, and gastric inhibitory peptide. VIP has a wide variety of actions which include vasodilation, inhibition of pentagastrin- and histamine stimulated gastric acid production and stimulation of insulin release The last two actions are shared with GIP. VIP also has actions similar to those of secretin and glucagon. Like secretin it causes an alkaline juice to flow from the pancreas and like glucagon it causes hyperglycaemia. In vitro experiments show that it stimulates juice flow in the small intestine and that it increases cyclic AMP concentration in the intestinal mucosa. The fact that it is rapidly cleared from the circulation after injection means that it has a high turnover rate and suggests that it does not act as a circulating hormone but that its role is probably that of a local tissue hormone or a neurotransmitter.

VIP is distributed throughout the gastrointestinal tract in large quantities from oesophagus to rectum.[22] The highest concentration is present in the intestinal wall, where immunocytochemistry has localised it to the innervation.[23] Immunoreactive VIP is present in nerves in the myenteric plexus, in the two muscle layers, and in the fine nerve fibres of the submucosa which form an intermingled mesh and run from the myenteric plexus to the inner parts of the submucosa, terminating in close contact with the mucosal epithelium (Fig. 5). VIP is also present in the mucosa in scattered endocrine cells. In the brain the largest quantities are found in the hypothalamus and cerebral cortex. Immunocytochemistry localises VIP to neurons of the cerebral cortex and to cell bodies and nerve fibres of the hypothalamus. It is also present in the fine innervation surrounding blood vessels in the brain and in many other areas. Using the technique of semithin/thin sections we have seen the ultrastructural localisation of VIP in P-type granules with a mean diameter of 1200 Å (Fig. 6). VIP may well be a regulator of the blood vessels and smooth musculature of the intestinal and genital tract walls and may be involved in control of vaginal or intestinal juice flow.

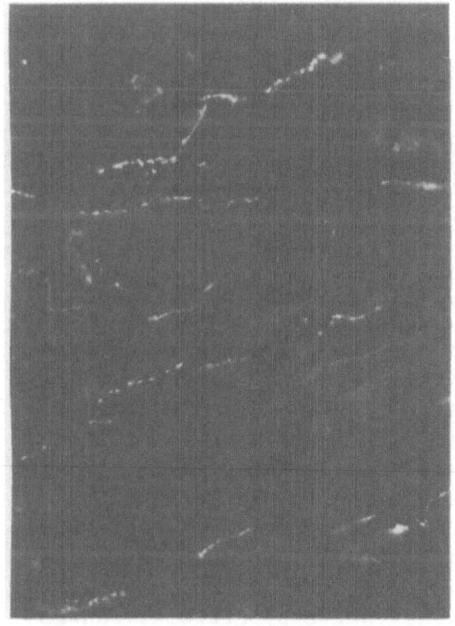

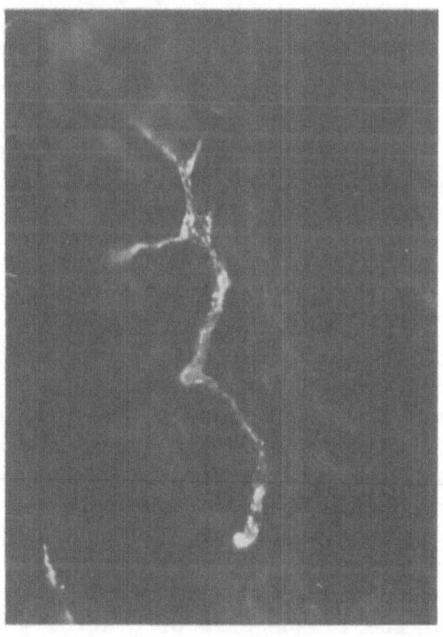

FIGURE 5. a. Human antrum ... numerous VIP nerves are seen in the muscle layer (x708).

b. Human colon. VIP nerves of the submucosa (x1134).

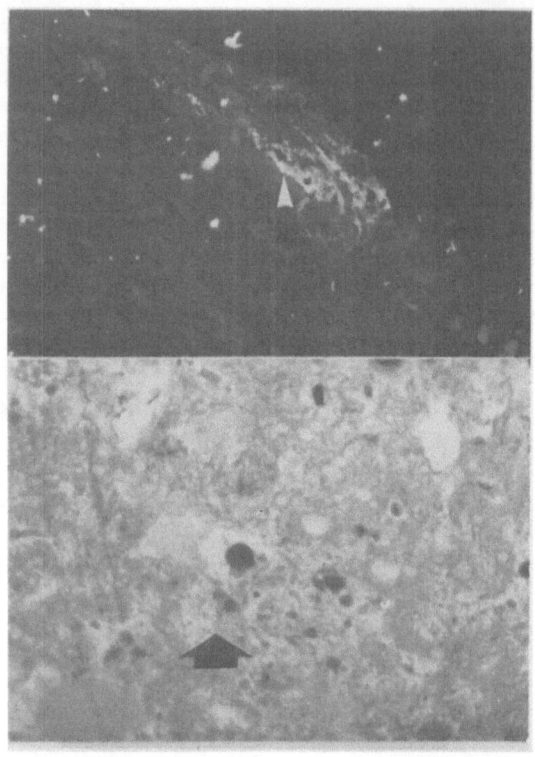

FIGURE 6. Upper - Semithin (1 μm) section showing VIP staining
of the Auerbach's plexus (X625).
            Lower - Consecutive (0.01 μm) ultrathin section showing
VIPergic neurosecretory granules (arrow) (X12,000).

        Abnormalities of the autonomic nerves taking the form of
thickened and split fibres have previously been noted in the area
of the gut affected by Crohn's disease. The overall distribution
of VIPergic nerves in the diseased area in Crohn's disease is
similar to the normal, but the number of nerves is increased, as
is the degree of immunostaining. The fibres are grossly distended
and they form disorganised but densely packed meshes (Fig. 7). The
myenteric plexuses are large, sprawling, and in places infiltrate
the muscle layers. The area covered by VIPergic nerves, estimated
by using an electronic scanning image analyser, is much increased,
occupying 4.15-4.75% of the total tissue area, compared with 1.26%
of the tissue area of control gut. This nerve hypertrophy can be
seen not only in the diseased part but also above and below the
centre of the lesion, indicating a probable primary involvement of
the peptidergic innervation in Crohn's disease.

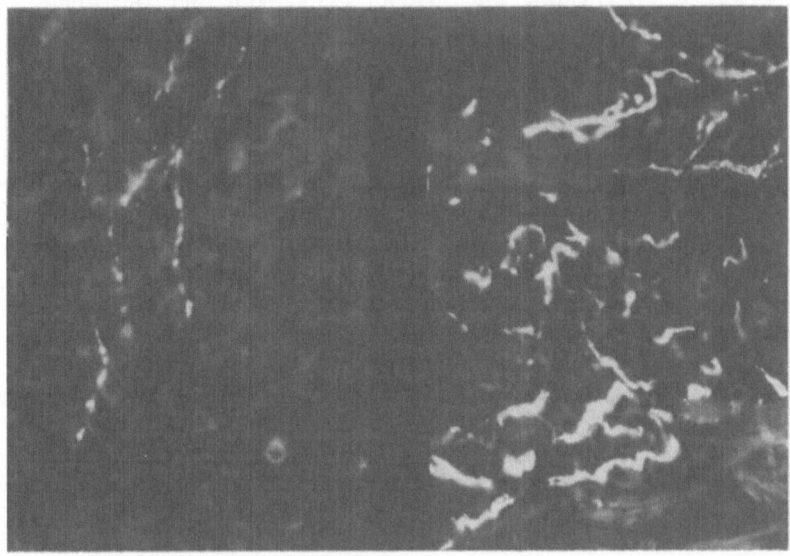

*FIGURE 7.   Left - Normal appearance of VIP nerves in colon (X1079).
              Right - Dramatic hyperplasia of VIP nerves in Crohn's
disease.   Nerve fibres are thick, arranged in disorganised meshes
(X1079).*

One of the most important pathological involvements of VIP is
in the Verner Morrison syndrome.   Verner and Morrison[24] drew attention to the association of severe, often fatal, watery diarrhoea
and islet cell adenoma of the pancreas.   An alternative name, WDHA,
has been given to the syndrome derived from its main features of
watery diarrhoea, hypokalaemia and achlorhydria.   The very severe
diarrhoea and pancreatic tumour association has also suggested the
term pancreatic cholera.   Remission of the symptoms was achieved
after removal of the tumour and it was postulated that the substance
which caused the diarrhoea must be released into the circulation.
The presence of large quantities of VIP in the plasma of patients
suffering from the Verner Morrison syndrome and in the causative
tumour was first shown in 1973.[25]   It is clear now that VIP may be
produced not only by pancreatic tumours but also by neural tumours.
Ganglioneuroblastoma is a common tumour in children and is often
accompanied by intractable diarrhoea.   These non-pancreatic tumours
show, with conventional histological staining, all the characteristics described for ganglioneuroblastomas, or ganglioneuromas with
a wide spectrum of cell differentiation.   Immunohistochemical studies
have shown that a number of cells are reactive to a specific antibody
to VIP.   The degree of reactivity closely correlates with the turnover rate of the tumour.   We have often seen the simultaneous
production of biogenic amines and VIP by ganglioneuroblastomas.

Neurotensin.   Carraway and Leeman extracted and isolated
neurotensin as a peptide from bovine hypothalamas which they
characterised, sequenced and synthesised.   It contains 13 amino
acids and is now widely available in synthetic form.   Neurotensin
stimulates contractions in the gastrointestinal tract.   It inhibits
pentagastrin- but not histamine-stimulated gastric acid secretion.
It appears that neurotensin is involved in the regulation of blood
glucose as it enhances glycogenolysis, stimulates the release of
glucagon, and inhibits the release of insulin.   Despite the fact
that neurotensin was first extracted from the brain, by far the
largest amount is found in the gut.   This is mostly localised in
the ileum, but smaller quantities are found in the jejunum with
minimal amounts in the stomach, duodenum and colon.   Immunocytochemistry, combined with electron microscopy, has localised neurotensin to the newly described large granulated cell of the human
ileum (the N cell) (Fig. 8).[15]   In the brain the highest concentrations of neurotensin are found in the hypothalamus and in the basal
ganglia.

The pathological involvement of neurotensin is difficult to
judge at present as the discovery of neurotensin is so recent.
Neurotensin cells are very easy to stain but as control tissue
from the normal ileal area is not easily available, comparative
studies have only just begun.   Neurotensin seems to be a powerful
hypotensive hormone and a highly significant rise in its serum
levels in the dumping syndrome has recently been shown.   It may

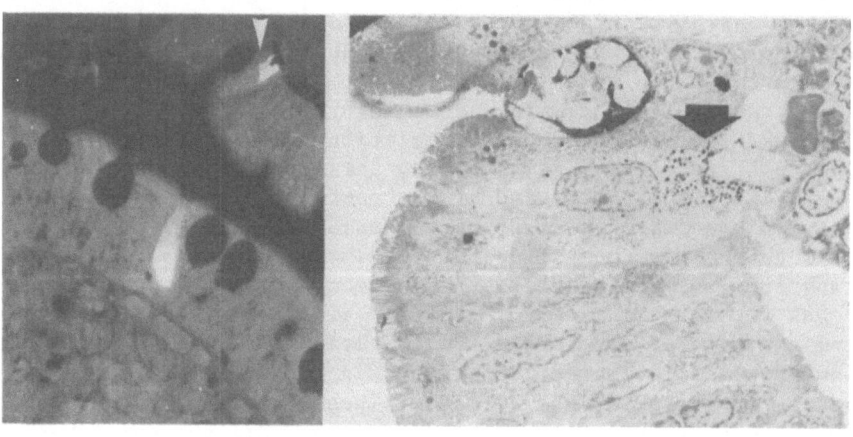

FIGURE 8.   *Left - Human ileum.   Semithin (1 μm) section stained
with antibodies to neurotensin.   Cell marked with an arrow is
compared with photo on the right (X675).*
          *Right - Ultrathin (0.01 μm) consecutive section.
Neurotensin (N) granules are seen in cell marked with arrow (X3000).*

thus be responsible for some of the previously unexplained aspects of this condition and perhaps some other similar 'functional' syndromes.

Cholecystokinin (CCK). The CCK purified by Jorpes et al was composed of 33 amino acids, the entire biological activity residing in the eight C-terminal residues. The last five amino acids of this part of the molecule are identical to the five C-terminal amino acids of gastrin so that these two peptides show a family resemblance. In 1968 Mutt and Jorpes also described a CCK variant of 39 amino acids. More recently a number of smaller forms of CCK have been found, of which a main component resembles the CCK C-terminal octapeptide. In the gut the most important actions of CCK are the promotion of gall bladder contraction and pancreatic enzyme secretion. In addition, CCK has powerful motor effects on the stomach and intestine, but it is not yet clear whether this is physiologically important. CCK greatly enhances the actions of secretin and in large pharmacological doses it stimulates gastric acid secretion in the same way as gastrin, again emphasising the similarities of these two hormones.

Vanderhaegen et al[26] described the presence of a peptide in the vertebrate central nervous system that reacted with antibodies against gastrin. This finding has been confirmed by Dockray,[27] who suggested on the basis of its immunoreactivity with different antisera and its chromatographic elution pattern on Sephadex G25 that the brain factor resembled cholecystokinin (CCK)-like peptides more closely than gastrin-like peptides. Immunocytochemistry using antibodies to CCK reveals the presence of numerous cell bodies throughout the cortical brain matter (3rd and 4th layer) and fewer cells in the subcortical white matter in the rat brain. In the gut the largest concentrations of CCK are found in the jejunum, where it is localised in a well defined secretory APUD cell, the I cell of the EM classification. The octapeptide of CCK has considerable biological potency and a short half-life and thus would match very well with a proposed neurotransmitter role of CCK in the brain. However at present no function has been proposed.

Substance P. A hypotensive and gut-contractile substance, quite distinct from acetylcholine, was extracted 40 years ago by Von Euler and Gaddum from equine brain and intestine. It was the first of the peptides to be considered common to the brain and gut. It was initially called Preparation P, for 'powder' and later referred to as substance P. Despite its early discovery it has only recently been purified, sequenced and synthesised, and has been shown to be an 11-amino acid polypeptide. There are a large number of papers reporting a wide range of pharmacological effects of substance P. In the spinal cord it appears to be involved in the transmission of pain whereas in the brain it may act as modulator of pain sensitivity. In other tissues, such as the salivary glands, substance P stimulates secretion, and in the intestine it increases motor activity. In pharmacological doses substance P inhibits the release of insulin from the pancreas.

Although widely distributed throughout the body, the highest concentrations of substance P occur in the nervous system and in the gut. In the brain the highest concentrations are present in the hypothalamus, pineal gland and substantia nigra, and in the grey matter of the spinal cord.[3] Substance P is localised to primary sensory neurons in a sub-population of small nerve cells with unmyelinated processes. In the spinal ganglia about 20% of the tissue is made up of these cells which are quite different from the somatostatin-containing cells of the corresponding areas. Substance P is also present in the dorsal root of the spinal horn. Substance P can be extracted not only from the gut, but also from almost all the tissues of the body including salivary glands, thyroid, trachea, pancreas, kidney, bladder, prostate, smooth muscle and skin. In the gut the largest concentrations are found in the duodenum and colon. Immunocytochemistry shows that substance P is present in nerve fibres surrounding the cell bodies of the myenteric plexus. It is not only present in the innervation but is also seen in one type of enterochromaffin (EC) cell of the gut mucosa.[17]

High levels of circulating substance P in plasma and tissue have been reported in patients with carcinoid tumours. The cytoplasmic granules of one such tumour were isolated and were shown to be argentaffin and to contain both substance P-like immunoreactivity and 5-HT. This supports the evidence that substance P is normally localised in a population of enterochromaffin cells where it is stored in the cytoplasmic granules together with 5 HT.

Bombesin. Bombesin is a 14 amino acid peptide which was originally extracted from amphibian skin by Erspamer's group in Italy. It has now been found to have powerful systemic, metabolic, gastrointestinal, and pulmonary effects in man and experimental animals. Bombesin stimulates the release of gastrin, and thus gastric acid secretion, independently of gastric pH. It stimulates pancreatic enzyme secretion and intestinal motor activity and enhances gall bladder emptying. These effects can also be seen in the isolated gall bladder and pancreas and thus the actions of bombesin cannot be solely explained as secondary to the release of CCK. Bombesin contracts uterine and urinary tract smooth muscle and also causes renal vasoconstriction.

Bombesin-like immunoreactive material has been found in man in the gastrointestinal tract, lung and brain.[28] In the gastrointestinal tract the largest quantities are found in the antrum and upper duodenum. Immunocytochemistry demonstrates bombesin to

be in fine nerve fibres of the submucosa throughout the entire
length of the intestine and also in endocrine APUD cells, especially
numerous in the avian gastrointestinal tract. In the brain the
largest quantities of bombesin are found in the hypothalamus, thalamus, and limbic cortex, where axons and terminals are specifically
stained by immunocytochemistry. Bombesin-like immunoreactivity
has also been detected in human foetal and neonatal lungs, in one
type of endocrine cell present in the small bronchial and bronchiolar
epithelium, as well as in the innervation.[29]

 Endorphin and enkephalin. One of the most exciting recent
developments in the neuropeptide field has been the discovery of
the endogenous opiates.[30] The first endogenous opiates to be
characterised were met- and leu-enkephalin. Many others have been
found subsequently and they have been shown to belong to a group
of naturally occurring peptides which have similar effects to those
produced by exogenous opiates. This group was called the endorphins,
named from ENDogenous mORPHINe. The amino acid sequences of most
endorphins are contained in the C terminal part of the 91 amino acid
peptide, gamma-lipotropin (gamma LPH). The largest member of the
class is β-endorphin, which corresponds to the sequence 61-91 of
gamma-LPH. Enkephalin corresponds to the sequence 61-65.

 All the endorphins have opiate-like effects.[31] Their central
nervous system distribution suggests that they are involved in the
control of pain and emotion. In the gut, enkephalins are probably
part of the in-built pain control mechanism as well as being, like
morphine, modulators of intestinal transit and muscle tone. The
association of the endorphins, especially of the small enkephalins,
with the brain synaptosomal fractions, their very short half-life
and the very small size of their molecules suggests that they may
play a neurotransmitter or neuromodulator role. In the gastrointestinal tract enkephalin may act physiologically in the same way as
the opiates. Morphine is a classical and very effective remedy in
the treatment of diarrhoea and acts by increasing intestinal and
gastric muscle tone and by slowing intestinal transit. However, it
is already obvious that the endorphins may serve many purposes in
addition to those suggested by our age-old knowledge of the opiates.

 Surprisingly, the distribution of the pentapeptide enkephalins
is completely different from that of the larger endorphins. The
enkephalins are present in the brain and their distribution closely
parallels that of the brain opiate receptors. Both are found in
particularly high concentration in areas associated with the sensory
input of pain signals. At present, enkephalin together with substance P appears to represent a major system, with cell bodies in
more than 20 areas of the central nervous system including the
spinal cord. Endorphin has a rather more limited distribution,
mainly within the hypothalamus but also extending into more rostral

areas, into the thalamus and the mesencephalon. α-endorphin positive cell bodies are not identical to any of the numerous enkephalin-positive cell groups and it appears that enkephalin and α-endorphin are present in different neurons. Enkephalins are also widely distributed in the gastrointestinal tract and are concentrated in the antrum and upper duodenum. By immuno-cytochemistry enkephalin-like immunoreactivity can be seen in fine nerve fibres in the submucosa and is associated with certain endocrine cells in the antral mucosa. The larger endorphins, on the other hand, are not present in the gastrointestinal tract in any of the species so far investigated. They are found particularly in the anterior and intermediate lobes of the pituitary, where they are produced by the corticotrophic cells.[30]

PATHOLOGY OF THE AUTONOMIC INNERVATION OF THE GUT

In the normal intestine stimulation of the parasympathetic nerves produces contraction and stimulation of the adrenergic nerves results in relaxation of the smooth muscle. Apart from these well-known opposing effectors a third, peptide mediated, mechanism has recently been attracting enormous interest. It has been termed the non-adrenergic inhibitory nervous system of the gut on the basis of in-vitro experiments. When adrenergic and cholinergic effects are prevented by specific blocking agents, electrical stimulation of intestinal nerves causes a dramatic muscle relaxation. This non-adrenergic, non-cholinergic system is thought to be responsible for the relaxation phase of peri-stalsis and for the relaxation of the internal anal sphincter. The neurons of the system are present in the myenteric plexus of Auerbach, together with the neurons of the parasympathetic system, and probably correspond with the P-type of nerves of the autonomic system to which we have already referred. Malfunction of this system is likely to produce many pathological features. For ex-ample, it has been demonstrated that after denervation, hyperplasia of the smooth muscle coat occurs. The hypertrophy is very striking and much greater than that which follows an organic obstruction. This hypertrophy, and possibly an additional obstruction element, stretches the muscle wall of the non-contracted segment, resulting in a myogenic movement known as segmentation. Many gastrointestinal diseases, manifested principally by defects in gut motility, are due to lesions of the autonomic innervation. Among these are Hirschsprung's disease, achalasia of the cardia, hypertrophic pyloric stenosis, small intestinal myenteric plexus deficiency (congenital or acquired) most commonly seen in the duodenum, Chagas' disease, and purgative abuse. Following denervation there is hypertrophy of the muscle of the sphincter regions which will in turn increase the obstructive element.

Hirschsprung's Disease. This is also called congenital mega-colon and is characterised by a total absence of neurons in the plexus at the level of the contracted segment. In the dilated part the neurons are abnormal in morphology and function. The muscle coat of both the distal contracted and the proximal dilated portions is extremely thick. The cause of the symptoms is not only the obstruction of the distal contracted segment but also the absence of normally coordinated muscle contractions above the lesion. The pathophysiology of Hirschsprung's disease is contro-versial. There is no accepted explanation as to why the absence of ganglion cells in the bowel wall results in the contraction of that segment of the bowel. An absence of the ganglion cells, which are thought to be part of the parasympathetic system, would be expected to leave the adrenergic system unopposed, resulting in relaxation of the smooth muscle. The diseased part of the bowel does not relax in response to electrical field stimulation, whereas the normal part of the bowel does. This finding indicates an ab-sence of, or a defect in, the non-adrenergic inhibitory autonomic nervous system in the spastic portion of the bowel. A fully func-tioning adrenergic system can be demonstrated in the abnormal colon with almost complete aganglionosis and no contraction following field stimulation. This indicates that the disease can probably exist in the presence of adrenergic innervation and does not support the theory of adrenergic deficiency. Electrical stimulation of tissue pre-incubated with radioactive adenosine has been shown to release radioactive ATP, and this has been taken as evidence that ATP is a mediator of the non-adrenergic inhibitory system. However Richardson[32], in experiments in mice suffering from congenital Hirschsprung's disease, has failed to find an increase in release of radioactive adenosine after electrical stimulation. His findings thus do not support the purinergic theory. Baumgarten et al[33] described a decreased number of P-type fibres in the contracted part of the bowel in Hirschsprung's disease and a possible increase or at least a normal number in the dilated portion. Two peptidergic hormones thought to be involved in the peristaltic movements of the gut (VIP and substance P) have been shown to be markedly decreased in the contracted segment of the diseased colon and possibly elevated or normal in the dilated portions. These studies are preliminary and further confirmation by the use of other methods is urgently needed.

Chagas' Disease. It is well established that in man and laboratory animals *Trypanasoma cruzi* causes lesions of the autonomic nervous system and reduction of the number of neurons, especially in the heart, oesophagus and colon. Peristalsis is greatly affected. There is very little morphological, physiological, and pharmacologic-al information on abnormalities of the peptidergic innervation (non-adrenergic inhibitory system of the gut) in Chagas' disease. There is some evidence which seems to indicate that the levels of substance P in the contracted segments are lower than in the dilated segments.

However, this has not been fully substantiated. No information
as to the levels of VIP is yet available.

CONCLUSION

It can be seen from the foregoing discussion that the recognition of this large new component of the autonomic nervous system
and its study by a combination of immunocytochemistry and radioimmunoassay has yielded some interesting information and is beginning to explain many previously puzzling aspects of gut function
and pathology. We have only mentioned a few instances of gut
pathological conditions which may involve the peptidergic nervous
supply. In many cases its role is still entirely speculative and
in others research is in progress. The large number of other
diseases which may be affected emphasises the importance of the
system.

Through advances in the techniques of radioimmunoassay and
immunocytochemistry physiologists have recently come to realise
that the function of the gut is greatly influenced by powerful
peptides such as VIP, somatostatin and substance P, which are
secreted from an extensive component of the autonomic nervous
system and are unaffected by cholinergic or adrenergic blockade.
These peptides apparently act locally and affect such functions
as gut motility. We may hope that future work will lead us to a
fuller understanding of the mechanisms of gut control.

Acknowledgements

This work was carried out with the help of grants from the
Medical Research Council, Cancer Research Campaign and Wellcome
Trust.

REFERENCES

1. Euler US von, Gaddum JH: An unidentified depressor substance
        in certain tissue extracts. J. Physiol. (Lond.) 192:
        74-87, 1931

2. Zetler G: The peptidergic neuron - a working hypothesis.
        Biochem. Pharmacol. 25: 1817-1818, 1976

3. Hökfelt T, Schultzberg M, Johansson O, Ljungdahl A, Elfvin L,
        Elde R, Terenius L, Nilsson G, Said S, Goldstein M:
        Central and peripheral peptide producing neurons. In:
        Gut Hormones, edited by SR Bloom, Churchill Livingstone,
        Edinburgh, 1978, p. 423-433

4. Dodgiel AS: Über der bau der ganglien in den geflechten des darmes under der gallenblase des menschen U. der Säugetiere. Arch. f. Anat. u. Phys. Anat. Abth., 1899

5. Baumgarten HG, Holstein AF, Owman Ch: Auerbach's plexus of mammals and man - electron microscopical identification of three different types of neuronal processes in myenteric ganglia of the large intestine from rhesus monkeys, guinea-pigs and man. Z. Zellsforsch 106: 376-397, 1970

6. Bishop A, Polak, JM, Buchan AMJ, Bloom SR, Pearse AGE: A third division of the autonomic nervous system. An important element in gut control. Gut 18: A416, 1977

7. Burnstock G: Purinergic nerves. Pharmacol. Rev. 24: 509-581, 1972

8. Bloom SR, Polak JM: Peptidergic versus purinergic. The Lancet I: 93, 1978

9. Pearse AGE: Common cytochemical properties of cells producing polypeptide hormones, with particular reference to calcitonin and the thyroid C cells. Vet. Rec. 79: 587-590, 1966

10. Pearse AGE, Takor Takor T: Neuroendocrine embryology and the APUD concept. Clin. Endocrinol. 5 (suppl.): 229-234, 1976

11. Feyrter F: Über diffuse endokrine epitheliale organe. J. A. Barth, Leipzig, 1938, p. 6-17

12. Pearse AGE, Polak JM: Bifunctional reagents as vapour- and liquid-phase fixatives for immunohistochemistry. Histochem. J. 7: 179-186, 1975

13. Bishop AE, Polak JR, Bloom SR, Pearse AGE: A new universal technique for the immunocytochemical localisation of peptidergic innervation. J. Endocrinol., in press

14. Sternberger L: The unlabelled antibody enzyme method. Prentice Hall Engelwood Cliff, 1974, p. 142-161

15. Polak JM, Sullivan SN, Bloom SR, Facer P, Pearse AGE: Enkephalin-like immunoreactivity in the human gastro-intestinal tract. The Lancet I: 972-974, 1977

16. Solcia E, Polak JM, Pearse AGE, Forssman WG, Larsson LI,
     Sundler F, Lechago J, Grimelius L, Fujita T, Creutzfeldt W,
     Gepts W, Falkmer S, LeFranc G, Heitz P, Hage E, Buchan AMJ,
     Bloom SR, Grossman MI: Lausanne 1977 classification of
     gastroenteropancreatic endocrine cells. In: Gut Hormones,
     edited by SR Bloom, Churchill Livingstone, Edinburgh,
     1978, p. 40-48

17. Pearse AGE, Polak JM: Immunocytochemical localisation of
     substance P in mammalian intestine. Histochemistry 41:
     373-375, 1975

18. Hokfelt T, Elfvin LG, Elde R, Schultzberg M, Goldstein M,
     Luft R: Occurrence of somatostatin-like immunoreactivity
     in some peripheral sympathetic noradrenergic neurons.
     Proc. Natl. Acad. Sci. USA 74: 3587-3591, 1977

19. Arimura A, Coy DH, Chihara M, Fernandez-Durango R, Samols E,
     Chihara K, Meyers CA, Schally AV: Somatostatin. In:
     Gut Hormones, edited by SR Bloom, Churchill Livingstone,
     Edinburgh, 1978, p. 437-445

20. Ganda OmP, Weir GC, Soeldner JS, Legg MA, Chick WL, Patel YC,
     Ebeid AM, Gabbay KH, Reichlin S: Somatostatinoma - A
     somatostatin-containing tumour of the endocrine pancreas.
     N. Engl. J. Med. 296: 963-967, 1977

21. Larsson LI, Hirsch MA, Holst JJ, Ingemansson S, Kuhl C,
     Jensen SL, Lundqvist G, Rehfeld JF: Pancreatic somato-
     statinoma - Clinical features and physiological implications
     Lancet I: 66-668, 1977

22. Polak JM, Bloom SR: The endocrine background. Scientific
     Foundations of Gastroenterology, edited by W Sizcus and
     AN Smith, Williams Heineman Medical Books, 1978

23. Bryant MG, Bloom SR, Polak JM, Albuquerque RH, Modlin I,
     Pearse AGE: Possible dual role for vasoactive intestinal
     peptide as gastrointestinal hormone and neurotransmitter
     substance. The Lancet I: 991-994, 1976

24. Verner JV, Morrison AB: Endocrine pancreatic islet disease
     with diarrhoea. Arch. Intern. Med. 133: 492-500, 1958

25. Bloom SR, Polak JM, Pearse AGE: Vasoactive intestinal peptide
     and watery-diarrhoea syndrome. The Lancet II: 14-16, 1973

26. Van der Haghan JJ, Signean J, Gepts W: New peptide in the
     central nervous system reacting with anti-gastrin
     antibodies. Nature 257: 604-605, 1975

27. Dockray GJ: Immunochemical evidence of cholecystokinin-like peptides in brain. Nature 264: 568-560, 1976

28. Polak JM, Buchan AMJ, Czykowska W, Solcia E, Bloom SR, Brown MI, Pearse AGE: Bombesin in the gut. In: Gut Hormones, edited by SR Bloom, Churchill Livingstone, Edinburgh, 1978, p. 540-542

29. Wharton J, Polak JM, Bloom SR, Ghatei MA, Solcia E, Brown MR, Pearse AGE: Bombesin-like immunoreactivity in the lung. Nature, in press

30. Polak JM, Sullivan SN, Buchan AMJ, Bloom SR, Facer P, Hudson D, Pearse AGE: Endorphins. In: Gut Hormones, ed. SR Bloom, Churchill Livingstone, Edinburgh, 1978, p. 501

31. Feldberg W: Pharmacology of the central actions of endorphins. In: Gut Hormones, ed. SR Bloom, Churchill Livingstone, Edinburgh, 1978, p. 495

32. Richardson J: Pharmacologic studies of Hirschsprung's disease on a murine model. J. Pediatr. Surg. 10: 875-884, 1975

33. Baumgarten HG, Holstein AF, Stelzner F: Nervous elements in the human colon of Hirschsprung's disease. Virchows Arch. Pathol. Anat. 358: 113-136, 1973

POLYPEPTIDES OF THE AMPHIBIAN SKIN ACTIVE ON THE GUT AND THEIR

MAMMALIAN COUNTERPARTS

V. Erspamer, P. Melchiorri, C. Falconieri Erspamer, and L. Negri

Institutes of Medical Pharmacology I and III

University of Rome, Italy

As repeatedly stated, the amphibian skin represents an un-exhaustible mine of biogenic amines and polypeptides. So far, more than ten different indolealkylamines, five imidazolealkylamines and more than twenty active peptides have been isolated in a pure form and, after elucidation of their structure, reproduced by synthesis. Another large group of amines and peptides (altogether more than twenty) has already been identified and is in part the object of studies now in progress.

However, in the context of this Symposium, our attention will be directed only to those few peptide families of the amphibian skin which have their counterpart in the mammalian gastrointestinal tract. It will be seen that the spectrum of activity of amphibian peptides completely covers that of the corresponding gut peptides and, as a consequence, that amphibian peptides may be conveniently used, in experimental and clinical work, in the place of mammalian peptides. Moreover, emphasis will be laid on the fact that studies carried out on amphibian skin may represent the premise for the discovery of unexpected gastrointestinal peptides in mammals and birds.

The three peptide families which will be discussed in this re-port are the bombesin-like peptides and, in less detail, the phy-salaemin-like peptides or tachykinins and the caerulein-like peptides.

BOMBESIN-LIKE PEPTIDES

The family of bombesin-like peptides will be considered first because the physiological and pharmacological actions of amphibian bombesin-like peptides, with their clinical implications, are presently the object of considerable attention and of fervid research and because data presented in this symposium on avian and mammalian gastrointestinal bombesin-like peptides are in part the first-hand result of research carried out by our group during the past few months.

Five polypeptides belonging to the bombesin-family have been so far isolated in a pure form, reproduced by synthesis, and submitted to a pharmacological study.

Bombesin
      Pyr-Gln-Arg-Leu-Gly-Asn-Gln-Trp-Ala-Val-Gly-His-Leu-Met-NH$_2$
Alytesin
      Pyr-Gly-Arg-Leu-Gly-Thr-Gln-Trp-Ala-Val-Gly-His-Leu-Met-NH$_2$
Ranatensin
      Pyr--------------Val-Pro-Gln-Trp-Ala-Val-Gly-His-Phe-Met-NH$_2$
Litorin
      Pyr$_2$--------------------Gln-Trp-Ala-Val-Gly-His-Phe-Met-NH$_2$
Glu(OMe)^2-Litorin
      Pyr----------------Glu(OMe)-Trp-Ala-Val-Gly-His-Phe-Met-NH$_2$

Additional bombesin-like peptides have been traced in amphibians of Australia, Papua, New Guinea, and Japan, but their complete amino-acid sequences still await elucidation, and no pharmacological studies have been carried out on them.

The bombesins represent a unique peptide family, possessing a very peculiar spectrum of biological activity. Several of their actions are indirect, as they are due to release and possibly inhibition of release of other active peptides. However, other actions, especially those on vascular and extravascular smooth muscle, are apparently direct.[1,2]

It has been unequivocally demonstrated that in the chicken, dog, and man the gastric secretagogue activity of bombesin is due to release of gastrin from the G-cells of the antral mucosa. As is the case with other gastrin releasers, the releasing effect of bombesin also was inhibited by somatostatin and occasionally by secretin

It would seem that gastrin-17, the gastrin form that appears to be predominant in the G-cells of the antral mucosa, is more promptly releasable than gastrin-34, which is mainly in extra-antral tissues[2].

Strong evidence suggests that in the dog bombesin releases also cholecystokinin (CCK) from the duodenal mucosa. In fact, the intravenous infusion of bombesin elicited contraction of the gall bladder, relaxation of the choledochoduodenal junction, and stimulation of flow of a pancreatic juice which was poor in bicarbonate but rich in protein and enzymes. Cholinergic mechanisms seem to play a permissive effect on CCK release by bombesin, as on CCK release by peptone in the intestine[2].

Following removal in the dog of the antrum and small intestine, bombesin failed to show not only any change in the serum gastrin level but also any stimulation of the pancreatic secretion. Somatostatin, in its turn, caused the inhibition of pancreatic protein secretion by bombesin but failed to affect the response to CCK-octapeptide[3].

Finally, Fender et al.[4] succeeded in demonstrating, by bioassay and radioimmunoassay, that in the dog release of gastrin by bombesin was accompanied by release of CCK, as shown by the increase in plasma levels of immunoreactive CCK from 107 ± 7 to 165 ± 16 pg/ml.

However, it should be stressed that caution is necessary before definitively accepting the above results, because reliable and simple methods for radioimmunoassay of CCK in blood and tissues are not yet available, in spite of the promising results obtained by Rehfeld[5] and by Rayford et al.[6]

At any rate, whereas the effects of bombesin on the dog gall bladder and pancreas seem to be mediated, at least to a great extent, by the release of CCK, an alternative mechanism of action of bombesin, direct and not mediated by CCK, may be postulated in other animal species.

In fact, while bombesin and litorin were virtually inactive on the isolated dog gall bladder, they showed 1 to 4% of the action of caerulein on the isolated guinea-pig gall bladder, and 5 to 20% of the action of caerulein on the in situ guinea-pig gall bladder. Upon intravenous injection this effect was immediate and tachyphylaxis was apparently lacking.

On the other hand, bombesin and litorin increased the output of amylase not only from fragments of rat pancreas but also from dispersed acinar cells of the guinea-pig pancreas. In this preparation, the peptides also caused increase in Ca^{45} outflux, increase in release of membrane-bound Ca^{45} and stimulation of cellular cyclic GMP. Cyclic AMP was not altered[7,8].

Also in keeping with a direct effect of bombesin on the rat acinar tissue is the finding by Linari[9] that enterectomy did not affect bombesin-stimulated pancreatic secretion.

The releasing or release inhibiting activity of bombesin does not seem to be limited to gastrin and CCK. In fact, the peptide proved to be a potent stimulant of pancreatic polypeptide release in the dog. A 2 hr infusion of 1 ng/kg-min of bombesin produced a mean increment in plasma pancreatic polypeptide concentration of 26.4 ± 17 pmol liter^{-1}, and an infusion of 4 ng/kg-min an increment of 214 ± 106 pmol liter^{-1}. Response to a meal in the same dogs over the same period was 170.6 ± 34 pmol liter^{-1}.[10]

However, it remains to be investigated to what extent release of pancreatic polypeptide is actually a direct effect of bombesin on the pancreatic F cells and not a consequence of CCK release. In fact, Bloom and Polak[11] found that pancreatic polypeptide release was greatly stimulated by small doses of the CCK analogue caerulein.

To complete the picture of the effects of bombesin on endocrine cells, it should be remembered that bombesin stimulated both insulin and glucagon release in man[12] and inhibited the release of VIP in dog, cat and man, as shown by a fall of plasma VIP immunoreactivity levels.[13]

Another important target of bombesin in the gastrointestinal tract is the musculature. Whether we are dealing with direct effects of the polypeptide on the smooth muscle or with effects again mediated or modulated by other agents is a problem which remains to be solved.

In conscious dogs provided with electrodes chronically implant-ed on the serosal surface of different gastrointestinal segments, bombesin produced a significant increase in the frequency of pace-setter potentials in antrum, duodenum, jejunum, and ileum. In the duodenum and jejunum, the increase in frequency showed linear cor-relation with the reduction of pacesetter potential amplitude. The propagation velocity of the pacesetter potentials was approxi-mately halved. At high infusion rates of bombesin, disappearance of rhythmicity of contractions in the duodenum and upper jejunum was observed, with appearance of an irregular sequence of slow and small potentials. This could be interpreted as due to the failure of coupling between relaxation oscillators over critical maximal frequencies. During recovery, electrical activity gradually return-ed towards the preinfusion pattern.[14]

Also of considerable interest was the suppressive effect of bombesin on the interdigestive myoelectric complexes of the dog intestine, which were reduced in frequency and duration.

The mechanical counterpart of the above changes of electrical activity was the disappearance of motility in the upper small intes-tine, sometimes followed by a post-bombesin phase of hyperactivity.

The human and the dog stomach responded to bombesin infusion (threshold in the dog 0.3 μg/kg-h) with contraction of the pylorus and the antrum accompanied by complete loss of basal motility and relaxation of the body and the fundus, as shown by disappearance of segmentation in man and sharp reduction of intragastric pressure in the dog. Discontinuing the infusion was immediately followed by relaxation of the pylorus and hyperperistalsis of the body and the fundus.[15,16,17]

Whereas, as shown below, the great majority of peptides found in amphibian skin have their counterparts in the mammalian organism, a mammalian duplicate of the amphibian bombesins was apparently lacking. This simply depended on the fact that bombesin-like peptides had never been sought in the gut.

Amphibian skin	Mammalian tissues or blood
Physalaemin-like peptides (Tachykinins) ————	Substance P
Caeruleins ————————————————————	CCK and Gastrins
Bradykinin-like peptides ————————————	Bradykinins
Angiotensin-II-like peptide —————————	Angiotensin-II
Pyr-His-Pro-NH$_2$ —————————————————	TRH
Xenopsin ————————————————————————	Neurotensin
Bombesins ————————————————————————	Bombesin-like intestinal (and brain) peptides

Strong evidence demonstrates that bombesin-like peptides occur in the gastrointestinal tract of mammals and birds, especially in the stomach and in the upper intestine. In fact:

a) extracts of gastric and intestinal mucosa of mammals and birds contain substances which react to an antibody to bombesin prepared in the rabbit, using bombesin conjugated with bovine serum albumin. A decapeptide of bombesin having at its N-terminus a tyrosyl residue labelled with radioactive iodine was used as a tracer. The antibody did not present cross reactions with human gastrin I, pentagastrin, caerulein, CCK, secretin, glucagon, VIP, GIP, or somatostatin. Limits of sensitivity of three antibombesin sera prepared in this laboratory were 20-30, 30-40 and 50 pg/ml respectively. In gastric extracts of different species, concentrations of immunoreactive bombesin-like activity, expressed as bombesin, ranged between 10-50 (dog stomach) and 1000-2000 ng/g (chicken proventriculus).[18]

The stomach of Rana pipiens contained 65 ng/g of bombesin-like immunoreactivity.[19]

b) purified extracts of chicken proventriculus were active on all the in vitro and in vivo test preparations on which bombesin and litorin were active. They stimulated isolated preparations of rat uterus, large intestine and urinary bladder, guinea-pig ileum and urinary bladder, kitten small intestine; they produced a moderate litorin-like rise of the dog blood pressure; they caused in the dog a release of gastrin with ensuing stimulation of gastric acid secretion; they contracted the in situ gall bladder of the guinea-pig and the dog and elicited in the dog and the rat a pancreatic secretion rich in amylase; finally, they provoked changes in the electrical activity of the dog stomach and intestine similar to those elicited by bombesin.

Concentrations of bombesin-like activity in some mammalian gastrointestinal tissues, as estimated on the rat uterus preparation and expressed as litorin, ranged between 10 and 100 ng/g. The bombesin-like activity of human gastric mucosa was of the order of 20-40 ng/g. So far, no bombesin-like activity has been detected either by bioassay or by radioimmunoassay outside the gastrointestinal tract, with the important exception of the central nervous system.

Preliminary experiments gave the following values of bombesin-like immunoreactivity in the CNS:

rat (5 animals)	whole brain	20-45 ng/g
	hypothalamus	150-250 ng/g
pig (3 mini-pigs, mean values)	cerebral cortex	15 ng/g
	cerebellum	23 ng/g
	hypothalamus	85 ng/g
rabbit (2 animals)	cerebral cortex	5-10 ng/g
	cerebellum	25-40 ng/g
	hypothalamus	105-130 ng/g

In our systematic search for bombesin-like activity in extracts of gastrointestinal tract, of particular importance was the finding that the avian proventriculus (chicken, turkey, pigeon) was exceptionally rich in this activity. Values ranged between 150-600 ng litorin per g fresh tissue, by bioassay, and between 500 and 2000 ng bombesin by radioimmunoassay.

c) bombesin-like immunoreactivity could be shown also in blood. In man it rose after a standard meal from very low fasting levels to levels comparable with those of gastrin. Release of the bombesin-like peptide preceded release of gastrin[18].

On the other hand Walsh and Holmquist[19]demonstrated that whereas bombesin-like immunoreactivity could not be measured in normal rabbit serum, boiled serum of an immunized rabbit contained 1.6 pmol, ml, indicating that circulating anti-bombesin was complexed with immumoreactive peptide and suggesting an endogenous source of such peptide in the rabbit.

d) the last important piece of evidence in favour of the occurrence of a bombesin-like peptide in the gut has been presented
by Polak et al.[20] who could see in all areas of the human gastrointestinal mucosa investigated cells specifically stained with antibodies to bombesin. They were predominantly localized in the basal
parts of the glands and were particularly bright and rather more
numerous in the duodenal mucosa. Full quenching was observed with
the synthetic tyrosine nonapeptide of bombesin, though none was
seen, even at high concentrations, with VIP, somatostatin, substance
P and gastrin.

Van Noorden et al.[21] have recently shown that bombesin-like
immunoreactivity may occur also in the myenteric plexus.

Our group is presently engaged in the difficult attempt
to isolate the avian bombesin-like peptide and to elucidate its
structure. The starting material will be the chicken proventriculus,
extracted with boiling 0.15 M HCl. Following passage through Elgalite
resin, precipitation with alcohol, chromatography on alumina columns
and gel filtration on Sephadex columns a good purification of the
peptide was obtained, with satisfactory yields. The presently available peptide preparation may be considered biologically pure, i.e.
free from active contaminants.

Moreover, extraction and purification procedures used in this
study demonstrated that the bombesin-like peptide is present in the
chicken proventriculus in at least two, and possibly three, forms,
differing from each other by their molecular size.

In fact, when the ground proventricular tissue was first extracted with methanol and then boiled with 0.15 M HCl, it could be
seen that after exhaustive extraction of the tissue with alcohol,
acid treatment was capable of releasing an additional amount of
bombesin-like activity, exceeding that extracted by the neutral
solvent. This would mean that in the chicken proventriculus most of
the bombesin-like peptide is in some way included in a protein pre-
cipitated by methanol. Upon acid hydrolysis (and possibly also
following trypsin digestion) of the big bombesin-like peptide, a
smaller, soluble and active peptide is liberated.

However, things may be even more complicated. In fact, whereas gel filtration through Sephadex G25 using 0.1 M acetic or formic
acid (pH 2.5-4) caused the appearance in the effluent of a single
activity peak, coinciding with that of litorin, gel filtration at
neutral pH, using 0.15 M phosphate buffer, caused the appearance of
two activity peaks. They were distinguishable from each other not
only by the elution profile, but also by their biological activity:
that of the first peak was more bombesin-like, that of the second
peak decidedly litorin-like.

It is evident that Sephadex experiments must be extended and other separation procedures must be tried before drawing definitive conclusions on the polymorphism of the bombesin-like peptides in avian proventriculus and on the relations between the different forms.

Similarly it is obvious that all the above results are presently valid only for the chicken proventriculus. Although preliminary experiments seem to indicate that even in mammals, including man, the intestinal bombesin-like peptide may occur in various forms, available data are exceedingly scanty to establish a comparison with the chicken proventriculus.

Concerning the possible physiological role of the bombesin-like intestinal peptide it is tempting to suggest, on the basis of pharmacological studies carried out with bombesin and litorin, that it may intervene:

a) in the regulation of gastrin release and hence of gastric acid secretion. Whether release of the bombesin-like peptide is an obligatory link in the chain of events which cause acid secretion following a meal remains to be established. In our few experiments the postprandial peak of bombesin-like radioimmunoactivity preceded the peak of gastrinaemia. Thompson et al.[22], in their turn, believe that release of antral gastrin elicited by infusion of liver extract into the small intestine of dogs might be due to release of a bombesin-like peptide.

b) in the control of gall bladder contraction and of pancreatic enzyme secretion. However, it remains to be ascertained whether the bombesin-like peptide acts directly on the effector cells or indirectly, through release of CCK. It is possible that the mechanism of action of the peptide is different in different species.

c) in the control of gastrointestinal motility. Bombesin and litorin have a tremendous effect on both the slow and the fast electrical activity of the gut, potently affecting pacesetter potentials, postprandial spike activity and interdigestive spike complexe with a general depression of motor activity. A thorough study of these bombesin effects and of the interactions coming into play between the bombesin-like peptide on the one side and the tachykinins (substance P), CCK (or caerulein) and motilin, on the other side, would be highly interesting and rewarding.

d) in eliciting all the secretory and motor effects which are attributed and will be attributed in the future to pancreatic polypeptide: augmentation of pancreatic bicarbonate and volume flow induced by secretin, relaxation of the gall bladder and increase of choledochal pressure, stimulation of gut motility, inhibition of pentagastrin-induced gastric acid secretion[23]. As previously stated bombesin was found to be a potent releaser of pancreatic polypeptide.

It is possible that stimulation of gastrin, CCK and pancreatic polypeptide release as well as inhibition of VIP release by the bombesin-like intestinal peptide may cause, in different cases, independent, additive or antagonistic effects.

To give a more complete panorama of the extraordinary versatility of the bombesin-like peptides it should be remembered that bombesin displays a formidable spasmogenic action on the afferent glomerular bed of the dog kidney with ensuing liberation of renin and activation of the renin-antgiotensin system, and that the same bombesin is one of the most potent peptides reported to affect the central nervous system. In fact the polypeptide caused, following injection into the cisterna magna, striking hypothermia in cold exposed rats, pointing to its possible involvement in producing the temperature changes observed in hibernating animals, and was a highly potent analgesic agent in some antinociceptive tests, pointing to a possible opiate dependent step in the mechanism of its action, not only in the mechanism of the central actions, but also in the peripheral actions of the peptide, especially on electrical and motor activity of the gut[24,25].

PHYSALAEMIN-LIKE PEPTIDES OR TACHYKININS

The amphibian (and molluscan) tachykinins, represented by the endecapeptides physalaemin, uperolein and eledoisin, the decapeptide phyllomedusin and the dodecapeptide kassinin, have aroused fresh attention after elucidation of the sequential structure of substance P, their counterpart in mammalian tissues.

Physalaemin
    Pyr-Ala-Asp-Pro-Asn-Lys-Phe-Tyr-Gly-Leu-Met-NH$_2$
Uperolein
    Pyr-Pro-Asp-Pro-Asn-Ala-Phe-Tyr-Gly-Leu-Met-NH$_2$
Phyllomedusin
    Pyr-----Asn-Pro-Asn-Arg-Phe-Ile-Gly-Leu-Met-NH$_2$
Kassinin
  Asp-Val-Pro-Lys-Ser-Asp-Gln-Phe-Val-Gly-Leu-Met-NH$_2$
Eledoisin
    Pyr-Pro-Ser-Lys-Asp-Ala-Phe-Ile-Gly-Leu-Met-NH$_2$
Substance P
    Arg-Pro-Lys-Pro-Gln-Gln-Phe-Phe-Gly-Leu-Met-NH$_2$

Substance P has a rather diffuse distribution in the central and peripheral nervous system, where it is localized both in epithelial cells (a particular type of enterochromaffin cells) and in nervous structures. It has a spectrum of biological activity very similar to that of the amphibian tachykinins[1].

In regard to the gastrointestinal tract the tachykinins have three main target structures: the salivary glands, the gastrointestinal smooth muscle and the splanchnic vascular bed.

Physalaemin, eledoisin and, to a lesser degree, substance P potently stimulated the secretion of salivary and lachrymal glands and moderately affected pancreatic secretion. Lachrymal stimulation has been profitably used in the topical treatment of Sjøegren's syndrome and other types of keratoconjunctivitis sicca. Recent research on dispersed acinar cells of the guinea-pig pancreas has shown that physalaemin and eledoisin moderately increased outflux of cellular ^{45}Ca, release of membrane-bound ^{45}Ca, both initial and steady-state uptake of ^{45}Ca, cellular cyclic GMP and amylase output[8]. On the other hand, Rudich and Butscher[26] could demonstrate that substance P and eledoisin caused a rapid, concentration-dependent increase in K^+ efflux and amylase release from slices of rat parotid glands. No effects were discerned on cyclic nucleotides. Ca^{++} was required for stimulation of efflux and amylase release.

On the electrical and mechanical activity of the gut, eledoisin and physalaemin displayed a striking stimulant effect manifested, at low dose levels, by an increase in frequency and duration of the interdigestive myoelectric complexes and an increase in coordinated mechanical activity and, at high dose levels, by the appearance of diffuse spike activity accompanied by intense local motor activity. Pacesetter potentials were not affected. The tachykinins seem to act directly on the effector muscular and glandular cells[27].

Finally, of considerable interest is the effect of amphibian tachykinins and substance P on the splanchnic vascular bed, with special reference to the hepatic area. The tachykinins produced, both by intravenous and by close arterial infusion, a clear-cut increase in hepatic arterial flow, with reduction of pressure and resistance in the hepatic artery, and increase of pressure in the portal vein. At the same time there was a decrease in the value of liver outflow resistance and an increase of pressure in sinusoid vessels, possibly caused by constriction of sinusoid sphincters and/ or of sphincters in the hepatic vein. On the splanchnic vascular bed and, more generally, on the vasculature and the systemic arterial pressure, substance P was the most active of the examined tachykinins especially when given by intravenous administration[28].

The above results seem to justify the conclusion that the tachykinins may intervene (physiologically in the case of substance P) in the control of gastrointestinal motility, as well as in the regulatio of hepatic blood flow, sinusoid pressure and, hence, sinusoid filtration.

CAERULEIN-LIKE PEPTIDES

The caerulein-like peptides, represented by the decapeptides caerulein and Asn[2], Leu[5]-caerulein and the nonapeptide phyllocaerulein, deserve only a few words because it is well known that they show a very close chemical analogy to the C-terminal portions of CCK and gastrin and possess a spectrum of activity on gastric secretion, pancreatic enzyme secretion, bile flow, gall bladder contraction and intestinal motility exactly superimposable, both in experimental animals and in man, on that of CCK and the C-terminal octapeptide of CCK.

Caerulein
     Pyr-Gln-Asp-Tyr(SO_3H)-Thr-Gly-Trp-Met-Asp-Phe-NH_2
Hylambates-caerulein
     Pyr-Asn-Asp-Tyr(SO_3H)-Leu-Gly-Trp-Met-Asp-Phe-NH_2
Phyllocaerulein
     Pyr-----Glu-Tyr(SO_3H)-Thr-Gly-Trp-Met-Asp-Phe-NH_2
Octa-CCK
         -Asp-Tyr(SO_3H)-Met-Gly-Trp-Met-Asp-Phe-NH_2
Hexagastrin II
            -Tyr(SO_3H)-Gly-Trp-Met-Asp-Phe-NH_2

On all tested preparations and parameters, caerulein showed a greater molar potency than CCK and a similar potency as the CCK-octapeptide[1].

Of interest, in this regard, is the statement by Rehfeld[29] that it is possible that the proper molecular form of CCK in the duodenum and jejunum is the C-terminal octapeptide of CCK, and the similar statement by Dockray[30] that the main brain CCK-component closely resembles the CCK-octapeptide.

This points once again to the validity of the model peptides found in amphibian skin and to the equivalence of caerulein to CCK.

Caerulein can be used: a) in cholecystography and cholangiography; b) in diagnostic testing of pancreatic function, in association with secretin; c) in the radiological study of the gastrointestinal tract; and finally d) in treatment of bowel hypomotility syndromes (adynamic ileus, chronic foecal stasis, megacolon).

MAMMALIAN PEPTIDES IN AMPHIBIANS

It has been stressed that amphibian peptides have generally their counterparts in the mammalian organism. But also the reverse seems to be true, probably to an unsuspected extent.

Van Norden et al.[21] have so far succeeded in proving the occurrence of a VIP-like immunoreactivity in the cutaneous glands of
Rana temporaria and Rana pipiens, but it is easy to foresee that
numerous peptides first found in the mammalian organism will be
traced in the amphibian skin.

Our group has started carrying out a systematic radioimmuno-
logical screening of our collection of extracts of amphibian skin.

Acknowledgement

Original research was supported throughout by grants from the
Consiglio Nazional delle Ricerche, Rome, Italy.

REFERENCES

1. Erspamer V, Melchiorri P: Active polypeptides of the amphibian
   skin and their synthetic analogues. Pure Appl. Chem. 35,
   463-494, 1973

2. Erspamer V, Melchiorri P: Actions of bombesin on secretions and
   motility of the gastrointestinal tract. In: Gastrointestinal
   Hormones, Thompson JC, ed., University of Texas Press, Austi
   and London 1975, p. 575-589

3. Konturek SJ, Krol R, Tasler J: Effect of bombesin and related
   peptides on the release and action of intestinal hormones
   on pancreatic secretion. J. Physiol. 267, 663-672, 1976

4. Fender HR, Curtis PJ, Rayford PL, Thompson JC: Effect of bombesi
   on serum gastrin and cholecystokinin in dogs. Surg. Forum 37
   414-416, 1976

5. Rehfeld JF: Radioimmunoassay of cholecystokinin. Intern. Symposi
   Gastrointest. Hormones and Dig. Pathol., Rome, June 13-15,
   1977, Abstr., p. 145

6. Rayford PL, Schafmayer P, Thompson JC: Secretin and CCK radioimmunoassay. Intern. Symposium Gastrointest. Hormones and Dig.
   Pathol., Rome, June 13-15, 1977. Abstr., p. 146-147

7. Deschodt-Lanckman P, Robberecht P, De Neef P, Lammers M, Christophe J: In vitro action of bombesin and bombesin-like peptides on amylase secretion, calcium efflux and adenylate
   cyclase activity in the rat pancreas. J. Clin. Invest. 58,
   891-898, 1976

8. May RJ, Conlon TP, Erspamer V, Gardner JD: Biochemical changes in pancreatic acinar cells produced by peptides isolated from amphibian skin. J. Clin. Invest. (in Press)

9. Linari G: Influence of enterectomy on bombesin stimulated pancreatic secretion. Joint Meet. German and Ital. Pharmacol., Venezia, October 4-6, 1977. Abstr., p. 284

10. Taylor IL, Walsh JH, Wood J, Chew P, Carter D: Bombesin is a potent stimulant of pancreatic polypeptide (PP) release. Clin. Research 25: 574A, 1977

11. Bloom SR, Polak JM: The entero-pancreatic axis. Symposium Gastrointest. Hormones and Dig. Pathol., Rome, June 13-15, 1977. Abstr., p. 76-77

12. Fallucca F, Delle Fave G, Gambardella S, Mirabella C, De Magistris l, Carratu' R: Glucagon secretion induced by bombesin in man. Internat. Symposium Gastrointest. Hormones. and Dig. Pathol., Rome, June 13-15, 1977. Abstr., p. 131

13. Melchiorri P, Improta G, Sopranzi N: Inibizione della secrezione di VIP da parte della bombesina nel cane, nel gatto e nell' uomo. Rendic. Gastroenterol. 7, Supl. 1, 57, 1975

14. Caprilli R, Melchiorri P, Improta G, Vernia P, Frieri G: Effects of bombesin and bombesin-like peptides on gastrointestinal myoelectric activity. Gastroenterology 68, 1228-1235, 1975

15. Corazziari E, Torsoli A, Delle Fave GF, Melchiorri P, Habib FI: Effects of bombesin on the mechanical activity of the human duodenum and jejunum. Rendic. Gastroenterol. 6, 55-59, 1974

16. Bertaccini G, Impicciatore M, Molina E, Zappia L: Action of bombesin on human gastrointestinal motility. Rendic. Gastroenterol. 6, 45-51, 1974

17. Broccardo M: Effect of bombesin and GH-RIH on intragastric pressure in the dog. Internat. Symposium Gastrointest. Hormones and Dig. Pathol., Rome, June 13-15, 1977. Abstr., p. 130

18. Erspamer V, Melchiorri P: Amphibian skin peptides active on the gut. J. Endocrinology 70, 12P-13P, 1976

19. Walsh JH, Holmquist AL: Radioimmunoassay of bombesin peptides: identification of bombesin-like immunoreactivity in vertebrate gut extracts. Gastroenterology 70, 948, 1976

20. Polak JM, Bloom SR, Hobbs S, Solcia E, Pearse AGE: Distribution
     of bombesin-like peptide in the gastrointestinal tract of
     man. Lancet I, 1109-1110, 1976

21. Van Noorden S, Polak JM, Negri L, Pearse AGE: Common peptides
     in brain, intestine, and skin: embryology, evolution, and
     significance. J Endocrinol 75: 33P-34P, 1977

22. Thompson MR, Debas HT, Walsh JH, Grossman MI: Release of antral
     gastrin by infusion of liver extract into the small intes-
     tine of dogs. Gut 17: 393, 1976

23. Schwartz TW: Pancreatic polypeptide (PP). Intern. Symposium
     Gastrointest. Hormones and Dig. Pathol., Rome, June 13-15,
     1977. Abstr., p. 78-79

24. Brown M, Rivier J, Vale W: Bombesin: potent effects on ther-
     moregulation in the rat. Science 196, 988-990, 1977

25. Brown M, Rivier J, Vale W: Actions of bombesin, thyrotropin
     releasing factor, prostaglandin E_2 and naloxone on thermo-
     regulation in the rat. Life Sciences 20, 1681-1688, 1977

26. Rudich L, Butscher FR: Effect of substance P and eledoisin on
     K^+ efflux, amylase release and cyclic nucleotide levels in
     slices of rat parotid gland. Biochim. Biophys. Acta 444,
     704-711, 1976

27. Caprilli R, Frieri G, Palla R, Broccardo M: Effects of eledoi-
     sin on gastrointestinal electrical activity. Proceed. In-
     ternat. Symposium on Physiol. and Pharmacol. of Smooth Mus-
     cle, Varna, Sept. 28-30, 1976, p. 12

28. Melchiorri P, Tonelli F, Negri L: Comparative circulatory ef-
     fects of eledoisin, physalamin and substance P in the dog.
     In: Substance P. Von Euler US, Pernow B, eds., Raven Press,
     New York 1977, p. 311-319

29. Rehfeld JF: Molecular forms of cholecystokinin in tissues. In-
     ternat. Symposium Gastrointest. Hormones and Dig. Pathol.,
     Rome, June 13-15, 1977. Abstr., p. 61

30. Dockray GJ: Polypeptides in brain and gut: cholecystokinin-
     like peptides. Internat. Symposium Gastrointest. Hormones
     and Dig. Pathol., Rome, June 13-15, 1977, Abstr., p. 132-
     133

PAIRED IMMUNOHISTOCHEMICAL STAINING OF GASTRIN-PRODUCING CELLS (G CELLS) AND PARIETAL CELLS IN PARAFFIN SECTIONS OF HUMAN GASTRIC MUCOSA

R. Stave, P. Brandtzaeg, J. Myren, K. Nygaard, and
E. Gjone

Histochemical Laboratory, Institute of Pathology, and
Medical Department A, National Hospital of Norway

Oslo, Norway

The purpose of this work was to develop a method to study directly the morphological relationship between G cells and parietal cells in parts of the gastric mucosa where both these cell types may occur - that is, in the transitional zone between the pyloric antrum and the body.

METHODOLOGY

Tissue Specimens

Stomach specimens obtained from patients subjected to partial gastric resection were fixed in paraformaldehyde or carbodiimide (CDI),[1] and small tissue blocks were systematically excised from the mucosa.[2]

Fluorochrome Conjugates

From human serum with an immunofluorescence parietal cell antibody titre of 1:512, fluorescein-labelled IgG was prepared; this "green" conjugate (molar fluorescein to protein ratio: 0.7) was used at a concentration of 4.2 g/l. Other reagents were rhodamine-labelled ("red") R-182 specific for gastrin, and fluorescein-labelled ("green") R-79 specific for human immunoglobulin light chains. These rabbit IgG conjugates have been characterized elsewhere.[2,3]

Immunohistochemistry

G cells were demonstrated by direct immunofluorescence,[2] whereas parietal cells were demonstrated either indirectly or directly. In the indirect method the tissue sections were first incubated at room temperature for 30 minutes with the diluted human serum containing parietal cell antibody; after a 15-minute rinse they were reincubated for 30 minutes with "green" anti-human light chains. In the direct method the fluorescein-labelled ("green") human IgG was used. Direct demonstration of G cells was combined in paired staining with indirect or direct method for parietal cells; "red" anti-gastrin was mixed with either "green" anti-human light chains or "green" human IgG in appropriate proportions.

The mounted sections were examined in a Leitz Orthoplan microscope equipped with a vertical Ploem-type illuminator. An HBO 200 W high pressure mercury lamp and an XBO 200 W Xenon lamp provided green and blue light for selective excitation of rhodamine and fluorescein fluorescence respectively. Photographic records were made on GAF 200 daylight colourslide film with selective filters for red or green fluorescence and by double exposures to show both colours simultaneously.

RESULTS

We found that the reactivity of the cytoplasmic parietal cell antigens was preserved after fixation and paraffin embedding. In our hands the intensity of the parietal cell immunofluorescence was similar for cryostat sections and for CDI-fixed material, whereas paraformaldehyde fixation resulted in slightly decreased intensity. The morphology of the paraffin section was much improved.

To secure distinct visualization of parietal cells in human sections by the indirect method, a serum dilution of 1:6 was selected since there was some background staining due to a direct anti-light chain conjugate reaction with intra- and extracellular immunoglobulins in the lamina propria. In CDI-fixed tissue this background staining was of relatively high intensity, and direct immunofluorescence method with the human IgG conjugate was therefore preferable.

Binding specificity of the indirect method was verified by the lack of parietal cell staining when pooled normal human serum diluted 1:5 was used in the first layer. Binding specificity of the direct method was ensured by showing that labelled normal human IgG with a fluorescein to protein ratio of 1.1 did not stain parietal cells when applied at 3.6 g/l. The specificity of the G-cell fluorescence has been defined elsewhere.[2]

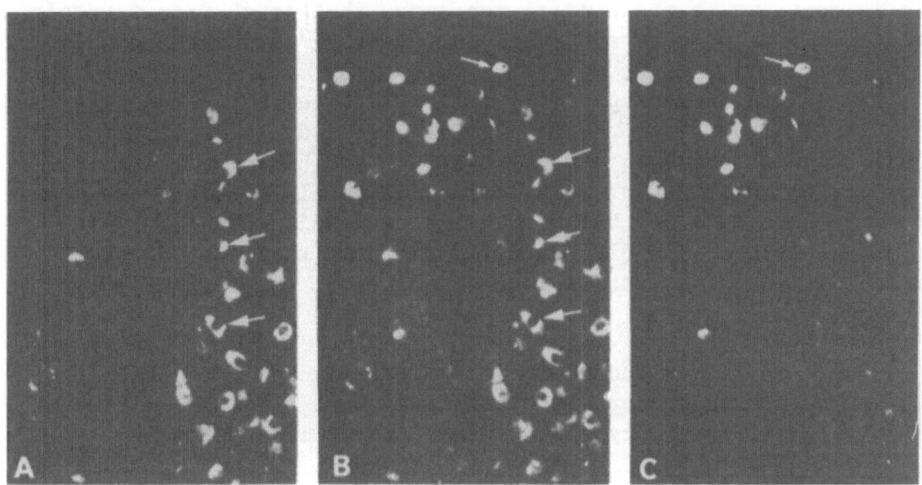

FIGURE 1. *Paired direct immunohistochemical staining of parietal cells (green) and G cells (red) in a section from the transitional body-antrum zone of a carbodiimide-fixed resected human stomach. The same field was photographed with green filtration (A), double exposure (B), and red filtration (C). Parietal cells and G cells are usually present in different glandular tubuli, but a single G cell (small arrow) can be seen in a typical pyloric gland tubulus along with several parietal cells (large arrows). Magnification: X280.*

In specimens from the transitional body-antrum zone, where both G cells and parietal cells are present, paired immunofluorescence staining gave a clear distinction between these two cell types (Fig. 1). In most sections from this area G cells and parietal cells were confined to different gland tubuli, but occasionally both cell types were present in the same tubulus.

DISCUSSION

Previous studies have demonstrated that human parietal cell antibodies do not react with other cell types in the gastric mucosa.[4] This was confirmed in the present investigation by comparing the immunohistochemical staining pattern with that obtained in adjacent sections by a combined Periodic Acid Schiff and Eosin fluorochrome method.[1] However, such antibodies have to our knowledge not been used before in a systematic morphological study of human parietal cells. We found that high-titred human serum was completely satisfactory both for indirect and direct immunofluorescence techniques on sections of fixed and paraffin-embedded human tissue specimens.

This is contrary to the general consensus that fresh-frozen sections have to be used as test substrate for parietal cell antibodies.

Paired staining of parietal cells and G cells is necessary for a precise evaluation of their relative distribution in the transitional body-antrum zone. We believe that the biological significance of the intimate morphological relationship of these two cell types in this region of the stomach needs elucidation. Our preliminary findings (unpublished data) indicate that the extent of the transitional zone varies considerably from one patient to another.

With a similar immunohistochemical approach it should be possible to obtain concurrent visualization of parietal cells and any other gastric cell type to which antibody is obtainable.

REFERENCES

1. Stave R, Brandtzaeg P: Fluorescence staining of gastric mucosa. A study with special reference to parietal cells. Scand J Gastroent (in press 1977)

2. Stave R, Brandtzaeg P: Immunohistochemical investigation of gastrin-producing cells (G cells). The distribution of G cells in resected human stomachs. Scand J Gastroent 11: 705-712, 1976

3. Brandtzaeg P: Conjugates of immunoglobulin G with different Fluorochromes II. Specific and non-specific binding properties. Scand J Immunol 2: 333-348, 1973

Taylor KB, Roitt IM, Doniach D, et al: Autoimmune phenomena in pernicious anemia: gastric antibodies. Brit Med J. 2: 1347-1352, 1962

HISTAMINE H$_2$-RECEPTORS AND GASTRIC SECRETION

G. Bertaccini

Institute of Pharmacology, University of Parma

Parma, Italy

In 1966 Ash and Schild[1] pointed out for the first time the existence of different types of histamine receptors. They defined as H$_1$ receptors those which could be blocked by the classical anti-histaminics and non-H$_1$ receptors those which could not be blocked. However, a real milestone in the history of histamine was represented by the paper of Black and coworkers[2] entitled "Definition and antagonism of histamine H$_2$ receptors" in which the first H$_2$ receptor antagonist, burimamide, was described. The subsequent availability of rather selective H$_2$ receptor stimulants and inhibitors, which served as a tool for characterizing H$_1$ and H$_2$ receptors, contributed noticeably to clarifying the role of histamine in gastric secretion.

A) H$_2$ RECEPTOR AGONISTS

As to the H$_2$ receptor agonists, compounds like 5-methylhistamine (5 MeHO), 5-methyl-N-methylhistamine (DMII), and dimaprit [S-(3-(N,N-dimethylamino)propyl)isothiourea] gave excellent results as stimulants of gastric secretion[3,4]: they were practically devoid of effects on H$_1$ receptors (bronchoconstriction, salivation, hypotension etc.) and they could be administered in doses which exceeded the maximum doses of histamine. In studies performed in cats[5] provided with gastric fistulas, 5 methyl-N-methyl histamine, which did not differ markedly from histamine in terms of threshold stimulant doses, was found significantly more active in terms of "efficacy". The maximal observed acid output to the di-methyl derivative (105 µmol min^{-1}) was produced by 2560 µg kg^{-1} h^{-1}, whereas the maximal response to histamine (65 mol min^{-1}) was produced by 320 g kg^{-1} h^{-1}. When histamine was combined with pyrilamine (an H$_1$ receptor antagonist) the maximal observed acid output did not differ significantly from

that of the di-methyl derivative. Moreover, when DMH was given together with histamine, the maximal response to this combination was
significantly lower than with DMH alone and similar to that elicited
by histamine alone. These results were quite similar to those observed following administration of dimaprit[6]. All these findings
suggested that in addition to its well known action on H_2 receptors
to stimulate acid secretion, histamine also acts on H_1 receptors to
inhibit gastric secretion. When the H_1 inhibitory effect is prevente
by using an H_1 blocker together with histamine or by employing an
analogue with little H_1 action, the full stimulant effect on H_2
receptors is exhibited.

Other experiments were performed in order to check whether the
site of the H_1 receptor for inhibition of acid secretion was in or
outside the stomach. Preparation of isolated guinea pig gastric
fundus (Impicciatore et al., unpublished observations) was tested
with different agonists and antagonists. In a first series of
experiments histamine, DMH and dimaprit were found to have the same
maximal responses; in a second series of experiments the observed
maximal response to histamine alone was not significantly different
from that obtained by using a combination of histamine plus H_1
blocker or by using histamine plus an H_1 receptor agonist. This
indicates that when the direct activity of histamine on the gastric
mucosa independent from that on cardiovascular and respiratory
systems was evaluated, the H_1 blocker failed to potentiate and the
H_1 agonist failed to inhibit the secretory action of the amine.
These findings are compatible with the hypothesis that the activity
of histamine in the isolated stomach fundus is only due to stimulation of H_2 receptors with no interference from H_1 receptors (at leas
in the guinea pig). Thus the inhibitory effect observed in the cat
was probably connected with excitation of H_1 receptors located outside the stomach.

Another H_2 receptor stimulant which seems particularly interesting is the compound 2(5 methyl-4-imidazolyl)-1-methyl ethylamine.
This substance was found to be approximately 10 times less potent
than histamine in stimulating H_2 receptors but it showed the highest
selectivity for these receptors so far observed (activity ratio H_2/
$H_1 \simeq 2000$). It is worth mentioning that the analogue of this compound lacking the methyl group in the imidazole ring was found[7] to
have an activity ratio H_2/H_1 of approximately 2.5. This once again
emphasizes the importance of the methyl group in position 5 of the
imidazole nucleus to enhance the activity on H_2 receptors.

B) H_2 RECEPTOR ANTAGONISTS

In the field of the H_2 receptor antagonists (burimamide,
metiamide and cimetidine) an enormous amount of experimental and
clinical work was reported in two International Symposia held in
London in 1973 and in 1976[8,9].

From a pharmacological point of view it is extremely interesting that H$_2$ antagonists inhibit gastric secretion stimulated not
only by histamine but also by a wide variety of other agents such
as gastrin or cholinergic agents. This indicates that histamine is
involved in every kind of gastric secretion. Several possibilities
were pointed out to explain the nature of this involvement: a) gastrin and acetylcholine release histamine, which represents the final
common mediator of acid secretion; b) there is an interaction among
the three receptors for histamine, acetylcholine and gastrin in the
oxyntic cells and blockade of one receptor changes the properties
of one or both of the other two receptors and c) histamine is being
constantly released and presented to the oxyntic cells as a tonic
background which sensitizes them to other stimuli. When this tonic
background is blocked by the H$_2$ antagonists, acid secretion in response to different stimulants is inhibited. This "permissive
hypothesis" received strong support from recent studies performed
on isolated oxyntic cells from dog gastric mucosa.[10] In this preparation gastrin had a very small effect which was not affected by
H$_2$ blockers. Conversely, administration of a threshold dose of histamine potentiated enormously the action of gastrin and this aug-
mentary action was counteracted by an H$_2$ antagonist.

Some general conclusions may be drawn from the recent studies
on histamine, histamine-like compounds and histamine antagonists:

1) Taking into account that histamine receptor stimulants and
blockers never show an absolute selectivity and that they can have
different side effects independent from the main effect on the histamine receptors, several criteria should be fulfilled to establish
that an observed pharmacological effect is surely connected with the
specific interaction histamine - H$_2$ receptors:
   a) The effect must be mimicked by specific H$_2$ receptor stimu-
      lants (5-methyl-N-methylhistamine and dimaprit);
   b) The effect must be blocked by at least two of the known H$_2$
      blockers (different nonspecific effects were demonstrated
      for each of these blockers);
   c) The effect must not be mimicked by specific H$_1$ receptor
      stimulants like 2-(2-aminoethyl) thiazole or 2-methylhista-
      mine and not be blocked by the H$_1$ antagonists.

Of course the situation is quite opposite if we want to establish
that an effect is connected with stimulation of H$_1$ receptors. In the
absence of some of these conditions the possibility exists that histamine may interact simultaneously with both types of receptors or
actually that histamine acts through excitation of other kinds of
receptors.

2) Another consideration concerns the distribution of the H$_2$ receptors outside the stomach, which is much wider than was suspected:
recent experiments indicate that besides the classical localizations

TABLE 1

EFFECTS OF HISTAMINE COMPLETELY OR PARTIALLY UNAFFECTED BY
ADMINISTRATION OF H_1 AND H_2 RECEPTORS BLOCKERS

ACTIVITY	REFERENCE
Excitation of hypothalamic cells (rat and cat)	Haas and Bucker, 1975[12]
Relaxation of rabbit trachea	Fleish and Calkins, 1976[13]
Behavioral changes in the rat (depression)	Calcutt and Reynolds, 1976[14]
Negative inotropic and chronotropic effect (isolated rat heart)	Dai, 1976[15]
Chemoattractant activity of histamine on eosinophils	Clark et. al., 1977[16]
Contraction of the rat pylorus	Bertaccini et. al., 1977[17]

suggested by Black and coworkers[2] (gastric mucosa, rat uterus and
guinea pig auricles), H_2 receptors are widely distributed throughout the body: in particular, vascular beds, gall bladder, sheep
bronchial muscle, cat trachea, guinea pig ileum and especially in
the central nervous system of different animal species. Thus the
physiopharmacological importance of the H_2 receptor agonists and
antagonists is certainly destined to increase and may transcend the
limits of gastric secretion.

3)   The ratio of activity between histamine and the specific H_2
receptor agonists varies profoundly in the different organs and
this suggests that the distribution of the H_2 receptors is extremely irregular in the various tissues and/or that they are mixed in
different degree with H_1 or other types of receptors or finally that
different types of H_2 receptors exist in the various tissues. This
heterogeneity makes it very important, when calculating the activity
ratio H_2/H_1 for agonistic compounds, to indicate the mean of the
values obtained in the different organs rather than giving values
obtained on only one parameter for each type of receptor.

4)   Finally a last consideration: recent observations concerning
different experimental conditions and different animal species seem
to suggest that the action of histamine can be mediated by other
types of receptors different from both H_1 and H_2 receptors. In
some cases the action is clearly of the muscarinic type and can be
easily blocked by atropine; in other cases it is the β-sympathetic
type and can be blocked by the common β-blocking agents; however,
there are cases in which none of the known inhibitors is effective
and this suggests the possible existence of still unknown histamine
receptors (see Table 1). It is easy to foresee that besides the
gastroenterologists, also biochemists, physiologists and pharmacolo-
gists will be stimulated to reconsider all the activities of histamine in the light of the new findings.

REFERENCES

1. Ash, ASF, Schild HO: Receptors mediating some actions of his-
    tamine. Br J Pharmac 27: 427-439, 1966

2. Black JW, Duncan WAM, Durant CJ, et al: Definition and antago-
    nism of histamine H$_2$-receptors. Nature 236: 385-390, 1972

3. Bertaccini G, Impicciatore M: Activity of 2-(5-methyl-4-imida-
    zolyl)-ethylamine (5-methylhistamine) on the gastric secre-
    tion of different laboratory animals and of man. Eur J
    Pharmacol 28: 360-367, 1974

4. Bertaccini G, Impicciatore M, Vitali T: Biological activities
    of a new specific stimulant of histamine H$_2$-receptors. Il
    Farmaco Ed Sc 31: 935-938, 1976

5. Impicciatore M, Bertaccini G, Mossini F, et al: N-methyl, 5-
    methyl histamine evokes higher maximal rate of gastric acid
    secretion than histamine. Clin Res 25: 109A, 1977

6. Carter DC, Grossman MI: Dimaprit evokes higher maximal rates of
    gastric secretion than histamine. Gastroenterology 72: 1036,
    1977

7. Durant GJ, Emmett JC, Ganellin CR, et al: Potential histamine
    H$_2$ receptor antagonists. 3 methyl histamines. J Med Chim 19:
    923-928, 1976

8. Wood CJ, Simkins MA: International symposium on histamine H$_2$-
    receptor antagonists. Smith Kline and French Laboratories
    Ltd. Welwyn Garden City, 1973

9. Burland WL, Simkins AM: Cimetidine. International symposium on
    histamine H$_2$-receptor antagonists. Excerpta Medica; Amster-
    dam-Oxford, 1976

10. Soll A, Isolated mammalian parietal cells: effects of atropine
    and metiamide on the actions and interactions of secretago-
    gues, Gastroenterology 72: 824, 1977

11. Ivey KJ, Baskin W, Jeffrey G: Effect of cimetidine on gastric
    potential difference in man. Lancet 2: 1072-1073, 1975

12. Haas HL, Bucker UM: Histamine H$_2$-receptors on single central
    neurones. Nature 255: 634-635, 1975

13. Fleish JH, Calkins PJ: Comparison of drug-induced responses of
    rabbit trachea and bronchus. J Appl Physiol 41: 62-66, 1976

14. Calcutt CR, Reynolds J: Some behavioral effects following
        intracerebroventricular (ICV) injection in rats of his-
        tamine H_1-receptor and H_2-receptor agonists and antagonists.
        Neurosci L. 3: 82-83, 1976

15. Dai S: A study of the actions of histamine on the isolated rat
        heart. Clin Exp Pharmacol Physiol 3: 359-367, 1976

16. Clark RAF, Sandler JA, Gallin JI, et al: Histamine modulation
        of eosinophil migration. J Immunol 118: 137-145, 1977

17. Bertaccini G, Coruzzi G, Molina E, et al: Action of histamine
        and related compounds on the pyloric sphincter of the rat.
        (unpublished)

THE GASTRINS: STRUCTURE AND HETEROGENEITY

R. A. Gregory

Physiological Laboratory, University of Liverpool

Liverpool L69 3BX, England

Since the isolation of the hormone gastrin as a pair of hepta-decapeptide amides from hog antral mucosa[1], a great deal more has been learned about the various forms of the hormone which are to be found in the tissues of origin (antral and upper intestinal mucosa and gastrinomas) and in the circulation in normal and pathological conditions. The heterogeneity of tissue gastrin raises the important question of metabolic relationships in terms of the mechanism of biosynthesis, while heterogeneity in the circulation give rise to the equally important problem of determining the various forms separately by radioimmunoassay and of evaluating their respective contributions to the stimulation of target organs.

The various forms of gastrin so far recognised are as follows:

1. Little gastrin (LG, G17) 17 aminoacid residues
2. Big gastrin (BG, G34) 34 aminoacid residues
3. Minigastrin (MG, G14) 14 aminoacid residues
4. Big big gastrin (BBG) molecular weight estimated as 20,000
5. Component I (C-I) size intermediate between G34 and BBG
6. NH_2-terminal tridecapeptide fragment of G17 (NT13).

Of these 1, 2, 3 and 6 have been isolated from antral mucosa or gastrinoma tissue and fully characterised chemically (Fig. 1). The remaining forms (BBG and C-I) have been recognised immunologically in the circulation and material which may resemble or represent them has been similarly identified in antral and gastrinoma extracts but not chemically characterised.

G34   Pyr-Leu-Gly-Pro-Gln-Gly-His-Pro-Ser-Leu-Val-Ala-Asp-Pro-Ser-Lys-Lys

      -Gln-Gly-Pro-Trp-Leu-Glu-Glu-Glu-Glu-Glu-Ala-Tyr-Gly-Trp-Met-Asp-Phe-NH$_2$
$$\qquad\qquad\qquad\qquad\qquad\qquad\qquad\qquad\qquad\qquad\qquad\qquad \overset{|}{R}$$

G17   Pyr-Gly-Pro-Trp-Leu-Glu-Glu-Glu-Glu-Glu-Ala-Tyr-Gly-Trp-Met-Asp-Phe-NH$_2$
$$\qquad\qquad\qquad\qquad\qquad\qquad\qquad\qquad\qquad\qquad\qquad\qquad \overset{|}{R}$$

G14   Trp-Leu-Glu-Glu-Glu-Glu-Glu-Ala-Tyr-Gly-Trp-Met-Asp-Phe-NH$_2$
$$\qquad\qquad\qquad\qquad\qquad\qquad\qquad\qquad\qquad\qquad\qquad\qquad \overset{|}{R}$$

NT-G17 Pyr-Gly-Pro-Trp-Leu-Glu-Glu-Glu-Glu-Glu-Ala-Tyr-Gly
$$\qquad\qquad\qquad\qquad\qquad\qquad\qquad\qquad\qquad\quad \overset{|}{R}$$

R = H(Type I) or SO$_3$H (Type II)

Figure 1. The aminoacid sequences of the chemically identified forms of human gastrin

LITTLE GASTRIN (G17)

This form of the hormone provided the first evidence of heterogeneity in a gastrointestinal hormone; it was isolated in the form
of a pair of heptadecapeptide amides having an identical aminoacid
constitution but differing in that in one member the single tyrosine
residue present was sulphated while in the other it was not. In
those species in which the heptadecapeptides have been chemically
characterised (hog, man, dog, cat, sheep and cow) trivial substitu-
tions are found to have occurred in the body of the molecule while
the C-terminal sequence, which contains the active region of the
molecule[2], remains unchanged. In a subsequent study on the forms of
gastrin present in a single large gastrinoma metastasis[3], it was
found that in addition to the types I and II (unsulphated and sulphated) previously isolated, there were present small amounts of
two further heptadecapeptides. These had the same aminoacid composition as types I and II and both had a sulphated tyrosine, so
that they could be described as types IIA and IIB. How they differ
chemically from types I and II is not yet understood and at present
the possibility that they are artefacts produced during extraction
or postmortem in the tissue cannot be excluded.

BIG GASTRIN (G34)

This form of the hormone was first identified immunologically
in the serum of patients with gastrinoma or pernicious anaemia[4,5].

Gregory and Tracy[6] subsequently isolated from hog antral mucosa
and gastrinoma tissue pairs of peptides corresponding exactly in
chromatographic and electrophoretic properties to the 'big gastrin'
immunoreactivity of Yalow and Berson. These peptides have been
sequenced and the human form has been synthesised. In each case
they consist of the heptadecapeptide amide (G17) covalently linked
at its N-terminus by two lysyl residues to a further heptadecapeptide, the N-terminus of which, as in free G17, is pyroglutamyl (Fig.
1). Tryptic digestion of G34 releases G17 by cleavage at the two
lysyl residues, together with one or two 'tryptic peptides', depending on the conditions of digestion.

MINIGASTRIN (G14)

Gregory and Tracy[7] isolated from gastrinoma tissue very small
amounts of a pair of peptides (sulphated and unsulphated) which are
now known to possess the C-terminal tetradecapeptide sequence of
G17. Owing to an error made in the aminoacid analysis of these
peptides at the time of isolation, which represented the tryptophan
content as one residue, it was at first believed that the minigastrins were the C-terminal tridecapeptide sequence of G17. This
error was discovered when the potency and chromatographic behaviour
of synthetic tridecapeptide proved to be different from that of
natural minigastrin. Harris (unpublished studies) has now established that the N-terminal residue of minigastrin is tryptophan; and
this together with the aminoacid composition proves that minigastrin
is the C-terminal tetradecapeptide fragment of G17.

In all species so far examined, the major gastrin component
present in G cells (and in most gastrinomas) is G17; it accounts for
about 95% of the total gastrin present in antral mucosa, most of the
remainder being G34;in duodenal mucosa the proportion of G34 is said
to be somewhat higher. The minigastrins have not been isolated from
antral mucosa, and the amount of them present in antral mucosa or
gastrinoma tissue is very small indeed compared with the amounts of
G17 and G34. In the circulation of man, dog and hog G34 commonly
predominates owing to its relatively long half-life[8]; the proportion
of G34 relative to G17 secreted by the antral mucosa appears to be
only a little higher than their proportions in the tissue. The very
small amount of minigastrin present in the circulation of man, dog
and hog makes it of little significance in the stimulation of acid
secretion; but it has recently been shown[9,10] that immunoreactive
material corresponding in size to G14 is a major component of cat
plasma, and there is evidence from the same sources that the minigastrin may be formed in the circulation from G17 by enzymatic
cleavage. We (Gregory and Tracy, unpublished studies) have recently confirmed that when substantial amounts of porcine G17 (5 mg or
more) are incubated with cat plasma or serum there is rapidly
formed a considerable amount of a single product which is either

G13 or G14. The peptide has been isolated from several preparations
and its exact composition is still under study. Similar experiments
made using human, dog and rat plasma showed only a slight degree of
cleavage, the amount of the product being too small to identify
chemically.

BIG, BIG GASTRIN (BBG)

    Yalow and Berson[11] reported the presence in serum and jejunal
extracts of an immunoreactive component which emerged in the void
volume on Sephadex G50; they termed it 'big, big gastrin'. The
amount of this component was not significantly increased by feeding,
and its amount, nature and functional significance has remained a
matter for speculation. It would now seem that further studies are
required to establish whether this component is truly a circulating
form of the hormone or whether it is an artefact, for recently both
McGuigan[12] and Rehfeld[13] have reported that the BBG in serum is not
taken up by immunoadsorption columns using antibodies which bind
the other forms of the hormone. It has been suggested[13] that circulating BBG is due mainly to interference in the radioimmunoassay
by serum proteins eluted in the void volume of small Sephadex G50
columns.

COMPONENT I (C-1)

    Rehfeld and Stadil[14] identified in serum an immunoreactive
component which emerges in advance of the G34 peak but later than
the void volume. On tryptic digestion, material of the size of G17
was liberated, suggesting that component I consists like G34 of G17
covalently joined by basic aminoacid residues at its N-terminus to
a larger peptide molecule. Owing to the very small amounts of C-I
present in antral tissue and in most gastrinomas, isolation of it
has not been achieved and nothing is known of its chemical nature.

THE N-TERMINAL TRIDECAPEPTIDE FRAGMENT OF G17 (NT13)

    In 1967 Tracy and I[15] isolated from hog antral mucosa small
amounts of a pair (sulphated and unsulphated) of inactive gastrin
fragments which proved to have the N-tridecapeptide sequence of G17.
Attempts to find in the extract the cleaved fragment, which would be
the active C-terminal tetrapeptide amide, were unsuccessful. Dockray
and Walsh[16] have identified in serum a component which appears to
correspond to NT13.

GASTRIN HETEROGENEITY: THE BIOSYNTHETIC SEQUENCE

It is becoming generally recognised that many, perhaps all, peptide hormones are formed as the product of genes which specify a molecule considerably larger than the major active form finally secreted, and that there occurs cleavage of this initial molecule in stages, commonly at points where basic aminoacids occur. This results in the formation of 'precursor' forms and cleaved fragments in addition to the final product, and all may find their way into the circulation or be detectable in the tissue.

The first hormone precursor to be discovered was proinsulin[17]; slices of an insulinoma were incubated with tritiated phenylalanine and leucine and an extract of the system fractionated by gel filtration. Besides labelled insulin there was also detected a larger labelled component which was transformed into insulin-like material by tryptic digestion. Proinsulin proved to be a straight-chain peptide of MW about 9000; trypsin cleaves it at the pairs of basic residues situated at either end of the 'C-peptide' fragment. Not only insulin, but also material corresponding to proinsulin and C-peptide, can be detected in the circulation.

In several other instances, similar trypsin-like cleavage of a 'prohormone' has been described; and by analogy it becomes highly probable that G34, which is rapidly cleaved by trypsin at the site of the two lysyl residues in the middle of the chain (Fig. 1), is the immediate precursor of G17, which is the main active product of the gastrin or gastrinoma cell.

The major difficulty of studying the biosynthetic pathway for gastrin directly lies in the fact that suitable preparations of antral G cells have not yet been made, and fresh gastrinoma tissue is now rarely obtainable because they are seldom resected, the preferred treatment for the disease being chemotherapy or total gastrectomy. There is however an alternative approach which has been used in several other instances and which is well exemplified by the brilliant studies of Habener and his colleagues[18,19] on the biosynthesis of parathyroid hormone (PTH). Messenger RNA was prepared from parathyroid adenomas (and from normal glands) and its translation studied in a cell-free system (wheat-germ). There was first formed a large peptide 'pre-pro-PTH' (115 residues) which was very rapidly cleaved at a basic aminoacid residue into a smaller peptide 'pro-PTH' (90 residues) which was in turn cleaved at a slower rate (again at a basic residue) into PTH (Fig. 2). Steiner[20], using the same method, has identified a 'pre-proinsulin' in the biosynthesis of that hormone.

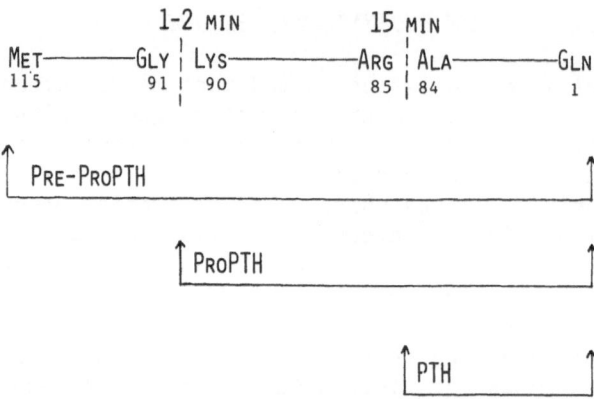

Figure 2.   Biosynthesis of parathyroid hormones (Harbener, 1977)

These results obviously suggest a place for component I in the
formation of gastrin; it may represent the 'pre-pro' form of the
hormone, as identified for PTH and insulin.   When component I is
digested in vitro with trypsin only immunoreactive material of the
size of G17 is liberated; but it may well be that G34, which is
cleaved very rapidly by trypsin, is in fact formed but is rapidly
converted to G17 and so does not accumulate.   Two facts support the
view that the biosynthetic sequence is: Component I → G34 → G17.
Firstly, both G34 and G17 have an N-terminal pyroglutamyl residue;
and it has been shown that in the tryptic cleavage of G34 the G17
at first liberated has N-terminal glutaminyl which then spontaneous-
ly changes to the pyroglutamyl form (Fig. 1).   Obviously if G34
were formed in a similar manner from a larger peptide, this would
account for its N-terminal pyroglutamyl residue.   Secondly, it has
recently been shown by Dr. G. J. Dockray in my laboratory (unpublished studies) that an antibody specific for the N-terminal region of
G34 detects in extracts of antral mucosa large amounts of a component with immunological and chromatographic properties corresponding
to the N-terminal 'tryptic peptide' of G34; and in collaboration
with Dr. Camille Vaillant of the Department of Histology, University
of Liverpool, immunocytological studies using the antibody have
localised G17 and the N-terminal fragment of G34 to the same gastrin
cell.   These observations thus provide strong indirect evidence that
G17 is formed from G34 in the gastrin cell and that the liberated N-
terminal fragment can also be found there.

No place in a biosynthetic scheme can yet be assigned with any confidence to minigastrin and the N-terminal fragment of G17. They might both be formed from G17 by an enzyme cleaving on the N-terminal side of the tryptophan residues, and the small amounts in which they occur suggest that this may be a trivial side-reaction.

Finally mention must be made of the fact that the C-terminus of the gastrins is phenylalanine amide, and the amide group is essential for physiological activity of the gastrin molecule[2]. There are a number of other active peptides in which the C-terminal residue is similarly masked (notably, of course, CCK and caerulein) which also depend for their activity on this feature. It has recently been suggested[21] (Fig. 3) that it results from cleavage by trans-amidation of a peptide longer at the C-terminal end, rather than by direct amidation of the free carboxyl group. It seems very doubtful whether such a peptide would be active and whether it would crossreact with the gastrin antibodies commonly in use, which are directed towards the C-terminus. The mechanism of C-terminal amidation remains an important unsolved problem in the biosynthesis of gastrin and similar peptides.

HOW HETEROGENOUS IS GASTRIN?

In our preparations of the gastrins from antral mucosa and gastrinoma tissue, Tracy and I[3] noted the presence of small amounts of two further forms of G17 and two further forms of G34; exactly how these differ from the well-known types I and II and whether they are artefactual in origin remains uncertain. However, the remarkable study of Rehfeld and his colleagues[22] reveals an hitherto unsuspected multiplicity of what might be termed 'subforms' of the various varieties of gastrin. They subjected serum containing a large amount of immunoreactive gastrin to repeated filtration on very large Sephadex columns and were able to identify no less than twenty

$$R^1CH_2.CO.NH.CHR^2.COOH + NH_3 \rightleftharpoons R^1CH_2.CO.NH_2 + NH_2.CHR^2.COOH$$

Figure 3. Formation of C-terminal amide by transamidation with simultaneous cleavage of a peptide bond (Bradbury, Smyth and Snell, 1975)

different gastrins - six component Is, six G34s, four G17s and four G14s. Probably the differences in structure of these multiple sub-forms will prove to be trivial; but the final account of the bio-synthesis of gastrin will need to include some explanation for their existence.

REFERENCES

1. Gregory RA, Tracy HJ: The constitution and properties of two gastrins extracted from hog antral mucosa. Gut 5: 103-117, 1964

2. Tracy HJ, Gregory RA: Physiological properties of a series of synthetic peptides structurally related to gastrin I. Nature (Lond.) 204: 935-938, 1964

3. Gregory RA, Tracy HJ: The chemistry of the gastrins: some recent advances. In Gastrointestinal Hormones, Edited by JC Thompson. University of Texas Press, Austin. p 13-24, 1975

4. Yalow RS, Berson SA: Size and charge distinctions between endogenous human plasma gastrin in peripheral blood and heptadecapeptide gastrins. Gastroenterology 58: 609-615, 1970

5. Berson SA, Yalow RS: Nature of immunoreactive gastrin extracted from tissues of gastrointestinal tract. Gastroenterology 60: 215-222, 1971

6. Gregory RA, Tracy HJ: Isolation of two 'big gastrins' from Zollinger-Ellison tumour tissue. Lancet ii: 797-799, 1972

7. Gregory RA, Tracy HJ: Isolation of two minigastrins from Zollinger-Ellison tumour tissue. Gut 15: 683-685, 1974

8. Walsh JH, Debas HT, Grossman MI: Pure human big gastrin: immuno-chemical properties, disappearance half-time and acid-stimulating properties in dogs. J Clin Invest 54: 477-485, 1974

9. Blair EL, Grund ER, Lund PK et al: Circulating gastrin variants in the cat. Gastroenterology 72: 812, 1977

10. Uvnäs-Wallensten K, Rehfeld JF: Type of gastrin released by vagal stimulation in anaesthetised cats. Gastroenterology 72: 825, 1977

11. Yalow RS, Berson SA: And now, 'big, big' gastrin. Biochem Biophys Res Commun 48: 391-395, 1972

12. McGuigan JE: The absence of big, big gastrin in human serum as determined by immune absorption techniques. In 1st International Symposium on Gastrointestinal Hormones, Asilomar, Oct 5-8, 1976, A096

13. Rehfeld JF: Circulating forms of gastrin in normals and DU. In 1st International Symposium on Gastrointestinal Hormones, Asilomar, Oct 5-8, 1976, A126

14. Rehfeld JF, Stadil F: Gel filtration studies on immunoreactive gastrin in serum from Zollinger-Ellison patients. Gut 14: 369-373, 1973

15. Gregory RA: The gastrointestinal hormones: a review of recent advances. J. Physiol (Lond) 241: 1-32, 1974

16. Dockray GJ, Walsh JH: Amino terminal gastrin fragment in serum of Zollinger-Ellison syndrome patients. Gastroenterology 68: 222-230, 1974

17. Steiner DF, Oyer PE: The biosynthesis of insulin and a probable precursor of insulin by a human islet cell adenoma. Proc Nat Acad Sci (Wash) 57: 473-380, 1967

18. Habener JF: New concepts in the formation, regulation of release, and metabolism of parathyroid hormone. In Polypeptide Hormones: Molecular and Cellular Aspects, Ciba Foundation Symposium 41 (new ser.), London, p 197-224, 1976

19. Habener JF: Biosynthesis of parathyroid hormone: an example of post-translational proteolytic processing of a polypeptide hormone. In EMBO Workshop on Hormone Fragments and Disease. Leiden, 15-18 May, 1977

20. Steiner DF: Biosynthesis of insulin and glucagon: a view of the current state of the art. In Polypeptide Hormones: Molecular and Cellular Aspects, Ciba Foundation Symposium 41 (new ser.), London, p 7-30, 1976

21. Bradbury AF, Smuth DG, Snell CR: Prohormones of β-melanotropin (β-melanocyte-stimulating hormone, β-MSH) and corticotropin (adreno-corticotropic hormone, ACTH): structure and activation. In Polypeptide Hormones: Molecular and Cellular Aspects, Ciba Foundation Symposium 41 (new ser.), London, p 61-76, 1976

22. Rehfeld JS, Stadil F, Malmstrom J et al: Gastrin heterogeneity in serum and tissue. A progress report. In Gastrointestinal Hormones. Edited by JC Thompson. University of Texas Press, Austin, p 43-58, 1975

BIOLOGICAL ACTIVITY AND CLEARANCE OF GASTRIN PEPTIDES IN DOG AND MAN: EFFECTS OF VARYING CHAIN LENGTH OF PEPTIDE FRAGMENTS

J. H. Walsh

Department of Medicine, U.C.L.A. Medical School

Los Angeles, California

Gastrin is known to exist in mammalian tissues and blood in multiple molecular forms.[1] Six of these forms have been characterized chemically and shown to be three pair of single chain peptides in which the single tyrosine residue either is nonsubstituted (Gastrin I) or is sulfated (Gastrin II). The three pairs have a common C-terminal tetradecapeptide sequence and differ only in the number of amino acids comprising the N-terminal portion of the molecule. The largest form contains 34 amino acid residues and is known as big gastrin or G-34. The other forms are the heptadecapeptide or little gastrin (G-17) and the tetradecapeptide or minigastrin (G-14). Other gastrin peptides which have been identified immunochemically but not characterized by chemical analysis include an amino terminal fragment of G-17, a molecule slightly larger than G-34 known as Component I, and a molecule with an apparent molecular weight similar to albumin known as big-big gastrin. None of these forms has apparent biological activity. In addition, synthetic preparations have been prepared of nonsulfated human G-34, G-17, and G-14, as well as shorter C-terminal fragments of gastrin which are not known to occur naturally. C-terminal peptides as short as the tetrapeptide (G-4) are known to be potent stimulants of gastric acid secretion and the tripeptide has slight activity.

The principal tissue which contains gastrin is the mucosa of the gastric antrum (pyloric gland area). Extracts of antral tissue reveal that the principal storage form of gastrin is G-17, which comprises 90% or more of extractable gastrin. The remainder consists of G-34 and small amounts of G-14. In contrast, the most abundant circulating form of gastrin is G-34. One possible explanation for the discrepancy between G-17/G-34 ratios in tissue

and blood is a difference in clearance rates from the circulation
of the two peptides. In order to explore this possibility, studies
were carried out to determine the clearance rates and disappearance
half time of G-17 and G-34 in dog[2] and in man[3]. In both species it
was found that G-17 was cleared 6-8 times more rapidly than G-34.
The estimated disappearance half times for G-17 in dog and man were
about 2.5 and 6 minutes, respectively, while the half times for G-34
were about 15 and 45 minutes, respectively. These differences were
reflected in much higher increments in circulating gastrin concentrations which were achieved during constant intravenous infusion
of G-34 compared with infusion of equimolar amounts of G-17. In
other studies it was found that sulfation had no effect on the
clearance rate of G-17 in dog and that G-14 had a clearance rate
similar to that of G-17. The differences in clearance rates could
account for most, but not all, of the discrepancy between tissue
and serum concentration of G-17 and G-34. The remaining difference
may be explained either by some selective release of G-34 from
tissue stores during stimulation or by differences in the efficiency
of extraction of these two molecules from tissue.

The proportions and absolute concentrations of G-17 and G-34
in blood differ over time during stimulation of gastrin release. In
order to estimate the biological activity of circulating gastrin it
is necessary to measure the biological effects produced by graded
increments in these circulating forms and to relate these effects to
changes in biological response obtained during endogenous stimulation
of hormone release. By administering pure peptides intravenously
as constant infusions over a sufficient period of time, it is possible to produce relatively steady state blood concentrations. At the
same time gastric acid secretion can be measured to monitor the principal biological activity of gastrin. When such studies were performed in man and in dog, it was found that equimolar exogenous doses
of G-17 and G-34 produced similar rates of gastric acid secretion.
However, when acid secretion was plotted as a function of increase
in circulating gastrin concentration, it was found that circulating
G-17 was 6-8 times as potent as circulating G-34. Since G-17 comprises about 20-50% of circulating gastrin, it can be estimated that
G-17 accounts for at least two-thirds of circulating gastrin biological activity. Therefore, in order to estimate the circulating gastrin which accounts for most of the acid-stimulating activity, it
would be desirable to use a radioimmunoassay specific for G-17, such
as the one described by Dockray and Taylor.[4]

Recent studies in man seem to confirm that G-17 may account for
most of the gastrin-stimulated acid secretion obtained during intragastric administration of amino acid solutions.[5] These studies were
performed with continuous perfusion of the stomach with either a
saline solution or with an isotonic solution of amino acids. Acid
secretion was measured by the technique of intragastric titration.[6]

In preliminary experiments it was shown that slight distention of
the stomach with saline caused a small increase in acid secretion
rate, compared with the secretion rate measured by gastric aspiration, and also increased gastric sensitivity to exogenous gastrin.
Distention alone did not increase serum gastrin concentration. Addition of amino acids to the gastric perfusate caused a further increase in acid secretion rate associated with an increase in serum
gastrin concentration. The increase in plasma G-17 concentration
was measured by radioimmunoassay, using an antibody (L-6) specific
for this molecular form. The increase in plasma G-17 and of acid
secretion which occurred during gastric perfusion with amino acids
were compared with the increase in plasma G-17 required to produce
the same rate of acid secretion when the stomach was perfused with
saline and synthetic human G-17 was given intravenously in graded
doses. It was found that the increase in G-17 which was obtained
during amino acid perfusion was sufficient to account for all of
the measured increase in acid secretion. Although this study did
not exclude the simultaneous occurrence of other stimulatory and
inhibitory factors regulating acid secretion, it could be concluded
that under these conditions of stimulation by amino acids G-17
could be a major physiological regulator of acid secretion.

All gastrin secreted into the portal blood from tissue gastrin
stores must traverse the liver before entering the general circulation. The liver thus could contribute to the proportion of various
forms of gastrin found in peripheral blood by selective removal of
one or more forms. The effects of liver passage were studied by
comparing circulating gastrin concentrations and acid secretory
responses to graded doses of various gastrin peptides in dogs with
gastric fistulas and portocaval transposition. In this preparation
peptide solutions could be administered intravenously either into
the portal or peripheral circulation by selective cannulation of the
front or hind limb.[7] It had been shown previously that the liver
removes more than 90% of gastrin tetrapeptide during a single passage. This finding was confirmed in the portocaval dogs. In contrast, intravenous systemic and portal infusions of G-34, G-17, and
G-14 produced similar increases in serum gastrin concentration and
similar acid secretory responses. It therefore is unlikely that
the liver plays a major role in the catabolism of gastrin or in the
regulation of the relative concentrations of the major biologically
active forms in the peripheral circulation. It also was found that
decreasing the peptide length of gastrin fragments from 13 to 10
resulted in considerable hepatic removal and that further shorten-
ing of the peptide to the heptapeptide resulted in nearly complete
hepatic removal. These studies suggest that it is unlikely that
unidentified short gastrin fragments could play any significant role
in total circulating gastrin activity. Even if such fragments were
synthesized and released into the portal circulation, they would be
removed or inactivated during hepatic transit.

Several groups have reported that the kidney removes gastrin from the circulation, and it has been postulated that the kidney is the major organ responsible for gastrin catabolism. Two types of observation make it unlikely that the kidney contributes more than a fraction of total body removal of circulating G-17. First, total renal blood flow is not sufficient to account for the short half life even if renal extraction of gastrin approaches 100%. Second, it has been shown that the renal arteriovenous difference in gastrin concentrations seldom exceeds 30%. Therefore it seems reasonable to presume that G-17 is removed at multiple sites in the body. In order to test this hypothesis, G-17 was administered by constant intravenous infusions to anesthetized dogs until a stable increment in blood concentration was produced.[8] Simultaneous blood samples then were obtained from the arterial circulation and from venous drainage from the kidney, intestine, head, and a limb. In a separate study, blood was obtained from the hepatic vein. It was found that venous concentrations were consistently lower than arterial concentrations by about 20-30% and that these differences were not significantly greater in the renal vein when compared with the other venous sites sampled. Hepatic venous gastrin concentration was lower than that found in other veins, but this finding could be explained by passage through two capillary beds (intestinal and hepatic). It was concluded that removal of circulating G-17 occurs at multiple sites, possibly all capillary beds, and that the kidney has no special role in G-17 catabolism. These results are consistent with reports that the disappearance half time of G-17 is not significantly prolonged in patients with severe renal failure.

It can be concluded that G-17, the most abundant form of gastrin in antral mucosa, also is the most important circulating form of gastrin and that G-17 is likely to play a major role in the regulation of acid secretion. G-34 is more abundant than G-17 in the circulation because of less rapid clearance, but the biological activity of G-34 is too low to account for a major effect on acid secretion. The liver and kidney appear to play no unique role in the removal of G-17 from the circulation. Instead, the rapid disappearance of G-17 is due to removal by organs throughout the body.

REFERENCES

1. Walsh JH, Grossman MI: Gastrin. New Eng J Med 292: 1324-1332, 1377-1384, 1975

2. Walsh JH, Debas HT, Grossman MI: Pure human big gastrin: Immunochemical properties, disappearance half time, and acid stimulating action in dogs. J Clin Invest 54: 477-485, 1974

3. Walsh JH, Isenberg JI, Ansfield J, et al: Clearance and acid-
       stimulating action of human big and little gastrins in duo-
       denal ulcer subjects. J Clin Invest 57: 1125-1131, 1976

4. Dockray GJ, Taylor IL: Heptadecapeptide gastrin: measurement
       in blood by specific radioimmunoassay. Gastroenterology 71:
       1114-1116, 1976

5. Feldman M, Richardson CT, Walsh JH: Mechanisms of acid secre-
       tory response to amino acid solutions in normal man. Clin
       Res 25: 467A, 1977

6. Fordtran JS, Walsh JH: Gastric acid secretion rate and buffer
       content of the stomach after eating: results in normal
       subjects and in patients with duodenal ulcer. J Clin Invest
       52: 645-657, 1973

7. Strunz UT, Walsh JH, Bloom SR, Thompson MR, Grossman MI: Lack
       of hepatic inactivation of VIP. Gastroenterology, 73: 768-
       771, 1977

8. Strunz UT, Walsh JH, Grossman MI: Removal of gastrin by various
       organs in dogs. Gastroenterology 74: 32-33, 1978

DIFFERENT FORMS OF GASTRIN IN PEPTIC ULCER

G. J. Dockray and I. L. Taylor

Physiological Laboratory, University of Liverpool

Liverpool L69 3BX England

The role of gastrin in the pathogenesis of peptic ulcer disease has received considerable attention in recent years[1]. It is now known that patients with duodenal ulcer are more sensitive to pentagastrin than normal subjects and have higher than normal postprandial serum gastrin concentrations, possibly due to impairment of inhibition of gastrin release by acid[1]. However, the significance of these observations remains difficult to interpret because gastrin is known to circulate in several different forms which differ in their biological activity. The main forms isolated from gastrinomas and antral mucosa are big gastrin, or G34, and little gastrin, or G17[2]. Other forms have also been identified but these either circulate in low concentrations or are not biologically active; they include minigastrin (G14), an NH_2-terminal fragment of G17, and two forms which are probably larger than G34 (big, big gastrin and Rehfeld's component I)[2]. In man and dog, Walsh has shown that the circulating concentrations of G34 required to stimulate half maximal rates of acid secretion are five-six times greater than those of G17, and that the metabolic clearance rates of G34 are about one fifth those of G17[3,4]. Clearly, any attempt to understand the physiological and pathophysiological significance of gastrin must take into account the relationships between endogenous circulating gastrin and acid secretion, and this can only be achieved through a knowledge of the relative proportions of the different forms in blood. As a first step in this direction we have compared the G17 and G34 responses to feeding in normal subjects and patients with gastric ulcer and duodenal ulcer.

Most previous attempts to estimate separate G17 and G34 in blood have involved the use of physico-chemical methods to separate the two forms[5]. These methods are relatively insensitive and laborious and

we have elected to use a different approach in which G17 and G34
are estimated by radioimmunoassay using two antisera which differ
in their specificity for G17 and G34[6]. One antiserum is specific
for the COOH-terminus of G17 and therefore estimates G17 and G34,
while in a separate assay an antibody is used which is almost ab-
solutely specific for G17, and which does not cross-react with G34
or with COOH-terminal or NH_2 terminal fragments of G17 (Fig. 1).
The difference between the two estimates is a measure of G34 concentration, although it should be noted that this fraction also includes the small quantities of G14 and Component I which may be
present. Dilution curves of serum from a Zollinger-Ellison patient
were parallel with those of G17, validating the use of this antiserum in the radioimmunoassay of gastrin in blood. Experiments in
which molecular forms were estimated by specific radioimmunoassay
and by separation of the forms by gel filtration showed good agreement between the two methods[6].

Gastrin responses to a standard meal of eggs, toast, and oxo
were studied in 25 normal subjects and an equal number of age and
sex matched patients with duodenal ulcer[7]. The increases in circulating G17 measured by specific radioimmunoassay rose to a peak of
about 18 pmol/l in the duodenal ulcer subjects compared to about
13 pmol/l in normal subjects, but the differences were not statistically significant. In both groups G17 immunoreactivity reached a
peak about 20 min after feeding and then declined towards basal.

In contrast, the increases in G34 were about 70% higher in
patients with duodenal ulcer than in normal subjects and these differences were significant ($p < 0.05$). In both duodenal ulcer and
normal subjects the peak G34 concentrations were higher than those
of G17 (Fig. 2), and occurred later (50 min in normals, 30 min in
duodenal ulcer). Several groups have reported that the total gastrin
responses to feeding are higher in duodenal ulcer than in normal subjects, and the present study therefore extends these findings by
showing that the elevated postprandial gastrin concentrations are
mainly attributable to increased G34. The differences between the
groups were most pronounced in the first hour after feeding, and
there were negligible differences in the second hour.

An indication of the biological activity of gastrin in blood in
the two groups can be obtained by relating the present data to that
obtained in studies of the biological activity of exogenously administered gastrins. A convenient expression for the comparison of circulating biologically active gastrin is obtained by describing the
gastrin concentrations as a multiple of the serum concentrations of
exogenously infused gastrin giving half maximal rates of acid secretion (D_{50}). In duodenal ulcer patients the D_{50} of G17 was reported
to be 25 pmol/l and that of G34 was 145 pmol/l.[4] Similar data obtained in our own laboratory (IL Taylor, unpublished observations)
indicates that in normal subjects the D_{50} of G17 is 70 pmol/l, or

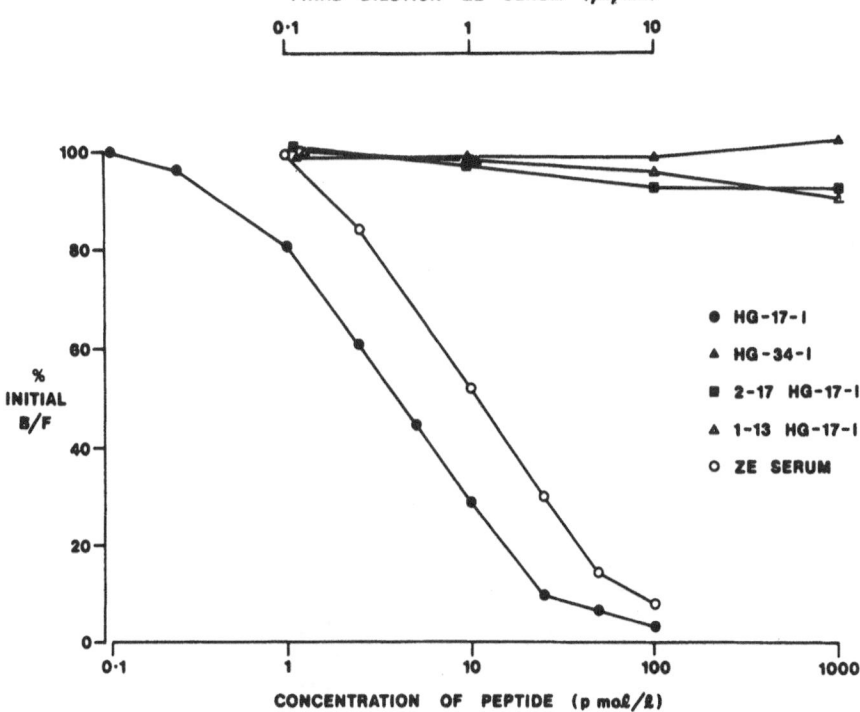

*Figure 1.   Inhibition of binding of^{125}I G17-I to antiserum L6 (1:
100,000) in the presence of graded amounts of human (H) G17-I, G34-
I, 2-17 G17-I, 1-13 G17-I and serum from a patient with Zollinger-
Ellison syndrome.   Inhibition of binding is described in terms of
the ratio of bound (B) to free (F)^{125}I G17-I expressed as a percentage of the ratio in the absence of unlabelled peptide or serum.
(Reproduced from Gastroenterology, ref 6).*

about three times that in duodenal ulcer subjects; these results
are in good agreement with the observation that duodenal ulcer subjects are about three times more sensitive to pentagastrin than
normal subjects[8]. If one assumes that in normal subjects, as in
dogs and duodenal ulcer patients[3,4], there is a six fold difference
in potency between circulating G17 and G34 then the D_{50} of G34 will
be about 420 pmol/l. Table 1 shows gastrin concentration 30 min
after the standard meal in normal and duodenal ulcer subjects expressed in 'D_{50} units'. The results indicate that G17 accounts for
over two thirds of circulating biological activity in both duodenal
ulcer and normal subjects even though G34 is present in higher molar
concentrations. In duodenal ulcer patients the total circulating
activity at peak gastrin concentrations is almost one D_{50} unit, in
other words if maintained these concentrations would produce about
half maximal acid secretion. In normal subjects on the other hand

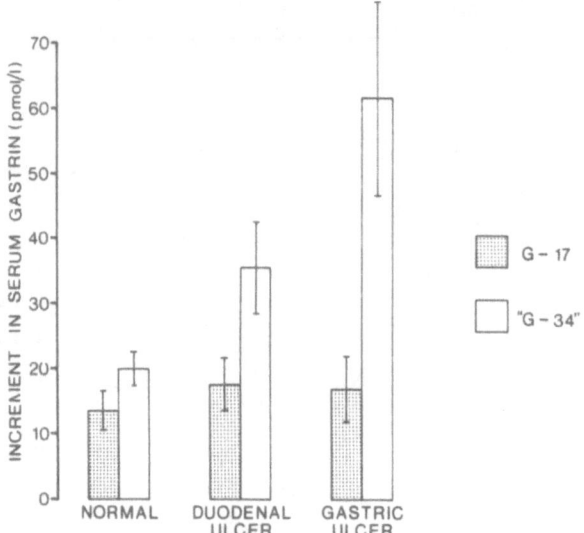

Figure 2. Peak increments in concentration of G17 and G34 measured by specific radioimmunoassay in normal (n = 25), duodenal ulcer (n = 25) and gastric ulcer (n = 5) subjects, in response to a standard meal. Vertical bars represent S.E.M. (Taylor IL and Dockray GJ, to be published).

the total circulating biological activity is only about 0.25 of a D_{50} unit, and from Michaelis-Menten kinetics this would correspond to less than 20% of maximal acid output. The differences in circulating gastrin concentrations in duodenal ulcer and normal subjects therefore become all the more striking when one takes into account the increased sensitivity of duodenal ulcer patients to gastrin.

In patients with gastric ulcer the G17 responses to a standard meal were similar to those in normal subjects. The peak increases in G17 were slightly higher than normal, but these differences were not significant. The time courses of the responses were also essentially similar. In contrast, the peak G34 responses to feeding in gastric ulcer were three-four times greater than those in normal subjects (Fig. 2). Moreover, the elevated G34 concentrations extended over the full two hour period after the meal and were therefore readily distinguishable from responses in duodenal ulcer subjects in which G34 concentrations were similar to normal in the second hour. It is possible, but unlikely, that these differences are caused by alterations in the metabolism of G34 in gastric ulcer. More likely is that in gastric ulcer there is enhanced secretion of G34 possibly due to increased tissue concentrations of G34. To examine this possiblity we have analysed molecular forms of gastrin in extracts of antral and duodenal biopsies obtained at endoscopy in

TABLE 1. Estimated contributions of G17 and G34 to biological activity of gastrin 30 min after a standard meal. Gastrin concentrations are expressed as a multiple of the serum concentration required for half maximal acid output.

	Normal	Duodenal Ulcer
G17	0.19	0.72
G34	0.05	0.25
Total	0.24	0.97

TABLE 2. Concentrations (nmol/g) of G17 and G34 in antral and duodenal biopsies obtained at endoscopy in patients with gastric ulcer (n = 5) and duodenal ulcer (n = 9). Values are means ± S.E.M.; asterisks indicate significant differences, p < 0.05 between the two groups. (Unpublished observations, Dockray GJ, Walker R, Owens D, Tracy HJ, Callam J).

	Antrum		Duodenum	
	G17	G34	G17	G34
Gastric ulcer	13.1 ± 4.5	2.9 ± 1.2	1.1 ± 0.6	1.8 ± 0.3
Duodenal ulcer	13.8 ± 4.1	1.2 ± 0.2*	0.3 ± 0.1	0.5 ± 0.1*

patients with gastric and duodenal ulcer. Preliminary results (Table 2) suggest that whereas the concentrations of G17 in antral and duodenal mucosa are similar in the two groups, the concentrations of G34 in both antral and duodenal mucosa are elevated in gastric ulcer subjects. In both groups the relative contribution of G34 to total gastrin concentrations was higher in the duodenum than in the antrum. These results point to functional differences in G-cell activity in gastric and duodenal ulcer. Whether these differences are secondary to other changes, for example in the patterns of acid secretion and inhibition of gastrin release, remains to be studied. The answers to these and other questions will depend on an understanding of the factors controlling the relative proportions of G17 and G34 produced and secreted by G-cells.

REFERENCES

1. Walsh JH, Grossman MI: Gastrin. New Engl J Med 292: 1324-1332,
    1975

2. Gregory RA: The gastrointestinal hormones: a review of recent
    advances. J Physiol 24: 1-32, 1974

3. Walsh JH, Debas HT, Grossman MI: Pure human big gastrin: immuno-
    chemical properties, half-life and acid stimulating activity
    in dogs. Gastroenterology 54: 477-485, 1974

4. Walsh JH, Isenberg JI, Arnfield J et al: Clearance and acid-
    stimulating action of human big and little gastrin in duo-
    denal ulcer subjects. J Clin Invest 57: 1125-1131, 1976

5. Yalow RS: Heterogeneity of peptide hormones with relation to
    gastrin. In, Gastrointestinal hormones, edited by JC Thompson
    Austin, Texas. University of Texas Press, 1975, pp 25-41

6. Dockray GJ, Taylor IL: Heptadecapeptide gastrin: measurement in
    blood by specific radioimmunoassay. Gastroenterology 71:
    971-977, 1976

7. Taylor IL, Dockray GJ: Little gastrin response to a meal; a
    comparison between patients with duodenal ulceration and
    normal subjects. Gut 17: 393, 1976

8. Isenberg JI, Grossman MI, Maxwell V et al: Increased sensitivity
    to stimulation of acid secretion by pentagastrin in duodenal
    ulcer. J Clin Invest 55: 330-337, 1975

ANTRAL G CELLS AND MUCOSAL GASTRIN CONCENTRATION IN NORMAL SUBJECTS AND IN PATIENTS WITH DUODENAL ULCER

L. Barbara, G. Biasco, M. Salera, F. Baldi, G. Di Febo and M. Miglioli

Cattedra di Gastroenterologia, Università di Bologna

Bologna, Italy

INTRODUCTION

The increase of serum gastrin after a protein rich meal,[1-2] insulin hypoglycaemia[3] and sham feeding[4] is higher in duodenal ulcer patients than in normal subjects. This may result from an increased number of antral G cells and/or from an increased antral gastrin content. Previous reports in this field have led to contradictory results. From a methodological point of view, it is still questionable if the antral gastrin content or the G cell count determined from a few antral bipsy specimens can be representative of the whole endocrine area of the antrum.

MATERIALS AND METHODS

We have examined 15 patients with duodenal ulcer and 12 normal subjects without known diseases of the gastrointestinal tract. In all cases consent was obtained.

Specimens of antral mucosa were excised during endoscopy from 2,3,4,5 and 6 cm from the pylorus along the greater curvature, using a small graduated tube previously introduced into the stomach. Samples were immediately cut vertically to the surface in two fragments under stereomicroscopic view: one was used for immunohistological examination and the other for estimation of mucosal gastrin concentration.

Immunohistochemistry

Tissue samples were fixed in Bouin's fluid and embedded in paraffin. Non-progressive sections 5 μ thick were incubated with antigastrin serum from a rabbit immunized with synthetic human gastrin (titre 1:10.000, Imperial Chemical Industries, kindly provided by Dr. G. Willems). A 1:50 dilution of the antiserum in phosphate-buffered saline (PBS) pH 7.4 was used for each incubation. After three PBS washings, the sections were allowed to react with sheep-antirabbit IgG peroxidase conjugated serum, diluted 1:50 in PBS. Diaminobenzidine was used to develop peroxidase activity. The sections were counterstained with Light Green.

Count of G cells

6 to 12 longitudinal sections have been examined in each biopsy with a Leitz microprojector (objective 25x, eyepiece 2x) at a distance of 30 cm from the screen. In agreement with Creutzfeldt,[5] the size of the area used for counting G cells (0.25x0.30 mm) encompassed the midzone of the antral mucosa, where antral G cells are predominantly situated in man.

Mucosal gastrin estimation

Each sample (weight: 1.8-5.6 mg) was immediately frozen on dry ice or in liquid nitrogen and stored at -20°C until assay. When examined, the frozen biopsy was placed into 2 ml 0.02 M Veronal buffer pH 8.4 in a test tube in a bath of boiling water for 10 min and then homogenized into a very fine suspension. After centrifugation at 4500rpm for 5 min, the gastrin content was measured by radioimmuno assay (Gastrin Radioimmunoassay Kit, Schwarz/Mann, Orangeburg, New York) in the supernatant in a series of dilutions in ratios from 1:10 to 1:1000. Gastrin mucosal concentration was expressed as μg-equivalent of the heptadecapeptide gastrin standard per g of tissue and was the mean of duplicate determinations of three homogenate dilutions.

Statistical analysis

All values are presented as means $\pm$ SD. Student's t test for unpaired values and the linear regression test were used for calculations of P values.

RESULTS

As shown in Figure 1, the mean gastrin concentration in antral mucosa of control subjects (35.9 $\pm$ 27.3 μg/g) was considerably higher than in antral mucosa of duodenal ulcer patients (20.7 $\pm$ 24.3 μg/g t = 2.25, P < 0.05). However, when the mean values of each site of biopsy were compared, no significant difference could be found between healthy subjects and duodenal ulcer patients (Fig. 2). The results of the G cell count are shown in Figures 3 and 4. The mean value of antral G cell concentration was 23.5 $\pm$ 9.8 G cells/area in

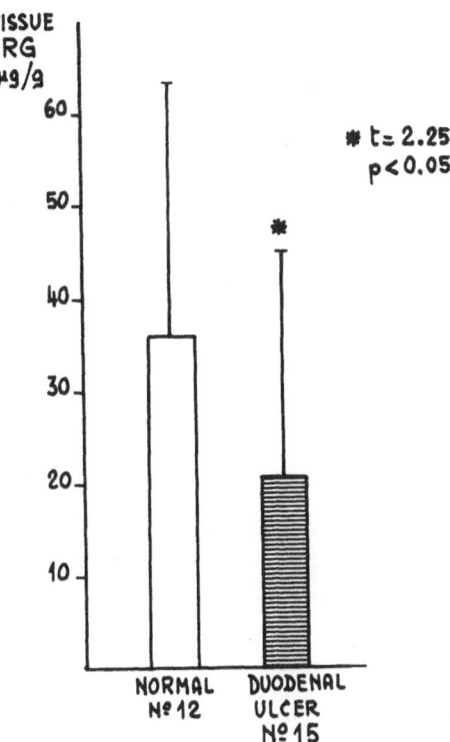

Figure 1. Immunoreactive gastrin (IRG) concentration in antral mucosa of duodenal ulcer patients and controls. Mean values ± SD of all sites of biopsy.

normal subjects and 15.9 ± 12.3 G cells/area in patients with duodenal ulcer (t = 2.48, $P < 0.02$). The comparison of the mean values of each site of biopsy shows a generally lower concentration of G cells per area in duodenal ulcer patients than in controls but the difference becomes significant only in biopsies taken at 6 cm from the pyloric ring (t = 2.98, $P < 0.02$).

A significant correlation was found between mucosal gastrin concentration and the number of G cells per area when all values obtained in normal subjects and in duodenal ulcer patients were considered (Fig. 5).

DISCUSSION

The relative importance of the various sites of gastrin release was insufficiently known until now. However, the gastrin content

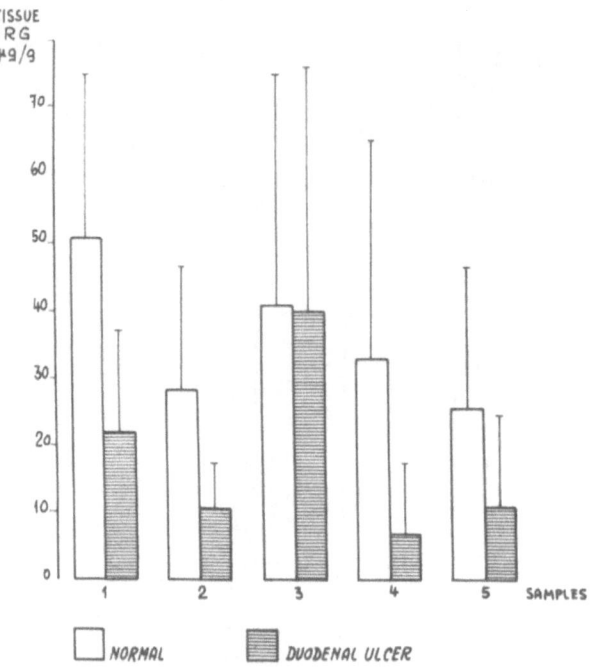

Figure 2.   IRG concentration in antral mucosa of duodenal ulcer patients and controls.   Mean values ± SD of each site of biopsy.

in the proximal and distal duodenum, in the jejunum and in the gastric body are only 0.5-10% of antral mucosa concentration,[6] so that it is generally accepted that the antrum is the major source of the increased serum gastrin levels after a protein rich meal.

The exaggerated gastrin release after various stimulants in duodenal ulcer patients has been considered to result from an increased G cell mass[7-10] or higher functional activity of the G cells.[5-6,11-16]

Absolute quantitative estimation of the total number of antral G cells can be made only in resected stomachs.[17] Similar studies have been carried out on endoscopic biopsies, presuming a homogeneou distribution of G cells in tissue and an equal extension of antral mucosa.[7-10]

In our study we aimed to encompass most of the antral mucosa using multiple biopsy specimens (5 samples taken from 2 cm proximal to the pylorus, 1 cm apart from one another, along the greater curvature). The greater curvature of the stomach was chosen as the site of biopsies because of technical reasons (ease in excising biopsies) and also because of lower incidence of inflammatory

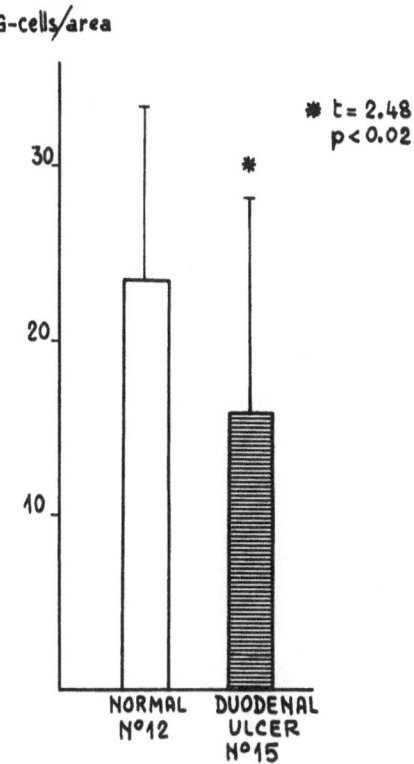

Figure 3. Number of G cells in antral mucosa of duodenal ulcer patients and normal subjects. Mean values ± SD of each site of biopsy.

damages in such a region and a more abrupt borderline zone between antral and fundic mucosa.

According to our previous results,[18] in the present study the G cell concentration appears to vary considerably even when adjacent areas in the same subject are compared or equivalent areas in different subjects are compared. This is true for mucosal gastrin concentration too. Furthermore, the extension of antral area varied considerably from one subject to another and appeared to be smaller in duodenal ulcer patients than in controls. Similar results have been obtained by Willems and Keuppens on resected stomachs[17] and by Creutzfeldt and his associates on biopsy specimens.[5,13] These authors point out that the analysis of gastrin content and G cell count upon a single or few biopsies may be misleading and does not give information on the total G cell number and the total gastrin content of the antrum.

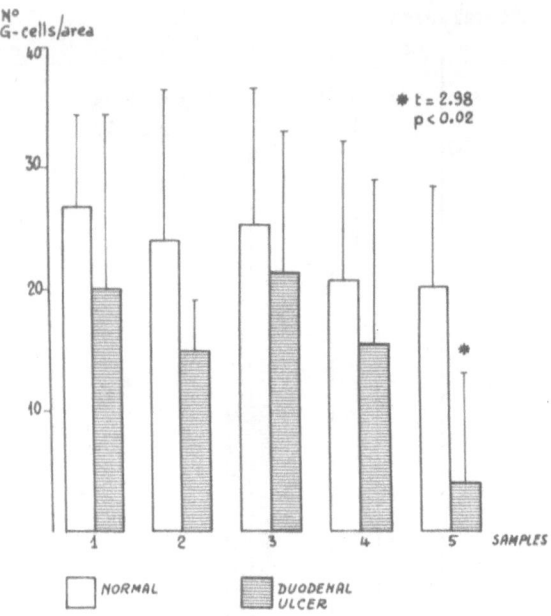

Figure 4. Number of G cells in antral mucosa of duodenal ulcer patients and normal subjects. Mean values ± SD of each site of biopsy.

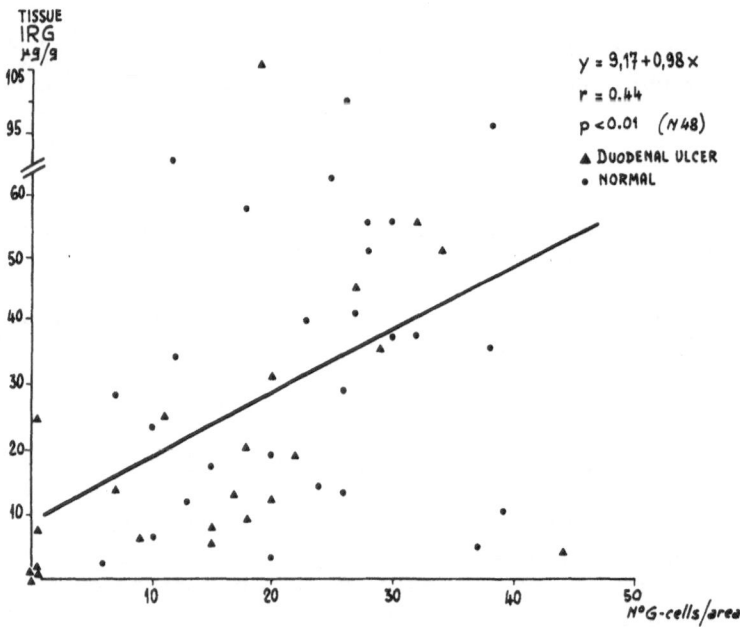

Figure 5. Relationship between mucosal IRG concentration and number of G cells in antral mucosa. All values from normal subjects (circles) and DU patients (triangles) were considered.

Our results do not confirm the suggestion of more abundant G cells in antral mucosa of duodenal ulcer patients nor the existence of a larger gastrin content. Both parameters were found to be lower in duodenal ulcer patients in each sample site and the difference was significant when the mean values from all sites were compared (P < 0.02). Moreover, the mean G cell number in sample 5 (6 cm from the pylorus) was significantly lower in patients with duodenal ulcer than in controls. The latter result is in agreement with the observations of Capper and his associates[19] and indicates a lesser extension of the antral area in duodenal ulcer patients. We cannot agree with the hypothesis of a G cell hyperplasia in duodenal ulcer patients;[7-10] the exaggerated gastrin response to different stimulants in patients with duodenal ulcer could be interpreted as the consequence of a higher functional activity of the G cells.

REFERENCES

1. Korman MG, Soveny C, Hansky J: Serum gastrin in duodenal ulcer. Part I: Basal levels and effect of food and atropine. Gut 12: 899-902, 1971

2. Mc Guigan JE, Trudeau WL: Differences in rates of gastrin release in normal persons and patients with duodenal ulcer disease. N Engl J Med 288: 64-66, 1973

3. Stadil F: Gastrin and insulin hypoglycaemia. A review of studies in gastrin determination and hypoglycaemic release of gastrin in man. Scand J Gastroenterol 9 (suppl 23): 1-49, 1974

4. Mayer G, Arnold R, Feurle G, et al: Influence of feeding upon serum gastrin and gastric acid secretion in control subjects and duodenal ulcer patients. Scand J Gastroenterol 9: 703-710, 1974

5. Creutzfeldt W, Arnold R, Creutzfeldt C, et al. Mucosal gastrin concentration, molecular forms of gastrin, number and ultrastructure of G-cells in patients with duodenal ulcer. Gut 17: 745-754, 1976

6. Malmström J, Stadil F, Rehfeld JF: Gastrins in tissue. Concentration and component pattern in gastric, duodenal and jejunal mucosa of normal human subjects and patients with duodenal ulcer. Gastroenterology 70: 697-703, 1976

7. Solcia E, Capella C, Vassallo G: Endocrine cells of the stomach and pancreas in states of gastric hypersecretion. Rendic Gastroenterol 2: 147-158, 1970

8. Pearse AGE, Bussolati G: Immunofluorescence studies of the dis-
    tribution of gastrin cells in different clinical states.
    Gut 11: 646-648, 1970

9. Pearse AGE, Coulling I, Weavers B, et al: The endocrine poly-
    peptide cells of the human stomach, duodenum and jejunum.
    Gut 11: 649-658, 1970

10. Polak JM, Staff B, Pearse AGE: Two types of Zollinger-Ellison
    syndrome: immunofluorescent, cytochemical and ultrastructural
    studies of the antral and pancreatic gastrin cells in dif-
    ferent clinical states. Gut 13: 501-512, 1972

11. Blair EL, Harper AA, Reed JD: An assay technique for gastrin.
    J Physiol (Lond) 163: 47-48, 1962

12. Emås S, Fyrö B: Antral gastrin activity in duodenal and gastric
    ulcers. Gastroenterology 46: 1-7, 1964

13. Creutzfeldt W, Creutzfeld C, Arnold R: The gastrin producing
    cells under normal and pathological condition. Rendic Gas-
    troenterol 7: 93-109, 1975

14. Emås S, Borg I, Fyrö B: Antral and duodenal gastrin activity in
    non-ulcer and ulcer patients. Scand J Gastroenterol 6: 39-
    43, 1971

15. Malmström J. Stadil F: Measurement of immunoreactive gastrin in
    gastric mucosa. Scand J Gastroenterol 10: 433-439, 1975

16. Malmström J, Stadil F: Gastrin content and gastrin release.
    Studies on the antral content of gastrin and its release to
    serum during stimulation by food. Scand J Gastroenterol
    suppl 37: 71-76, 1976

17. Willems G, Keuppens F: Total number of gastrin cells in the
    normal human stomach and in peptic ulcer patients (abstr).
    Rendic Gastroenterol 9: 57, 1977

18. Biasco G, Salera M, Santini D, et al: On the possible relation-
    ship between number of G-cells, mucosal gastrin and plasma
    gastrin levels in normal subjects (abstr). Rendic Gastroen-
    terol 9: 72, 1977

19. Capper WM, Butler TJ, Buckler KG, et al: Variation in size of
    the gastric antrum: measurement of alkaline area associated
    with ulceration and pyloric stenosis. Ann Surg 163: 281-291,
    1966

ROLE OF THE SMALL BOWEL IN REGULATING SERUM GASTRIN AND GASTRIC INHIBITORY POLYPEPTIDE (GIP) LEVELS AND GASTRIC ACID SECRETION

H. D. Becker, N. J. Smith, H. W. Börger and A. Schafmayer

Department of Surgery, University of Goettingen

Germany

An increase in gastric acid secretion after resection or by-pass of the small bowel has been described by several authors in man and dogs[1,2,3]. Furthermore, a high incidence of peptic ulcer disease has been observed in patients with small bowel resection[4,5]. The mechanism which induces gastric hypersecretion after exclusion of large parts of the small bowel from food passage is still poorly understood. Several authors have described hypergastrinemia after small bowel exclusion or bypass, which may be caused by either a decrease in catabolism or a diminished release of inhibitors from the small intestine[6,7,8]. One of the physiologically important inhibitors of gastric acid secretion may be the gastric inhibitory polypeptide (GIP) which is released from the duodenum and jejunum after food intake and shows besides its insulinotropic effect a strong inhibition of stimulated acid secretion in dogs[9].

The present experiments in dogs were undertaken to examine the effect of small bowel exclusion (SBE) and subsequent resection of the excluded part (SBR) on basal and food-stimulated serum gastrin and serum GIP levels and Heidenhain pouch (HP) acid secretion.

MATERIAL AND METHODS

Four adult mongrel dogs were prepared with Heidenhain-pouches. After a recovery period of 6 weeks a standardized feeding study was performed: After a 12-hour fasting period the dogs were fed a standard 400 g meat meal. During the basal period and during the experiment blood samples for measurement of serum gastrin and serum GIP concentrations were collected at regular intervals from a peripheral vein. Acid secretion from Heidenhain-pouch was measured

every 15 minutes during the basal period and for 150 minutes after
food intake. Acid concentration was determined by titration with
0.1 N NaOH using phenol red as indicator.

After completing 2 studies in each dog we performed a bypass
of the majority of the small bowel: The jejunum was transsected
20 cm distal from the ligament of Treitz and after closing the distal part of the transsected jejunum the proximal jejunal stump was
anastomosed end – to – side to the distal ileum 20 cm proximal of
the ileocoecal valve (SBE). After a recovery of at least 4 weeks
2 food studies were repeated as described above. After completing
this series of studies the four surviving dogs underwent a third
operation: the jejunum and ileum which had been excluded in the
second operation was resected (SBR). Finally, after a recovery
period of 4 weeks 2 food studies were repeated.

RESULTS

The results are tabulated in Table 1.

GASTRIN

The basal serum gastrin concentration in the control group was
36 + 6 pg/ml; this was significantly increased after SBE to 59 + 7
pg/ml (p < 0.01). After SBR basal serum gastrin showed a further
significant increase to 95 + 17 pg/ml (p < 0.01). After food serum
gastrin increased to a maximum of 135 + 16 pg/ml in the control group
to 212 + 27 pg/ml in the SBE group and to 325 + 59 pg/ml in the SBR
group. The integrated postprandial gastrin output for the 150 minute:
after food intake was 9.03 + 1.84 ng/ml in the control group; this
increased significantly to 20.00 + 4.28 ng/ml after SBE. After SBR
integrated postprandial gastrin output increased further to 29.92
+ 6.76 ng/ml.

GIP

The basal GIP concentration was 665 + 72 pg/ml; this decreased
to 357 + 46 pg/ml after SBE. The resection of the excluded small
bowel did not change basal serum GIP levels (374 + 49 pg/ml after
SBR). After food GIP rose to a peak of 1297 + 135 pg/ml in the control group, whereas after SBE GIP only increased to 847 + 66 pg/ml
and after SBR to 821 + 88 pg/ml. The integrated postprandial GIP
output was significantly higher in the control group (73.93 + 16.14
ng X 150 min/ml) than after SBE (49.66 + 12.07 ng X 150 min/ml) or
after SBR (50.97 + 11.02 ng X 150 min/ml).

TABLE 1. Serum gastrin, serum GIP and HP acid secretion after small bowel exclusion (SBE) or small bowel resection (SBR) compared to controls.

		Controls	SBE	SBR
GASTRIN	basal serum gastrin pg/ml	36 ± 6	59 ± 7	95 ± 11
	peak post-prandial serum gastrin pg/ml	135 ± 16	212 ± 27	325 ± 59
	integrat. post-prand. gastrin output ng. 150 min/ml	9.03 ± 1.84	20.00 ± 4.28	29.92 ± 6.76
GIP	basal serum GIP pg/ml	665 ± 135	357 ± 46	374 ± 49
	peak postprand. serum GIP pg/ml	1297 ± 135	847 ± 66	821 ± 88
	integrated postprand. GIP output ng. 150 min/ml	73.93 ± 16.14	49.66 ± 12.07	50.97 ± 11.02
HP acid secretion	postprandial acid output mEq/150 min.	4.93 ± 1.08	4.50 ± 0.97	6.34 ± 1.09

HP ACID SECRETION

HP acid output after food did not differ significantly in the
control group compared to SBE as shown in figure 1. However, after
resection of the excluded small bowel the postprandial HP acid output increased significantly in the SBR group.

DISCUSSION

Gastric acid hypersecretion has been documented after small
bowel resection in man and dogs[1,3,4]. After exclusion of large
parts of small intestine, however, the results are controversial:
whereas some authors found an increase in basal[12] or stimulated[13]
acid secretion, others could not detect a change in gastric secretion[2]. In the present studies exclusion of approximately 80% of
the small intestine did not change significantly the gastric acid
secretion from a denervated Heidenhain fundic pouch. However, after
resection of the excluded small bowel HP acid secretion increased
significantly.

The mechanisms for these changes in acid secretion are still
poorly understood. Since gastrin is the most potent stimulator of
acid secretion, several authors have suggested that an increase in
gastrin release may be responsible for the observed hypersecretion
[7,8,14,15]. In the present studies exclusion of the small bowel
caused a significant increase in basal and postprandial serum gastrin
levels. After resection of the excluded small intestine serum gastrin showed a further increase. These results are in agreement with
findings of Solhaug and Schrumpf[12] and Lennon and coworkers[16] who
observed an increase in basal and food stimulated serum gastrin
levels after jejunal-ileal bypass, whereas Coyle et al.[13] found
no change in serum gastrin in human patients with small bowel bypass.
Our findings after small bowel resection are in agreement with several other groups who observed hypergastrinemia in the short bowel
syndrome[12,14,15].

The increase in basal or postprandial serum gastrin concentrations after small bowel resection is not understood. In earlier
studies we have observed an uptake of elevated serum gastrin levels
by the small bowel[7]. However, Strunz and Grossman[17] have described
a diffuse uptake of gastrin in the body, which may explain the relative short half - life of gastrin in the circulation. Straus and
Yalow[8] speculated that the time course of appearance and disappearance of plasma gastrin militates against the hypothesis of a significant role of decreased catabolism after small bowel resection. However, our present results show that after resection of the excluded
small bowel there is a further significant increase in basal and
postprandial serum gastrin levels which might be explained by the

loss of a catabolic site of gastrin which cannot be taken over by
other sites.

Another explanation for the changes in gastric acid secretion
and in gastrin release is the decreased release of an intestinal
inhibitor after small bowel bypass or resection. Gastric inhibitory
polypeptide (GIP) has a strong inhibiting effect on basal and postprandial gastric acid secretion in dogs[9]. In the present studies
basal and postprandial serum GIP levels are decreased after exclusion of the majority of the small bowel. However, gastric acid
secretion from the denervated fundic pouch was not changed. After
resection of the bypassed small bowel serum GIP levels were not
further changed. These results indicate that GIP may not be of major importance in regulating gastric acid secretion after small
bowel bypass or resection.

Several other peptides have been isolated from the mucosa of
the small intestine which may be important in the regulation of
serum gastrin levels and acid secretion after small bowel resection.
Our studies seem to indicate that at least two mechanisms may be
responsible for the hypergastrinemia and changes in acid secretion
after small bowel bypass or resection: besides the decrease in
catabolic site, an intestinal factor may be postulated which regulates
the release of gastrin and which is decreased after small bowel bypass. The nature of this intestinal factor remains to be elucidated.

REFERENCES

1. Osborne MP, Frederich PL, Sizer JS, et al: Mechanism of gastric
       hypersecretion following massive small bowel resection. Ann.
       Surg. 164; 622, 1966

2. Wise L, Vanghan R, Stein TH: Studies on the effect of small bowel
       bypass for massive obesity on gastric secretory function.
       Ann. Surg. 183; 259, 1976

3. Buchwald H, Coyle JJ, Varco RL: Effect of small bowel bypass on
       gastric secretory function: Post-intestinal exclusion
       hypersecretion. Surgery 75; 821, 1974

4. Frederich PL, Sizer JS, Osborne MP: Relation of massive small
       bowel resection to gastric secretion. N. Engl. J. Med. 272;
       509, 1965

5. Winawer SJ, Broitman SA, Osborne MP, Zamchetzn: Successful manage-
       ment of massive small bowel resection based on assessment of
       absorption defects and nutritional needs. N. Engl. J. Med.
       274; 72, 1966

6.   Straus E, Gerson CD, Yalow RS: Hypersecretion of gastrin asso-
     ciated with the short bowel syndrome. Gastroenterology 66;
     175, 1974

7.   Becker HD, Reeder DD, Thompson JC: Extraction of circulating
     endogenous gastrin by the small bowel. Gastroenterology 65;
     903, 1973

8.   Straus E, Yalow RS: Differential diagnosis of hypergastrinemia.
     In: Thompson JC: Gastrointestinal hormones. University of
     Texas Press, Austin, 1975

9.   Brown JC, Dryburgh JR, Moccia P, Pederson RA: The current status
     of GIP. In: Thompson JC: Gastrointestinal hormones. Univer-
     sity of Texas Press, Austin, 1975

10. Mayer G, Arnold R, Feurle G, Fuchs K, Ketterer H, Track NS,
     Creutzfeldt W: Influence of feeding and sham feeding upon
     serum gastrin and gastric acid secretion in control subjects
     and duodenal ulcer patients. Scand. J. Gastroenterol. 9;
     703, 1974

11. Kuzio M, Dryburgh JR, Malloy KM, Brown JC: Radioimmunoassay for
     gastric inhibitory polypeptide. Gastroenterology 66; 357,
     1974

12. Solhaug JH, Schrumpf E: Effect of small bowel bypass on serum
     gastrin levels and gastric acid secretion in man. Scand. J.
     Gastroent. 11; 329, 1976

13. Coyle JJ, Varco RL, Buchwald H: Gastric secretion and serum gas-
     trin in human small bowel bypass. Arch. Surg. 110; 1036, 19'

14. Wickbom G, Landor JH, Bushkin FL, Mc Guigan JE: Changes in canil
     gastric acid output and serum gastrin levels following mass:
     small intestinal resection. Gastroenterology 69; 448, 1975

15. Junghanns K, Kaess H, Dörner M, Encke A: The influence of resec·
     tion of the small intestine on gastrin levels. SGO 140; 27,
     1975

16. Lennon J, Lidgard GP, Tarry SH, Sircus W: Gastrin response to a
     standard meal after jejuno-ileal bypass in morbid obese
     patients. Gut 16; 407, 1975

17. Strunz U, Grossman MI: Removal of gastrin by various organs.
     Cure Symposium: Hormones and Peptic Ulcer. Los Angeles,
     October 1976.

CIMETIDINE TREATMENT IN ZOLLINGER-ELLISON SYNDROME

S. Bonfils, M. Mignon and G. Kloeti

Unité de Recherches de Gastroénterologie (INSERM U.10)
et Service de Gastroénterologie B; Hospital Bichat

170 Bd Ney 75877 Paris Cedex 18, France

Total gastrectomy is the accepted method of symptomatic treatment for the Zollinger-Ellison syndrome (ZES) but has disabling side-effects. However, treatment with H_2-receptor antagonists has proved promising. The usefulness of this series of drug was first demonstrated with metiamide[2,3] but despite the fact that only a small number of patients was studied, some practical and theoretical problems appeared: drug toxicity, assessment of efficacy of long term treatment, with variations consisting of an escape phenomenon or, conversely, prolonged secretory inhibition.[1,2]

We report 8 cases of ZES treated with cimetidine for 15 to 195 days.

PATIENTS AND CONDITIONS OF TREATMENT

Information on the 8 treated cases is summarized in Table I. Cases 1 to 5 are considered as "chronic ZES", followed in a medical ward and, from time to time, were treated as outpatients. The other cases ("acute ZES") were hospitalized in an intensive care unit because of postoperative complications making gastric aspiration necessary (cases 6 and 7), and (case 8) because of acute dehydration.

Cimetidine was administered orally, the dose depending upon the clinical and biological information. For instance the dose was originally 800 mg for cases 1 to 4, increasing up to 1,000 mg for cases 1 and 2, and up to 2,000 mg for case 3; as for case 5, still

TABLE I: ZES CASES TREATED WITH CIMETIDINE

No	Ulcer	Diarrhea	Previous Surgery	BAO mEq/h	Serum gastrin Basal pg/ml	Increase (%) under secretin	Tumor
1	0	+	0	38	820	27.2	Pancreas (head)
2	0	+	0	35	300	33.3	Duod. wall
3	Bulb	+	0	13.6	210	42.2	Duod. wall
4	Stomach (L.C.)	0	Intestinal resection for jej. ulcer	21.4	450	100	Islet cell hyperplasia
5	Bulb	0	Vagotomy + pyloroplasty	36.3	415	7	Pancreas (head)
6	Stomach (multiple)	+	Partial gastrectomy Intest. resection for jej. ulcer	15	200	–	Islet cell hyperplasia
7	Anastomot.	0	Vagotomy, partial gastrectomy	–	1500	–	Not found.
8	Bulb	++ dehydration	0	15	200	–	Lymph node metastases.

TABLE II: ORAL CIMETIDINE TREATMENT IN ZES

No	CIMETIDINE		THERAPEUTIC RESULTS				SURGERY DURING OR AFTER CIMETIDINE
	Total dose	day	Ulcer healing	Diarrhea improvement	Secretion inhibition	Conclusion	
1	115 g	117 d	–	No	Acute	Unsuccessful	Total gastrectomy
2	173 g	195 d	–	Yes	Prolonged	Successful	Duod. tumor excision
3	102 g	74 d	Yes	Yes	Prolonged	Successful	Duod. tumor excision
4	40 g	56 d	Ulc. under treatment	–	Acute	Unsuccessful	Total gastrectomy
5	240 g	120 d	Yes	Yes	Acute	Successful	No
6	9 g	15 d	Yes	Yes	Acute	Successful, + escape	Total gastrectomy
7	36 g	45 d	Yes	–	Acute	Successful	Total gastrectomy
8	44 g	22 d	No	No	Acute	Unsuccessful	Total gastrectomy

under treatment, adequate control was obtained only with 2,000 mg,
the original dose of 1,200 mg having been ineffective.

Efficacy criteria were 1) ulcer healing (endoscopically controlled), 2) suppression of diarrhea and 3) inhibition of gastric
secretion. In relation to the last criterion, we considered separately a) acute inhibition observed during the time of the drug's
pharmacological action (within 4 hours after administration) either
on basal or stimulated secretion, and b) prolonged secretory inhibition of basal secretion assessed later than 12 hours after the
drug's final administration.[5]

RESULTS (Table II)

1. Cimetidine was a useful therapeutic agent in 3 out of 5
chronic ZES and in 2 out of 3 acute cases. In one case, an escape
phenomenon was observed after 8 days of treatment with a low dosage
(600 mg/day) resulting in relapse of gastric hypersecretion and
ulcers.

2. Prolonged secretory inhibition was observed in 2 cases.
In each case at laparotomy an isolated duodenal tumor was found.
Following its excision a definite symptomatic cure was obtained.

3. They were no signs of toxicity related to either kidney,
liver or blood. No subjective functional side effects were observed.

4. In the only case still under treatment, the tumor located
in the head of the pancreas could not be surgically removed. The
patient's duodenal ulcer healed with 2,000 mg/day cimetidine. After
a 4 month treatment, oral administration of 400 mg cimetidine (in
one dose) resulted in an acute inhibition of the gastric basal
acid secretion of only 41% (2nd hour after administration).

DISCUSSION

Although good results were obviously obtained in ZES treatment
with cimetidine a number of problems are so far unresolved.[4]

1. Dose: is it possible to define a standard dose or should
the dose be adapted for individual cases and, within each case, for
various circumstances[7]?

In other words, to what extent are we allowed to increase the
dosage in case of therapeutic resistance?

2. Efficacy: which are the best efficacy criteria for a prolonged follow-up? After healing, ulcers may relapse and acute

relapse could occur without ulcer pain. We observed this with cimetidine as well as with metiamide. Diarrhea improvement or re-appearance is indicative of gastric secretion changes; but high acid secretion can be observed with normal stools.

Secretory tests are obviously necessary for quantifying the drug's efficacy and for defining a prolonged inhibition.[5] But how often should (and can) the test be repeated?

3. Indications and duration of the treatment: according to our experience this should be correlated with anatomical findings and with the individual sensitivity to the drug.

Even if evidence is obtained that cimetidine strongly reduces acid secretion, a surgical exploration is required: about 25% of the patients have an isolated benign tumor,[6] either in the tail of the pancreas or within the duodenal wall, that could be operated on and removed. This chance of definitive cure must be offered to the patients. Cimetidine gives the possibility to act on the tumor ex-clusively while leaving the stomach in place;[7] if surgery is un-successful, cimetidine is useful to control acid hypersecretion in the postoperative period.

If metastases are noted either clinically or during surgery, cimetidine treatment might be chosen instead of total gastrectomy.

Prolonged medical treatment is acceptable providing that clini-cal and biological controls are frequently performed. A greater margin of security is obtained in the cases with prolonged secretory inhibition.

As for "acute ZES", cimetidine should only be considered as a preoperative treatment, allowing the patient to improve while await-ing surgery.

REFERENCES

1. Bonfils S, Bernier JJ, Mignon M, Hautefeuille R, Corbic M, Marteau J: Syndrome de Zollinger-Ellison traité médicale-ment par un inhibiteur des récepteurs H_2 á l'histamine. Nouv Presse Med 1975, 4: 2377-2381

2. Bonfils S, Bernier JJ, Mignon M, Lambert R, Bernades P, L'Hirondel C, Accary JP, Kloeti G: Metiamide treatment in five patients with Zollinger-Ellison syndrome. Digestion 1977, 15: 43-52

3. Bonfils S, Mignon M, Accary JP: Inhibiteurs des récepteurs H_2 á l'histamine et syndrome de Zollinger-Ellison. Nouv Presse

Med 1974, 30: 1883 (lettre)

4. Bonfils S, Mignon M, Kloeti G, Jian R: Therapeutic assessment of histamine H_2 blockers in 10 cases of Zollinger-Ellison syndrome. Gastroenterology 1977, 72: Ad/813 (abst.)

5. Bonfils S, Mignon M, Jian R, Kloeti G: Biological studies during long-term cimetidine administration in the Zollinger-Ellison syndrome. In: Cimetidine. Edited by Burland WL, Simkins MA, Excerpta Medica, Amsterdam, Oxford. 1977, chapter 29, 311-321

6. Ellison EH, Wilson SD: The Zollinger-Ellison syndrome: reappraisal and evaluation of 260 registered cases. Ann Surg 1964, 160: 214-227

7. Stage JG, Rune SJ, Stadil F, Worning H: Treatment of Zollinger-Ellison patients with cimetidine. In: Cimetidine. Edited by Burland WL, Simkins MA, Excerpta Medica, Amsterdam, Oxford. 1977, chapter 28, 306-310

CALCITONIN, PARATHYROID HORMONE AND INSULIN CONCENTRATIONS IN SERA FROM PATIENTS WITH GASTRINOMA

M. Cecchettin, A. Albertini, G. Bonora and P. Vezzadini

Instituto di Chimica, Facoltà Medica E.U.L.O. di Brescia
I Clinica Medica Generale e Terapia Medica, Università
di Bologna

Bologna, Italy

The most recent studies of endocrinology have noted a close relationship between gastro-enteric hormones and phosphorus-calcium balance regulator hormones in different physiological conditions. The existence of one or more substances (hormones or electrolytes) capable of completing the last links of this chain is evident. The disorders of gastro-enteric hormones and those of phosphorus-calcium metabolism are always mutually interesting; in fact, hypercalcemia accompanies the highest levels of gastrinaemia in some patients; calcitonin inhibits gastric secretion, and vice-versa gastrin stimulates calcitonin release. Previous studies led us, in agreement with data of other authors, to suggest the hypothesis of the existence of a gastrin-calcitonin system in the regulation of phosphoremia.

On the basis of these experiences, we have studied the regulating factors of phosphorus-calcium metabolism and of some gastro-enteric hormones in 31 in-patients of the Clinica Medica I (University of Bologna), in whom a diagnosis of gastrinoma was made on the basis of clinical evidence and of laboratory tests[1]. The presence of a tumor was found in only 21 patients. In addition, no patient proved to have medullary cancer of the thyroid or parathyroid adenoma. Only five patients presented clinical and biochemical signs of hyperparathyroidism.

In our research we have determined the serum levels of parathyroid-hormone (PTH)[2], gastrin[3], insulin and C-peptide[4] and calcitonin (CT)[5] with radioimmunoassay.

TABLE 1

HORMONE	PATIENTS WITH GASTRINOMA	CONTROLS
Parathyroid-hormone pg/ml	2563 ± 617	555 ± 37
Gastrin pg/ml	951 ± 305	43 ± 2.65
Calcitonin pg/ml	1110 ± 144	350 ± 25
Insulin µU/ml	7.5 ± 2.35	10.9 ± 0.98
C-peptide ng/ml	2.7 ± 0.4	2.5 ± 0.5

In particular, the assay method of PTH permits the determination of the 1-84 fragment COOH terminal group, while that of gastrin measures G-17 and G-34 molecular forms.

All 31 patients studied presented serum values of PTH, CT and gastrin significantly higher (p < 0.01) than in control subjects. No significant differences were found in serum levels of insulin and of C-peptide (Table 1).

The hormone content of the tumors was also evaluated. In particular, PTH and C-peptide were found in 5 patients, in addition to the presence of gastrin.

In the global evaluation of the data, there is a significant correlation between PTH and CT and between PTH and gastrin. The data resulting from this study indicate that generally there is a hypersecretion of PTH and CT in patients with gastrinoma.

In addition, gastrin-secreting tumors also produce PTH or PTH-like substances along with hormones such as insulin and glucagon, as already indicated by various authors. As PTH has been determined using an antibody directed against the 1-84 fragment, the weak biological activity of this fragment could explain the absence of clinical and biochemical signs of hyperparathyroidism. Some authors have excluded the presence of PTH in gastrin-producing tumors; a possible explanation is that the assay included the I-34 NH_2-terminal fragment with short half-life and was capable of pointing out only notable parathyroid hyperfunction.

On the basis of these preliminary data, we can conclude that in gastrinoma the alterations of the hormones regulating the phosphorus-calcium metabolism, in the absence of neoplastic pathology in the structures that secrete them, could depend on the continuous hypersecretion of gastrin, capable of chronically stimulating CT secretion. Calcitonin, through variations induced in the levels of phosphoremia and calcemia, would thus be able to maintain an increased PTH secretion.

REFERENCES

1. Vezzadini P, Rendic. Gastroenterol. 9, 12, 1977

2. Masson ED, New York, 1976

3. Vezzadini P, Tomassetti P, Cipollini F, Atti del II Convegno
   Metodi Radioimmunologici in Endocrinologia, 1975, pg. 291

4. Grodsky GM, Forsham PH, J. Clin. Invest., 39, 1070, 1960

5. Heinen, Eur. J. Clin. Invest., 4, 213, 1974

HORMONAL CONTROL OF THE LOWER ESOPHAGEAL SPHINCTER IN MAN AND DOG: REEVALUATION OF THE PRESENT MANOMETRIC METHOD FOR DIAGNOSIS OF GE REFLUX

Z. Itoh, R. Honda, K. Hiwatashi and I. Takahashi

G. I. Laboratories, Department of Surgery, Gunma University School of Medicine

Maebashi, Japan

Previously, we reported that the contractile activity of the lower esophageal sphincter (LES) and stomach were coordinated during the interdigestive state[1,2]. Both the LES and stomach cycled between periods of strong contractions and longer periods of quiescence; these data were gathered by long-term measurements made by chronically implanted force transducers in conscious dogs. In the present study, we will report on the effect of gut hormones on LES interdigestive contractile activity in the dog and human.

A. DOG EXPERIMENTS

Materials and Methods

Five healthy male mongrel dogs weighing 9-13 kg were used in the present study. A silastic tube (Medical Grade, 602-205, Dow Corning Corporation, Midland, Mich.) was introduced into the jugular vein via the common cephalic branch so that the tip of the tube was placed in the superior vena cava. The other end of this tube was pulled out through a stab wound on the back through a subcutaneous tunnel. This tube was used for postoperative fluid transfusion or i.v. administration of test materials[3]. Transducers were sutured onto the serosal surface of the LES, the gastric body, and antrum to measure the contractile force of the respective circular muscle. In order to suture the transducer over the LES, a 3 cm longitudinal gastrotomy was made on the anterior wall of the stomach, and by

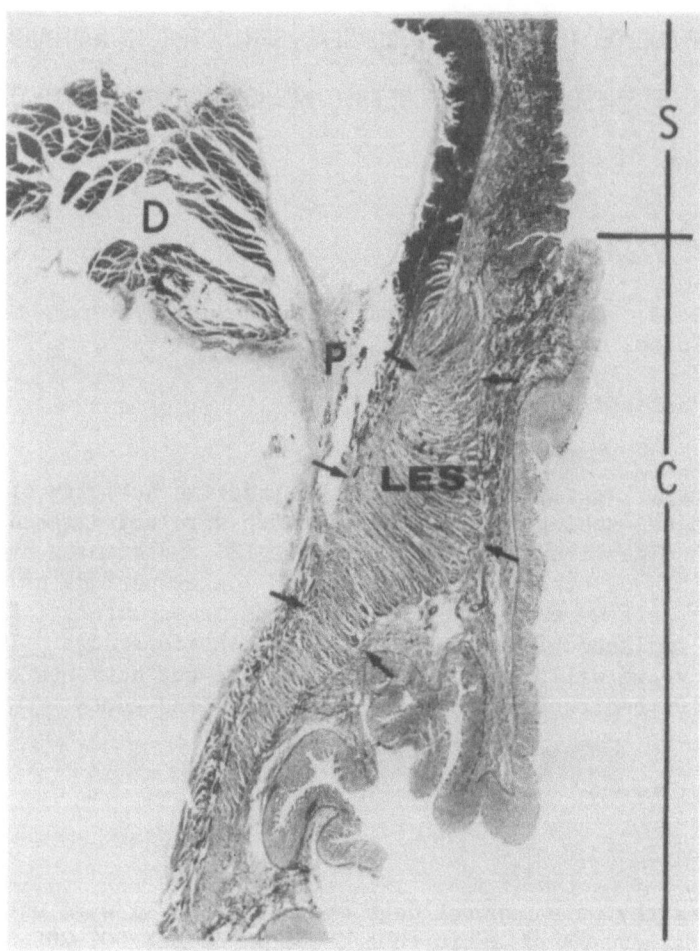

Figure 1. Histologic section of the gastroesophageal junction of the dog. LES, lower esophageal sphincter; D, diaphragm; P, phreno-esophageal membrane; S, squamous epithelium; C, columnar epithelium. (Reprint from: Scand J Gastroent 11 Suppl 39: 93-110, 1976

using a finger inserted through this gastrotomy up into the gastroesophageal junction as a guide, the transducer was accurately sutured
directly over the muscle thickening of the LES. As reported previously[1,2], the LES was found to be located immediately inferior to the
attachment of the phreno-esophageal membrane to the abdominal esophagus (Figure 1). A transducer was then sutured on the gastric body
just opposite the splenic hilus and another on the gastric antrum
3 cm proximal to the gastroduodenal junction. All the lead wires
were drawn out of the abdominal cavity through a stab wound on the
left abdominal wall, pulled through a subcutaneous tunnel on the
left costal flank and brought out through a stab wound between the
scapulas.

The experiments were begun two weeks after surgery. Contractile
activity was recorded on a three channel pen-writing oscillograph at
a paper speed of 1 mm/min when observing diurnal changes and at a
paper speed of 10-30 mm/min when observations of detailed changes
were desired. During all the experiments, the dogs were allowed to
walk freely in a 5-meter diameter circle and drink water as they
wished.

Gut hormones were dissolved in 0.9% saline to the desired concentration and infused i.v. through the implanted silastic tube by
a Harvard infusion pump. All data were processed by statistical
analysis with the Student's t test.

RESULTS

It was found that the episodes of contractile activity were
completely synchronized with those of the stomach during the interdigestive state[4]. These bursts of motor activity were found to be
followed by long periods of motor quiescence. The timing of these
bursts and episodes of quiescence were quite regular. Figure 2 shows
the interdigestive contractile activity recorded at a paper speed of
10 mm/min to show detailed changes. The synchronous contractions of
the LES and the gastric body can be seen.

In the present study, the effect of GI peptides on LES contractile changes together with those of the stomach was investigated
mainly during the interdigestive state. During the digestive state,
when pentagastrin was given in a bolus, transient contractions of
the LES were observed, but when infused continuously in a dose at
0.1-2.7 µg/kg-hr it did not increase contractile activity of the
LES. However, when it was infused during the period of interdigestive contractions, the contractions in the LES and the stomach were
simultaneously suppressed by pentagastrin infusion as shown in Figure 3. The secretin family of peptides,(secretin, VIP, GIP and glucagon), had little effect on LES contractile activity during the interdigestive state. These peptides slightly lowered the contractile

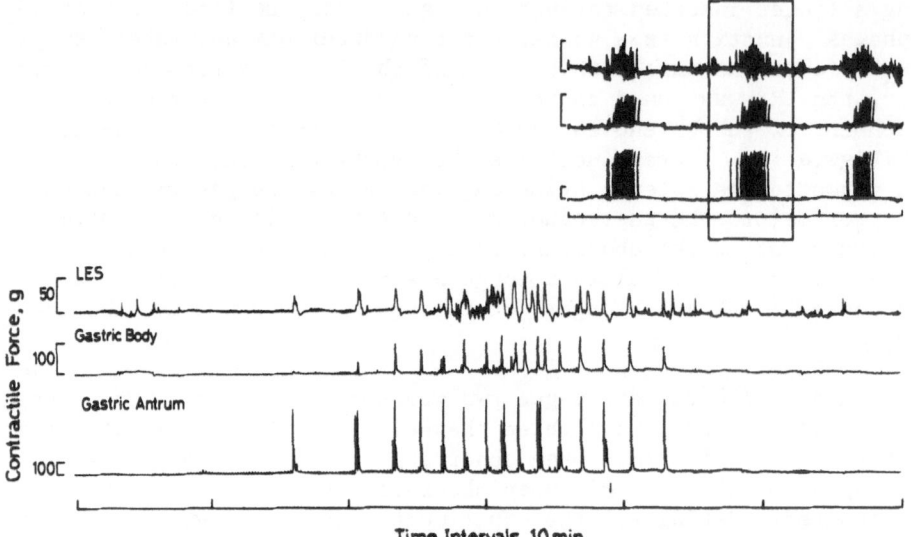

Figure 2.   Detailed changes in LES and gastric contractile activity in the interdigestive state, recorded at paper speed of 10 mm/min. The inset in this and following figures shows the same record and its immediately previous and subsequent changes in each figure taken at a slower paper speed of 1 mm/min.

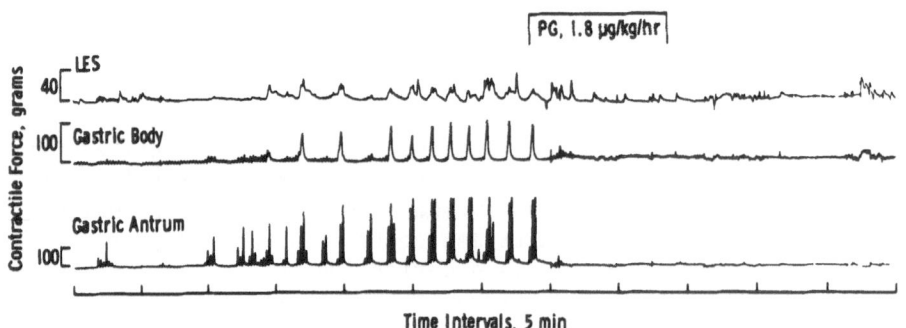

Figure 3.   Inhibitory effect of pentagastrin on the naturally-occurring interdigestive contractions in the LES and the stomach.

force of the LES during the digestive state in a large doses.   As
reported previously[1,5] we first demonstrated that motilin induced
a motor pattern of the gut precisely similar to that of the naturally occurring contractions during the interdigestive state in the
dog.   A similar effect of motilin was also demonstrated in the LES;
when synthetic motilin in a dose at 0.3 μg/kg-hr was infused intravenously 10 min after the termination of the natural contractions,
a synchronous increase in contractile activity of the LES and the
stomach was induced as shown in Figure 4.   However, when motilin
was given during the digestive state, no significant changes were
observed in either the LES or the stomach.   Motilin-induced contractions in the LES and the stomach were instantly inhibited by feeding
or an i.v. infusion of pentagastrin in a dose at 0.2-1.9 μg/kg-hr
as shown in Figure 5.   Furthermore, the secretin family peptides, as
mentioned above, did not influence motilin-induced contractions in
the LES and the stomach.

B.   CLINICAL STUDIES

Materials and Methods

      In the clinical studies, LES pressure was measured on twenty
healthy, non-obese adult volunteers of both sexes.   The manometric
tube assembly consisted of five polyvinyl catheters (Argyle feeding
tube).   The four distalorifices were arranged radially in the same
horizontal plane.   The recording system was standardized and calibrated taking atmospheric pressure as zero.   Recordings were made
by perfusing the catheters with   tap water (2.5 ml/min).   For
manometry the recording assembly was passed through either the mouth
or the nose into the stomach.   LES pressure was measured by a rapid
pull-through technique according to the method reported by Dodds et
al[6].   LES pressure was calculated by subtracting gastric pressure
from the peak pressure at the LES.   LES pressure values determined
for each of the four radial catheter tips were averaged.   All subjects
were fasted for 16 hr and resting LES pressure was monitored before
and after intravenous pulse dose or continuous infusion of GI peptides.   In the present study, pentagastrin (ICI), secretin (Karolinska Institute) and glucagon (Eli Lilly) were used.   All data obtained were analysed statistically with Student's t-test and p values
less than 0.05 were considered as significantly different.

RESULTS

      An i.v. infusion of pentagastrin did not change LES pressure
in a dose of 0.1-2.7 μg/kg-hr; however, when it was given as a bolus

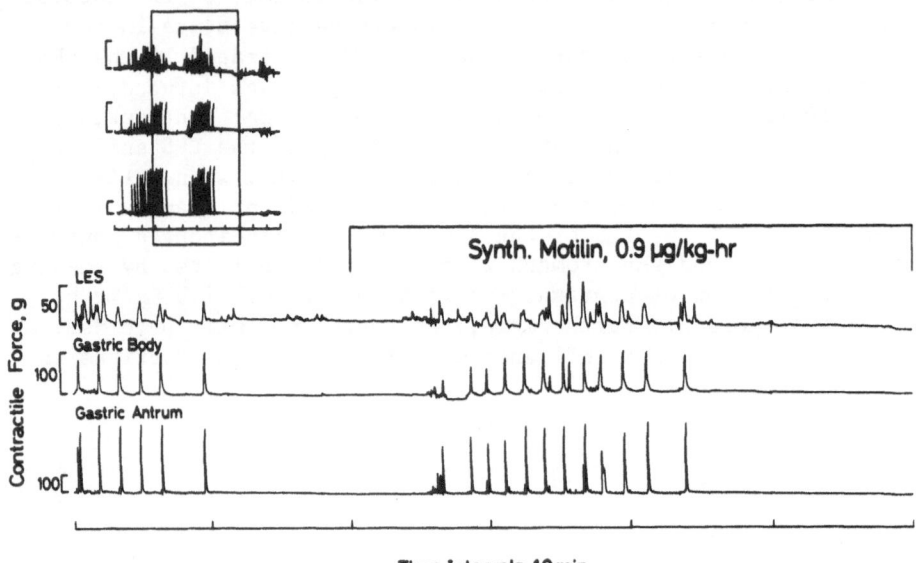

Figure 4. *Effect of motilin on motor activity of the LES and stomach during the interdigestive state.*

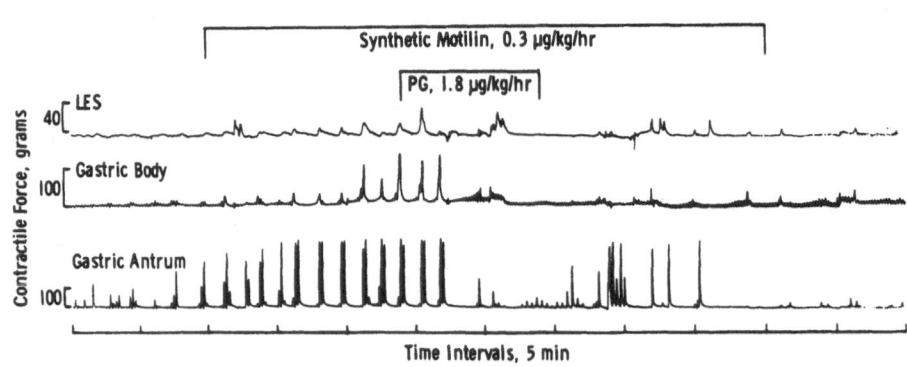

Figure 5. *Inhibitory effect of pentagastrin on motilin-induced contractions during the interdigestive state.*

injection in a dose of 0.9 μg/kg, LES pressure increased 62.4 + 8.6%
from the control values as shown in Figure 6. Secretin and glucagon
had no definite influence upon the resting LES pressure when they
were given by a continuous i.v. infusion. However, these two peptides caused transient decrease of the LES pressure after bolus doses.
The effect of motilin could not be tested because it is not approved
for use in human studies. However, in our recent study in man, we
found a concomitant rise in LES and stomach pressure in man which
lasted for approximately 30 min and was interrupted by longer periods
of quiescence.

DISCUSSION

     According to the present study, it was clearly shown that the
function of the LES in the dog was to seal the oral orifice of the
stomach during the interdigestive contractions. Therefore, a synchronous increase in motor activity of the LES with that of the
gastric body would meet the need to prevent reflux into the esophagus. We have previously reported that the diurnal pattern of gastric motility is divided into two major patterns: the digestive and
the interdigestive pattern[4]. The digestive pattern consists mostly
of the steady and low amplitude contractions which are confined to
the gastric antrum alone; no significant changes are observed in
the body during this period. On the other hand, the interdigestive
pattern is characterised by a series of high-amplitude contractions
lasting for 24 + 2.6 min separated by long periods (89 + 4.6 min)

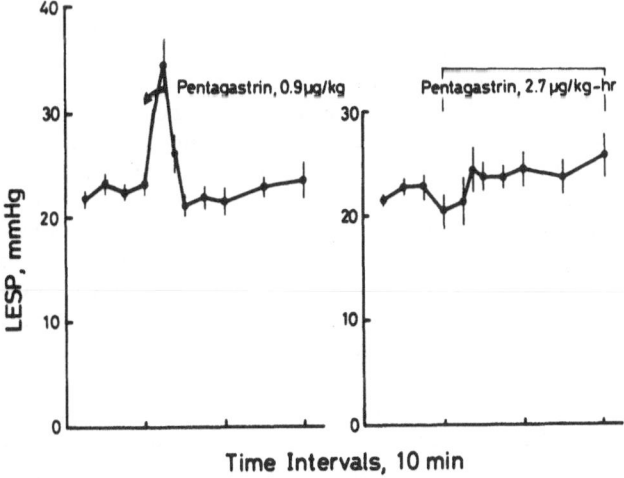

*Figure 6. Effect of pentagastrin give as a pulse dose (left) or
an i.v. continuous infusion (right) on LES pressure in 5 human
subjects.*

of motor quiescence; these contractions, in contrast to the digestive
pattern, occur simultaneously in both the gastric body and antrum.
In the present study we showed that during the interdigestive state,
the motor activity of the LES also increased; this increase in the
LES was in complete association with the increase in the gastric
body. Therefore, it is concluded that at least in dog a great difference exists between the function and control mechanism of LES
motor activity in the digestive and interdigestive state. This concept is of utmost important to the clinician when asking patients
about their symptoms. Reflux esophagitis is a clinical problem of
great importance at present. To understand its pathophysiology,
it is essential to know the function and control mechanism of the
LES. In the present study, therefore, the effect of GI peptides
was examined in both the digestive and interdigestive state. In
dog experiments, it was clearly shown that pentagastrin inhibited the
interdigestive contractions in the LES. In general, it has been
believed that gastrin increases LES pressure in man and animal[7,8].
However, as demonstrated in the present study, it is in the digestive state that gastrin increases LES pressure, and when it was
given during the interdigestive state, gastrin promptly inhibited
LES contractions. But if gastrin is given during the period of
quiescence of the interdigestive state, it will be found that gastrin has no significant influence on the LES motor activity in both
dog and man as shown by others[9]. We consider at present that gastrin may increase LES pressure slightly when the animal is in the
digestive state; however, it is not involved in the control of the
interdigestive movements of the LES. Furthermore, the secretin
family peptides are not involved directly in the regulation of LES
motor activity. The present report is not the first to indicate
that motilin increases LES pressure, but no other study had demonstrated that the increase in LES motor activity evoked by motilin
is a component of the caudad-moving interdigestive contractions of
the gastrointestinal tract, which is also known to be under control
of exogenous motilin[1,2,10]. In fact, there are no other hormones
except motilin that induce such strong contractions and reproduce
the naturally-occurring contractile pattern in the LES. Recent
studies have indicated that motilin may play a role in the control
of the LES. Jennewein et al[11] studied LES pressure activity by a
rapid pull-through method in conscious dogs and reported that highly
purified natural motilin (1.0 µg/kg-hr) induced phasic contractions
of the LES. They also confirmed the occurrence of a caudad-moving
band of strong contractions elicited by the i.v. injection of motilin (50 ng/kg) which resembles closely the motor pattern seen in
the interdigestive state. Lux et al[12] investigated the effect of
motilin on LES pressure in normal subjects. They found that i.v.
infusion of various doses of 13-norleucine-motilin caused significant and dose-dependent increase in LES pressure. Domschke et al[13]
studied LES pressure and plasma motilin concentration in the same
group of normal volunteers. They reported the LES pressure

and plasma motilin levels rose concomitantly after duodenal acidification; they exceeded the basal level by 80 and 90%, respectively,
after 3-4 min. However, Helleman et al[14], using the same technique
also measured the changes in LES pressure and motilin concentrations
in plasma in normal volunteers. They introduced acid and alkaline
solution into the gastric antrum and the duodenum, and LES pressure
was measured by means of a Honeywell probe. They could not obtain
consistent results; they concluded, therefore, that the role of endogenous motilin in the regulation of LES pressure is either non-
existent or frequently overwhelmed by other factors. Similar studies
were reported by Eckardt and Grace[15]: They could not obtain correlation between fasting motilin level and resting LES pressure. More
recently, Meissner et al[16] studied the effect of motilin on LES
pressure in trained dogs using an infusion manometric technique and
reported that motilin produced significant rises in resting pressure
and contractions of the LES. However, they did not point out what
type of contractions are induced by motilin. These studies, however, suggest the existence of a motilin control mechanism of LES
motor activity; but these workers did not clearly demonstrate the
state of motor activity when their measurements were taken.

In the present study, we also demonstrated the existence of
interdigestive contractions in the human stomach precisely similar
to those observed in dogs during the interdigestive state; it seems
most likely that in the human LES cyclic recurring increase of motor
activity may exist during the fasted state. We believe there may be
some reservations about the data obtained by the present pull-through
method. The interdigestive changes in contractile activity of the
LES are divided into two phases: an active phase of strong contractions, and a long-lasting quiescent period. If GE reflux occurs
during the interdigestive state, it must be during the phase of strong
contractions in the stomach. However, the LES pressure measured by
a rapid pull-through technique in general represents momentary resting pressure of the LES when the stomach and the LES are both in the
period of motor quiescence, because 75% of the interdigestive state
consists of the period of motor quiescence. What is necessary for
diagnosis is the pressure of the sphincter when intragastric
pressure rises and intragastric content refluxes into the esophagus
due to the strong gastric contractions. Further studies of this are
needed.

In conclusion, the LES contractile activity is under control of
GI peptides. However, it is suggested that gastrin is not such a
major factor in regulation of the motor activity of the LES as it
has been considered. The secretin family of peptides also has no
significant influence upon LES motor activity in both human and dog.
On the contrary, motilin is the most important substance in regulating the interdigestive contractions in the LES and stomach and is
closely related to GE reflux or heart burn which occurs in the fasted
state.

ACKNOWLEDGEMENTS

This work was in part supported by a Grant for Cancer Research to Z. Itoh in 1975-77 from the Ministry of Public Welfare and Health of Japan. Excellent animal care was provided by Mr. T. Koganezawa, Supervisor of the Animal House, Gunma University School of Medicine. Manuscript was prepared by Miss M. Koike, G. I. Laboratory. Photographic materials were prepared by Mr. F. Ohshima, Chief, Photo Center, Gunma University Hospital.

REFERENCES

1. Itoh Z, Honda R, Hiwatash K, Takeuchi S, Aizawa I, Takayanagi R, Couch EF: Motilin-induced mechanical activity in the canine alimentary tract. Scand J Gastroent 11 (Suppl 39): 93-110, 197

2. Itoh Z, Honda R, Aizawa I, Takeuchi S, Hiwatashi K and Couch EF: Interdigestive motor activity of the lower esophageal sphincter in the conscious dogs. Submitted to Am J Dig Dis.

3. Itoh Z, Carlton N, Lucien HW, Schally AV: Long-term tubing implantation into the external jugular vein for injection or infusion in the dog. Surgery 66: 768-770, 1969

4. Itoh Z, Aizawa I, Takeuchi S, Takayanagi R: Diurnal changes in gastric motor activity in conscious dogs. Am J Dig Dis 22: 117-124, 1977

5. Itoh Z, Aizawa I, Takeuchi S, Couch EF: Hunger contractions and motilin. In Proceedings of 5th International Symposium on Gastrointestinal Motility, edited by G Vantrappen. Typoff-Press, Herentals, Belgium, 1975. p. 48-55

6. Dodds WJ, Hogan WJ, Stef JJ, Miller WN, Lydon SB and Arndorfer RC: A rapid pull-through technique for measuring lower esophageal sphincter pressure. Gastroenterology 68: 437-443, 1975

7. Cohen S and Harris LD: The lower esophageal sphincter. Gastroenterology 63: 1066-1073, 1972

8. Pope CE, II: Pathophysiology and diagnosis of reflux esophagitis. Gastroenterology 70: 445-454, 1976

9. Walker CO, Frank S, Manton J and Fordtran JS: Effect of continuou infusion of pentagastrin on lower esophageal sphincter pressure and gastric acid secretion in normal subject. J Clin Invest 56: 218-225, 1975

10. Wingate DL, Ruppin H, Green WER, Thompson HH: Motilin-induced
     electrical activity in the canine gastrointestinal tract.
     Scand J Gastroent 11 (Suppl 39): 111-118, 1976

11. Jennewein HM, Bauer R, Hummelt H, Lepsin G, Siewert R: Motilin
     effects on gastrointestinal motility and lower esophageal
     sphincter (LES) pressure in dogs. Scand J Gastroent 11 (Suppl
     39): 63-65, 1976

12. Lux G, Rösch W, Domschke S, Domschke W, Wünsch E, Jaeger E and
     Demling L: Intravenous 13-nle-motilin increases the human
     lower esophageal sphincter pressure. Scand J Gastroent 11
     (Suppl 39): 75-79, 1976

13. Domschke W, Lux G, Mitznegg P, Rösch W, Domschke S, Bloom SR,
     Wunsch E, Demling L: Relationship of plasma motilin re-
     sponse to lower esophageal sphincter pressure in man. Scand
     J Gastroent 11 (Suppl 39): 81-84, 1976

14. Hellemans J, Vantrappen G and Bloom SR: Endogenous motilin and
     the LES pressure. Scand J Gastroent 11 (Suppl 39): 67-73,
     1976

15. Eckardt V, Grace ND: Lower esophageal sphincter pressure and
     serum motilin levels. Am J Dig Dis 21: 1008-1011, 1976

16. Meissner AJ, Bowes KL, Zwick R and Daniel EE: Effect of moti-
     lin on the lower esophageal sphincter. Gut 17: 925-932,
     1976

PROGRESS IN INTESTINAL HORMONE RESEARCH

V. Mutt

Department of Biochemistry II, Karolinska Institute

Stockholm, Sweden

I greatly appreciate the opportunity given to me by the organizing committee to participate in this conference.

A conference like this one demands much work from its organizers. In a certain sense work on it started here in Italy almost two centuries ago. It was in the early 1780's that Luigo Spallanzani allowed various kinds of large birds to swallow perforated metal capsules containing pieces of meat and found that as the digestive juices entered the capsules the meat was dissolved[1]. This, and similar experiments carried out by Réaumur, proved that digestion was a chemical process and initiated the investigations of its various aspects. Later came the study of the control mechanisms involved - the topic of this conference.

All intestinal hormones that have been isolated hitherto are peptides and as such derive their individuality from the type, number and sequence of their amino acid residues. There are no other specific chemical characteristics that have been exclusively found only among these substances although there are structures that are rather uncommon elsewhere. The phenolic group of the tyrosine residue of the cholecystokinins is sulphated, as it is in fibrinogen, where this type of structure was first discovered in a peptide[2], and in the sulphated forms of the gastrins[3], and in caerulein[4]. The pyroglutamyl structure has not yet unequivocally been shown to occur in peptides of intestinal origin, but it occurs in gastrin and gastrin has by both bioassay[5,6] and radioimmunoassay[7] been found to be present in the proximal intestine, and also a peptide apparently identical with bovine hypothalamic neurotensin has recently been isolated from bovine intestinal tissue[8]. In the intestinal calcium binding protein the N-terminal amino group is acetylated[9], as it was found to be in the tobacco mosaic virus protein[10] and subsequently in many other proteins. Then there are the C-terminal alpha amide structures. This type of structure, first found in oxytocin[11,12], has later been found in gastrin, secretin, cholecystokinin, Substance P and VIP, as well as numerous other peptides of hormonal or toxic type. Recently it has been shown by Suchanek and Kreil[13] that when the messenger RNA for mellitin is translated in a cell free system from wheat germ a peptide terminating at its C-terminus with glutaminylglycine instead of the glutaminamide of mellitin is formed. This peptide is evidently a biosynthetic precursor of mellitin giving rise to the latter by the exchange of glycine for ammonia. Shoul this prove to be a general mechanism for the synthesis of peptides with C-terminal alpha amides we may expect the discovery of the corresponding precursors of CCK, gastrin, secretin, VIP and Substance P.

The peptides that have hitherto been isolated from intestinal tissue fall into two categories: such as have had their complete amino acid sequences disclosed and such as have not. The former group is slowly growing at the expense of the latter, which, however, is in some kind of steady state; as peptides leave it for the first group others enter it from the unknown. At the time of writing (June 1977) the first group consists of the bovine calcium binding intestinal protein[9], which, however, is (probably) not a hormone; chicken VIP (vasoactive intestinal polypeptide)[14]; equine substance p[15]; porcine CCK-33 and CCK-39 (cholecystokinin-pancreozymin)[16,17]; GIP (gastric inhibitory peptide)[18]; motilin[19]; secretin[20]; and VIP[21]. The amino acid sequences of these peptides are given in Table 1. With the exception of those given for CCK, the above references are to publications giving detailed accounts of the completed sequence work. For CCK they are to publications where the sequences of CCK-33 and CCK-39 have been disclosed, but details of this work have been described only in part[22]. The reason for this is that although we believe the sequences to be correct we would nevertheless like to confirm some points. However, we have had a large number of requests for the essentially pure CCK preparations from physiologists and, especially, radioimmunoassayists from many laboratories, and we have considered it more reasonable for the time being to try to meet these requests rather than use the material for the checking of sequences where we do not expect revisions. Another reason for waiting is that Dr. K. Tatemoto has in our laboratory obtained evidence for the occurrence in one of our peptide side-fractions from the isolation of CCK-33 and CCK-39 of still another, highly basic, CCK variant.

In the second category are: 1.) Bovine intestinal neurotensin[8] identical in amino acid composition with the neurotensin first isolated from bovine hypothalami by Carraway and Leeman[23]. 2.) Porcine enteroglucagon I, a polypeptide of 100 amino acid residues, the N- and C-terminal sequences of which have been disclosed[24]. 3.) Chicke

Table 1.   Amino acid sequence of:

1. bovine calcium binding protein
2. chicken VIP
3. equine substance P
4. porcine CCK-39 (CCK-33 lacks the 6
   N-terminal residues of CCK-39)
5. porcine GIP
6. porcine motilin
7. porcine secretin
8. porcine VIP

+ acetylated, ° amide, · sulphated

1. +K Q S P L E K Y A A E K S I Q K E I E K G F F K Q L L V
   S V Q K A G D K E S L Q P L F T L L K S G P E E N L K E
   S Q N G P D L K S G P Q N D L E E K G T D F V L F S L K Q

2. H S D A V F T D N Y S R F R K Q M A V K K Y L N S V L T °

3. R P K P Q Q F F G L M °

4. Y I Q Q A R K A P S G R V S M I K N L Q S L D P S H R I S
   D R D Y M G W M D F °

5. Y A E G T F I S D Y S I A M D K I R Q Q D F V N W L L A
   Q Q K G K K S D W K H N I T Q

6. F V P I F T Y G E L Q R M Q E K E R N K G Q

7. H S D G T F T S E L S R L R D S A R L Q R L L Q G L V °

8. H S D A V F T D N Y T R L R K Q M A V K K Y L N S I L N °

The amino acid residues are referred to by the one-letter
symbols, recommended by the IUPAC-IUB Commission on Biochemical nomenclature (Eur J Biochem 5 (1968) 151-153)

secretin, small quantities of which have recently been obtained in
apparently pure form in our laboratory by A. Nilsson and M. Carl-
quist. The amino acid composition of this material is markedly
different from that of porcine secretin, which is somewhat unexpected
since chicken glucagon is very similar to porcine glucagon[25], and
porcine glucagon and secretin are chemically closely related. 4.)
Porcine chymodenin, a polypeptide with a molecular weight of about
5000 found by Adelson and Rothman[26] to selectively stimulate the
secretion of chymotrypsinogen from the pancreas. This peptide
seems to contain a disulphide bridge which would make it chemically
different from all the other polypeptides that have hitherto been
obtained in pure form from intestinal tissue[27]. 5.) PI-HIA-27
(Porcine intestinal heptacosapeptide with N-terminal histidine and
C-terminal isoleucine amide) recently isolated in our laboratory by
Dr. K. Tatemoto. There is still some doubt as to the exact number
of its amino acid residues. Besides these isolated peptides there
are - unknown as to their chemical composition - a large number of
physiological principles that have been postulated to occur in the
intestinal wall. There is the group of enterogastrone[28], incretin[29],
and enterocrinin[30]. Physiological work suggested their existence
but chemical work has complicated matters by showing that they may
be concepts rather than individual substances. This conceptual
change was suggested for enterogastrone by Gregory[31] and by Johnson
and Grossman[31]. It is now known that CCK, GIP, secretin, and VIP
may, at least in some species, function as enterogastrones[31]. GIP
has incretin properties[31].   Stimulation of intestinal secretion,
a function of enterocrinin, has been found by Barbezat and Grossman[32]
in dogs to be exerted by pancreatic glucagon, GIP and VIP.

There is almost no doubt that additional substances with enterogastrone, incretin and enterocrinin properties will be discovered.
Indeed for incretin there is already one such report[33]. The question
is rather whether or not in this group some substance will be discovered with so selective and potent an action that it could lay
claim to the original designation.

Ugolev and his colleagues have found that in cats and dogs
duodenectomy with transplantation of the papilla Vateri into the
jejunum results in decreased secretory activity of the thyroid gland
and the hypothalamus. This does not occur if the duodenum is isolate
but preserved. They postulate the existence of specific "enterines"
with systemic functions[34] and describe the partial purification of
one such substance,appetite regulating enterin (arenterin).[35]

Harper and coworkers[36] have described the presence in extracts
of the terminal ileal and colonic mucous membrane of pigs and cats
of a factor that inhibits exocrine pancreatic secretion and suggest
the name pancreotone for it. Sarles and coworkers have made similar
observations[37]. Recently Wilson and Boden found that somatostatin

is, in dogs, a potent inhibitor of meal-stimulated exocrine pancreatic
secretion[38]. Orloff and coworkers have described the partial purification from pig intestinal mucosa of a stimulant of gastric acid
secretion[39]. We have recently achieved some purification of a
factor[40] that might be the gastrozymin of Blair and coworkers[41].

Recently Krumdieck and Ho discussed the possible existence of
an intestinal factor regulating hepatic cholesterol synthesis[42], and
Felber and coworkers have described experiments suggesting the
existence of intestinal hormones, in addition to chymodenin, that
regulate the selective secretion of pancreatic enzymes, and of insulin[43]. Still other candidate hormones have been discussed by
Grossman[44]. A special case is the release of gastrin. The possibility that intestinal hormonal factors might be involved in gastrin release was first pointed out to me by Morton Grossman in 1974.
This possibility has gained support by the discovery of gastrin releasing effects of bombesin and the finding of bombesin immunoreactivity in duodenal and jejunal mucosa[45].

Recently Dr. T. J. McDonald from the University of Western
Ontario has, during a stay in our laboratory - in collaboration with
Dr. Göran Nilsson (Dept. of Pharmacology, Karolinska Institute),
Dr. Stephen Bloom (Hammersmith Hospital, London) and Dr. Monique
Vagne, (Inserm, Lyon) - found that a peptide fraction from the non-
antral part of pig gastric tissue, on administration to dogs and
cats, increases the concentrations of immunoreactive gastrin in
plasma and stimulates gastric acid and pepsin secretion in cats.
There is some preliminary chemical evidence that the material might
be rather different from amphibian bombesin.

The methods by which intestinal hormones have been, and are
being, isolated, as well as the techniques that have been used for
the determination of the amino acid sequences of the isolated peptides, will, however important, not be discussed here.

Some sequence similarities
There is a fascination in the finding of sequence similarities
between peptides of natural origin. Sometimes there are different
similarities depending on how the peptides are aligned in respect to
each other and the finding of the maximal similarity, important for
an understanding of evolutionary relationships, may obscure other
similarities that may have possible functional significance. For
instance in the case of bombesin and VIP it has been pointed out by
Track[46] that if bombesin is aligned with porcine VIP so that the N-
terminal pyroglutamyl residue of bombesin corresponds to Tyr-10 of
the VIP then a striking similarity between the two peptides is revealed.

(Amino acid notations as in table 1. except for *Q
which signifies pyroglutamyl):

```
Bombesin              *QQRLGNQWAVGHLM°
Porcine VIP      HSDAVFTDNYTRLRKQMAVKKYLNSILN°
```

However, with this alignment there will be no amino acid residues
in identical positions in bombesin and glucagon if glucagon is aligned with the VIP with the N-terminal amino acids of the two in corresponding positions. (There will be two marked similarities: Arg-
3 of bombesin and Lys-12 of glucagon, and asparagine-6 of bombesin
and aspartic-15 of glucagon).

If bombesin is now aligned with glucagon so that its pyroglutamyl residue corresponds to leucine-14 of the latter four identities will be revealed in the amino acid sequences.

```
Bombesin                           *QQRLGNQWAVGHLM°
Porcine glucagon      HSQGTFTSDYSKYLDSRRAQDFVQWLMNT
```

This similarity draws attention to some structural similarity to
glucagon of, besides bombesin, Substance P[47], eledoisin[48], physalaemin[49], and chicken secretin (sequence work in progress, the C-terminal leucylmethioninamide structure found by Dr. K. Tatemoto):

```
Physalaemin                              *QADPNKFYGLM°
Eledoisin                                *QPSKDAFIGLM°
Substance P                               RPKPQQFFGLM°
Bombesin                                 *QQRLGNQWAVGHLM
Porcine glucagon          HSQGTFTSDYSKYLDSRRAQDFVQWLMNT
Chicken secretin          H-----------------------LM°
```

Another case is motilin. A marked sequence similarity between porcine motilin and porcine gastrin-34 has been pointed out by, again,
Track[50].

```
Porcine motilin           FVPIFTYGELQRMQEKERNKGQ
Porcine G-34       *QLGLQGHPPLVADLAKKEGPWMEEEEEAYGWMDF°
```

However, an admittedly somewhat less impressive similarity may be
visualized between motilin and secretin.

```
Porcine motilin           FVPIFTYGELQRMQEKERNKGQ
Porcine secretin          HSDGTFTSELSRLRDSARLQRLLQGLV°
```

To what extent similarities in sequence indicate similarities in
function varies from case to case. Some similarities are almost
certainly purely coincidental, and without any importance for the
understanding of either functional or evolutionary relationships.

The difficulty is to know in a specific case whether this is so or
whether the importance is only dormant, e.g. for the similarity between the amino acid sequences KYLDS (residues 12-16) in glucagon
and KYLNS (residues 21-25) in VIP or the sequences LDPSHRIS (residues 16-23) of CCK-33 and the corresponding LNNFHRFS of porcine
calcitonin[51,52]. In other cases it is almost impossible to believe
that the sequence similarities could be without functional and/or
evolutionary significance.

For several years one of the best examples has been the well-
known gastrin II - caerulein - CCK relationship[53]. It is still so,
although the finding that the structurally very different bombesin[45]
exhibits cholecystokinetic and pancreozyminic activities in mammalian
systems does complicate the picture.

The finding that the C-terminal octapeptide amide of CCK is, on
a molar basis, more potent in eliciting gallbladder contraction than
the whole molecule[54] has found a parallel in the finding that the
C-terminal octapeptides of Substance P and of physalaemin are in at
least one assay system more potent than the whole molecules[55].

The dose of the octapeptide of CCK necessary for obtaining
maximal gallbladder contraction in man is 20 ng/kg[56] or approximately 10,538,090,000,000 molecules/kg.

In the glucagon - VIP - secretin group of substances there is
an excellent example of the necessity of taking not only identical
but also similar acids into consideration when comparing amino acid
sequences in peptides. In these peptides the N-terminal parts of
the molecules have a larger number of identical residues than the
C-terminal parts, suggesting an important functional role for the
former. The importance of the N-terminal histidine for the stimulatory action of secretin on pancreatic secretion in vivo was clear
already from the first synthesis of secretin by Bodanszky and coworkers[57]. It was found that the 2-27 sequence had only a slight
activity[58]. In dispersed pancreatic acinar cells, from the guinea
pig, secretin 1-14 was found to increase cellular cyclic AMP[59].
However, Christophe et al.[60] found that the porcine secretin sequence
14-27 which has only 2 identities with the corresponding sequence in
porcine VIP, was more potent in inhibiting the binding of[125] I-VIP
to dispersed pancreatic cells than was the 1-14 secretin sequence
which has 8 identities:

Porcine secretin   HSDGTFTSELSRLRDSARLQRLLQGLV°
Porcine VIP   HSDAVFTDNYTRLRKQMAVKKYLNSILN°

It is however evident that although the C-terminal half of VIP shows
only one sequence identity with the corresponding part of secretin
there is a high degree of similarity in the types of amino acid
residues. This is seen on comparing the C-terminal heptapeptide

amide of secretin with the C-terminal octapeptide amide of VIP
(fig. 1). As pointed out by Christophe et al.[60] these similar,
although not identical, amino acid residues in the C-terminal parts
of the molecules probably explain the apparently anomalous inhibitory effect on VIP binding produced by secretin 1-14 and 14-27.

A comparison of the amino acid sequences of VIP, secretin and
glucagon has given some indications concerning which amino acid
residues might be of importance for the biological activities of
these hormones. For instance the finding that secretin and VIP,
both of which stimulate pancreatic bicarbonate secretion, have as-
partic acid in position 3 whereas glucagon,which does not have this
stimulatory activity,has a residue of glutamine in this position
suggests an important role for Asp-3 for secretin function. In
this connection we have some expectations that the amino acid sequence of PI-HIA-27 when elucidated, might give additional information on the structures necessary for secretin and glucagon activity. PI-HIA-27 does not have secretin-like activity. Of course we
hope that it will anyhow be found to exhibit some sort of hormonal
activity. However,even if it proves to be hormonally inert it might
prove to be of interest. Many publications now describe the postprandial increases in plasma levels of various hormones and hormone
candidates. There is no doubt that a remarkable specificity exists
in some cases for the release of hormones by food components as
clearly shown for pancreozymin already by Wang and Grossman in 1951.
However, there may be a temptation to assume that if the plasma concentration of a polypeptide increases during a meal then that peptide

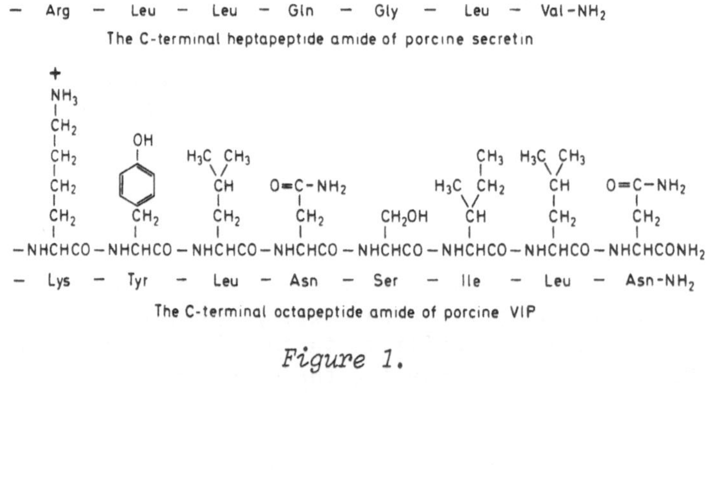

The C-terminal heptapeptide amide of porcine secretin

The C-terminal octapeptide amide of porcine VIP

Figure 1.

is a hormone. After all, eating initiates rather profound physiological activity in the splanchnic area, with, for instance, a great increase in blood flow through the intestine and the pancreas as beautifully illustrated by Claude Bernard in his Mémoire sur le pancréas[62] 120 years ago.

It might be worth knowing whether or not such changes in activity will lead to the appearance in blood of hormonally inert peptides from intestinal or pancreatic tissue, or increases in their fasting concentrations if there are such.

The finding of VIP in neural tissue by several groups of workers[63,64,65] is of course not the first instance of a peptide being implicated in neural activity. However, in cases like that of substance P[66] it could be a matter of specific substances evolved for the purpose. Should a functional role for VIP in neural activity be demonstrated there will be a direct structural link between a substance with neural activity and the typical metabolic hormone glucagon.

REFERENCES

1. Spallanzani L: Expériences sur la digestion de l'homme et de différentes espèces d'animaux. B Chirol Genève, 1783

2. Bettelheim FR: Tyrosine-o-sulfate in a peptide from fibrinogen. J Am Chem Soc 76: 2838-2839, 1954

3. Gregory H, Hardy PM, Jones DS, Kenner GW, Sheppard RC: The antral hormone gastrin. Structure of gastrin. Nature 204: 931-933, 1964

4. Anastasi A, Erspamer V, Endean R: Isolation and structure of caerulein, an active decapeptide from the skin of Hyla caerulea. Experientia 23: 699-700, 1967

5. Emås S, Fyrö B: Gastrin-like activity in different parts of the gastro-intestinal tract of the cat. Acta Physol Scand 74: 359-367, 1968

6. Uvnäs B: The presence of a gastric secretory excitant in the human gastric and duodenal mucosa. Acta Physiol Scand 10: 97-101, 1945

7. Malmstrom J, Stadil F, Rehfeld JF: Gastrins in Tissue. Concentration and component pattern in gastric, duodenal, and jejunal mucosa of normal human subjects and patients with duodenal ulcer. Gastroenterology 70: 697-703, 1976

8. Kitabgi P, Carraway R, Leeman SE: Isolation of a tridecapeptide
    from bovine intestinal tissue and its partial characteriza-
    tion as neurotensin. J Biol Chem 251: 7053-7058, 1976

9. Huang W-Y, Cohn DV, Hamilton JW: Calcium-binding protein of
    bovine intestine. The complete amino acid sequence. J Biol
    Chem 250: 7647-7655, 1975

10. Narita K: Isolation of acetylpeptide from enzymic digests of
    TMV-Protein. Biochim Biophys Acta 28: 184-191, 1958

11. Tuppy H: The amino-acid sequence in oxytocin. Biochim Biophys
    Acta 11: 449-450, 1953

12. Du Vigneaud V, Ressler C, Trippett S: The sequence of amino-
    acids in oxytocin with a proposal for the structure of
    oxytocin. J Biol Chem 205: 949-957, 1953

13. Suchanek G, Kreil G: Translation of melittin messenger RNA in
    vitro yields a product terminating with glutaminylglycine
    rather than with glutaminamide. Proc. Natl Acad Sci USA
    74: 975-978, 1977

14. Nilsson A: Structure of the vasoactive intestinal octacosapeptid
    from chicken intestine. The amino acid sequence. FEBS Lett
    60: 322-325, 1975

15. Studer RO, Trzeciak A, Lergier W: Isolierung und Aminosäurese-
    quenz von Substanz P aus Pferdedarm. Helv Chim Acta 56: 860-
    866, 1973

16. Mutt V, Jorpes E: Hormonal polypeptides of the upper intestine.
    Biochem J 125: 57p-58p, 1971

17. Mutt V: Further investigations on intestinal hormonal polypep-
    tides. Clin Endocrinol 5 (Suppl) 175s-183s, 1976

18. Brown JC, Dryburgh JR: A gastric inhibitory polypeptide II: the
    complete amino acid sequence. Can J Biochem 49: 867-872, 197

19. Schubert H, Brown JC: Correction to the amino acid sequence of
    porcine motilin. Can J Biochem 52: 7-8, 1974

20. Mutt V, Jorpes JE, Magnusson S: Structure of porcine secretin.
    The amino acid sequence. Eur J Biochem 15: 513-519, 1970

21. Mutt V, Said SI: Structure of the porcine vasoactive intestinal
    octacosapeptide. The amino-acid sequence. Use of kallikrein
    in its determination. Eur J Biochem 42: 581-589, 1974

22. Mutt V, Jorpes JE: Structure of porcine cholecystokinin- pancre-
     ozymin. I. Cleavage with thrombin and with trypsin. Eur J
     Biochem 6: 156-162, 1968

23. Carraway R, Leeman SE: The isolation of a new hypotensive pep-
     tide, neurotensin, from bovine hypothalami. J Biol Chem 248:
     6854-6861, 1973

24. Sundby F, Jacobson H, Moody AJ: Purification and characteriza-
     tion of a protein from porcine gut with glucagon-like im-
     munoreactivity. Horm Metab Res 8: 366-371, 1976

25. Pollock HG, Kimmel JR: Chicken glucagon. Isolation and amino
     acid sequence studies. J Biol Chem 250: 9377-9380, 1975

26. Adelson JW, Rothman SS: Chymodenin, a duodenal peptide: speci-
     fic stimulation of chymotrypsinogen secretion. Am J Physiol
     229: 1680-1686, 1975

27. Adelson JW: Chymodenin: An overview. In: Gastrointestinal Hor-
     mones. Ed James C. Thompson, Univ of Texas Press, Austin:
     pp 563-574, 1975

28. Kosaka T, Lim RKS: Demonstration of the humoral agent in fat
     inhibition of gastric secretion. Proc Soc Exp Biol Med 27:
     890-891, 1930

29. La Barre MJ: Sur les possibilités d´un traitement du diabète
     par l´incrétine. Bull Acad Roy Med Belg Ser 5, 12: 620-634,
     1932

30. Nasset ES: Enterocrinin, a hormone which excites the glands of
     the small intestine. Am J Physiol 12: 481-487, 1938

31. Brown JC: "Enterogastrone" and other new gut peptides. Med
     Clin North Am 58: 1347-1358, 1974

32. Barbezat GO, Grossman MI: Intestinal secretion: Stimulation by
     peptides. Science 174: 422-424, 1971

33. Moody AJ, Agerbak GS, Sundby F: The isolation of incretin. VIII
     Congress of the International Diabetes Federation, Brussels,
     Excerta Medica, Amsterdam. International Congress Series No
     280 Abstracts 118: 51-52, 1973

34. Ugolev AM: Non-digestive functions of the intestinal hormones
     (enterines). New data and hypotheses based on experimental
     duodenectomy (Short Review) Acta-Hepato-Gastroenterol 22:
     320-326, 1975

35. Skvortzova NB, Volkonskaya VA, Tuljaganova EH, Zabolotnykh VA, Ugolev AM, Khokhlov AS: on the existence of a special appetite-regulatin intestinal hormone arenterin. Doklady Akad Nauk SSSR 220: 493-495, 1975

36. Harper AA, Hood AJC, Mushens J, Smy JR: Inhibition of external pancreatic secretion by extracts of ileal and colonic mucosa. Gut 15: 825, 1974

37. Hage G, Tiscornia O, Palasciano G, Sarles H: Inhibition of pancreatic exocrine secretion by intra-colonic oleic acid infusion in the dog. Biomedicine 21: 263-267, 1974

38. Wilson RM, Boden G, Shore LS, Essa-Koumar N: Effect of somatostatin on meal-stimulated pancreatic exocrine secretions in dogs. Diabetes 26: 7-10, 1977

39. Chandler JG, Rosen H, Kester RC, Orloff MJ: Extraction of the hormone responsible for the intestinal phase of gastric secretion from the intestinal mucosa of the pig. Gastroenterology 64: A-24/707, 1973

40. Vagne M, Mutt V, Perret G, Lemaitre R: A fraction isolated from porcine upper small intestine stimulating pepsin secretion in the cat. Digestion 14: 89-93, 1976

41. Blair EL, Harper AA, Lake HJ: The pepsin-stimulating effects of gastric and intestinal extracts in cats. J Physiol 121: 20p-21p, 1953

42. Krumdieck CL, Ho K-J: Intestinal regulation of hepatic cholesterol synthesis: an hypothesis. Am J Clin Nutr 30: 255-261, 1977

43. Felber JP, Zermatten A, Dick J: Modulation, by food, of hormonal system regulating rat pancreatic secretion. Lancet II 185-188, 1974

44. Grossman MI, and others. Progress in gastroenterology. Candidate hormones of the gut. Gastroenterology 67: 730-755, 1974

45. Erspamer V, Melchiorri P: Actions of bombesin on secretions and motility of the gastrointestinal tract. In: Gastrointestinal Hormones, Ed James C. Thompson, Univ of Texas Press, Austin: pp 575-589, 1975

46. Track NS: Bombesin and the human gastrointestinal tract. Lancet II: 148, 1976

47. Chang MM, Leeman SE, Niall HD: Amino-acid sequence of substance
     P. Nature New Biology 232: 86-87, 1971

48. Erspamer V, Anastasi A: Structure and pharmacological actions
     of eledoisin, the active endecapeptide of the posterior
     salivary glands of Eledone. Experientia 18: 58-59, 1962

49. Erspamer V, Anastasi A, Bertaccini G, Cei JM: Structure and
     pharmacological actions of physalaemin, the main active
     polypeptide of the skin of Physalaemus fuscumaculatus.
     Experientia 20: 489-490, 1964

50. Track NS: Evolution of the gastrointestinal hormones. In:
     Endocrinology of the Gut. McMaster Univ Symposium, Hamilton,
     Canada. Ed Norman S Track, pp 126-139, 1976

51. Potts JT jr, Niall HD, Keutmann HT, Brewer HB Jr, Deftos IJ:
     The amino acid sequence of porcine thyrocalcitonin. Proc
     Natl Acad Sci USA 59: 1321-1328, 1968

52. Neher R, Riniker B, Zuber H, Rittel W, Kahnt FW: Thyrocalcitonin.
     II. Struktur von α-Thyrocalcitonin. Helv Chim Acta 51: 917-
     924, 1968

53. Jorpes JE: Memorial lecture. The isolation and chemistry of
     secretin and cholecystokinin. Gastroenterology 55: 157-
     164, 1968

54. Ondetti MA, Rubin B, Engel SL, Pluscec J, Sheehan JT: Cholecysto-
     kinin-pancreozymin. Recent developments. Am J Dig Dis 15:
     149-156, 1970

55. Bergmann J, Bienert M, Niedrich H, Mehlis B, Oehme P: Uber den
     Einfluss der Kettenlänge bei C-terminalen Sequenzen der
     Substanz P - in Vergleich mit analogen Physalaemin - und
     Eledoisin-Peptiden - auf der Wirksamkeit am Meerschweinchen-
     Ileum. Experientia 30: 401-403, 1974

56. Sturdevant RAL, Stern DH, Resin H, Isenberg JI: Effect of graded
     doses of octapeptide of cholecystokinin on gallbladder size
     in man. Gastroenterology 64: 452-456, 1973

57. Bodanszky M, Ondetti MA, Levine SD, Narayanan VL, von Saltza M,
     Sheehan JT, Williams NJ, Sabo EF: Synthesis of a heptacosa-
     peptide amide with the hormonal activity of secretin. Chem-
     istry and Industry 42: 1757-1758, 1966

58. Mutt V, Jorpes JE: Contemporary developments in the biochemistry
     of the gastrointestinal hormones. Rec Progr Hormone Res 23:
     483-503, 1967

59. Robberecht P, Conlon TP, Gardner JD: Interaction of porcine
     vasoactive intestinal peptide with dispersed pancreatic
     acinar cells from the guinea pig. Structural requirements
     for effects of vasoactive intestinal peptide and secretin
     on cellular adnosine 3´:5´-monophosphate. J Biol Chem 251:
     4635-4639, 1976

60. Christophe JP, Conlon TP, Gardner JD: Interaction of porcine
     vasoactive intestinal peptide with dispersed pancreatic
     acinar cells from the guinea pig. Binding of radioiodinated
     peptide. J Biol Chem 251: 4629-4634, 1976

61. Wang CC, Grossman MI: Physiological determination of release of
     secretin and pancreozymin from intestine of dogs with trans-
     planted pancreas. Am J Physiol 164: 527-545, 1951

62. Bernard MC. Mémoire sur le pancréas. Mallet-Bachelier Paris,
     1856

63. Said SI, Rosenberg RN: Vasoactive intestinal polypeptide:
     Abundant immunoreactivity in neural cell lines and normal
     nervous tissue. Science 192: 907-908, 1976

64. Larsson L-I, Fahrenkrug J, Schaffalitzky de Muckadell O,
     Sundler F, Håkanson R, Rehfeld JF: Localization of vaso-
     active intestinal polypeptide (VIP) to central and peri-
     pheral neurons. Proc Natl Acad Aci USA 73: 3197-3200, 1976

65. Bryant MG, Bloom SR, Polak JM, Albuquerque RH, Modlin I, Pearse
     AGE: Possible dual role for vasoactive intestinal peptide
     as gastrointestinal hormone and neurotransmitter substance.
     Lancet I: 991-993, 1976

66. Otsuka M, Konishi S, Takahashi T: Hypothalamic substance P as
     a candidate for transmitter of primary afferent neurons.
     Fed Proc 34: 1922-1928, 1975

TROPHIC EFFECTS OF ENDOGENOUS AND EXOGENOUS PANCREOZYMIN UPON THE EXOCRINE AND ENDOCRINE PANCREAS

T. Fujita, Y. Matsunari, Y. Koga, K. Sato and M. Hayashi

Department of Anatomy Niigota University School of Medicine

Niigota, Japan

Pancreozymin is known to have a trophic effect on the exocrine pancreas. Feeding rats or chicks with soybeans causes hypertrophy of the pancreas and this effect is ascribed to trypsin inhibitor contained in soybeans. Furthermore, it is now accepted by many authors that the pancreatotrophic effect of trypsin inhibitor is mediated by CCK-PZ released from the gut mucosa.[1,2]

Although there had been a report that [14]C – thymidine incorporation into DNA was increased in rats repeatedly injected with CCK-PZ,[2] our previous reports first demonstrated increased mitotic figures in various portions of exocrine pancreas of rats fed soybean or trypsin inhibitor or repeatedly injected with CCK-PZ or caerulein.[3,4] In these papers of ours, we also revealed that endocrine pancreas is enlarged by these treatments and B cells, thereby, show increased mitotic figures.

The present paper aims to confirm and extend our previous findings, especially by measuring the insulin content of the pancreas under the effect of endogenous and exogenous CCK-PZ and by the use of its synthetic analog, caerulein.

MATERIAL AND METHODS

In male rats weighing 170-200 g, peroral administration of trypsin inhibitor or repeated injections of CCK-PZ or caerulein were performed for 7 days. "Trypsin inhibitor soybean" (Miles), 400 mg, was dissolved in 100 ml water and the pH value of the solution was

adjusted to 7.0 with dilute NaOH. This solution was given to a group
of rats as drinking water. CCK-PZ (Karolinska Institute), 10 I.D.
U./kg, was subcutaneously injected 4 times a day (7:00, 12:00, 17:00,
21:00), thus 40 I.D.U./kg-day. Caerulein (Pharmitalia; ampuled and
provided by Kyowa Hakko Co.) was injected 0.8 µg/kg-day, divided into 4 doses in the same way as above described. Control rats received
injections of physiological saline.

The pancreases of rats of each group were fixed in GPA fixative[5]
and paraffin sections were examined by hematoxylin-eosin, aldehyde
fuchsin and Hellman-Hellerström's silver impregnation.

In six animals from each group treated with trypsin inhibitor
and CCK-PZ, and control animals, the whole pancreas was removed,
weighed and frozen. Insulin was extracted from each pancreas and
measured by radioimmunoassay.

RESULTS

1. Pancreas Hypertrophy
As partly reported previously, oral administration of trypsin
inhibitor, repeated injections of CCK-PZ or caerulein produced
markedly hypertrophic pancreas which was usually clear at a glance.
The weight increase in the treated animals was significant in many
groups analyzed statistically (though not in the CCK-PZ group in
Table 1, incidentally). The weight increase was usually more con-
spicuous in the cases of oral inhibitor administration than in repeated hormone injections, presumably because more continuous release of endogenous CCK-PZ is induced by the inhibitor.

2. Histological Study
In the exocrine pancreas of treated animals, acinar cells were
markedly enlarged with increased zymogen granules. Mitotic figures
were frequent in the acinar, centroacinar and duct epithelial cells.

The islets of Langerhans in most cases of treated animals were
increased in size and sometimes unusually large islets (about 1 mm
diameter) were encountered. Cell hypertrophy was not clear, while
mitotic figures were markedly increased. Mitosis was found only in
the B cells; this was clear after aldehyde fuchsin staining silver
impregnation.

A and D cells of islets were apparently indifferent to the
trophic effect of CCK-PZ.

3. Insulin Content
Insulin content in the pancreas was measured in six individuals
from each of the inhibitor administered, CCK-PZ injected and control
groups.

TABLE 1 - *Effects on pancreas of trypsin inhibitor and CCK-PZ in rats*

	Body Wt. (g)	Pancreas Wt. (g)	Insulin Cont. (mU)	Insulin (mU) / B.Wt. (g)	Insulin (mU) / P.Wt. (g)
Trypsin Inhibitor in Drink Water (400mg/ml)	214.3 ± 6.98	1.42 ± 0.13 $P<0.01$	743.7 ± 118.1 $P<0.005$	3.481 ± 0.620 $P<0.005$	525.5 ± 68.5 $P<0.05$
CCK-PZ Subcutan. (10⋅4 IDU/Kg/day)	212.7 ± 11.4	1.25 ± 0.12	671.3 ± 155.1 $P<0.02$	3.155 ± 0.707 $P<0.01$	535.9 ± 99.4 $P<0.05$
Control	212.2 ± 7.56	1.14 ± 1.66	448.3 ± 107.5	2.064 ± 0.436	398.6 ± 100.0

7 Day Treatment, N = 6 for Each Group

Table 1 shows: (1) body weight after 7 days of experiment, (2) weight of the pancreas, (3) insulin content in the pancreas, and (4) insulin content relative to body weight and (5) to pancreatic weight. All the absolute and relative values of insulin content in treated animals are significantly increased, even to 150 to 170% of the control values.

DISCUSSION

Our previous papers demonstrated hypertrophy and hyperplasia of exocrine cells of the pancreas, as well as hyperplasia of endocrine B cells by the continuous action of endogenous CCK-PZ by orally given trypsin inhibitors and by exogenous CCK-PZ and its analog, caerulein. The present paper confirms those results and adds the new finding that insulin content is markedly increased in the pancreas by the treatments.

Recently Ihse, Lundquist and their associates[6,7,8] reported that oral administration of trypsin inhibitor to rats improved the exocrine and endocrine functions of the pancreas as well as the impaired carbohydrate metabolism in alloxan diabetic rats, although these authors could not demonstrate   clear morphological or metabolic changes in normal rats given trypsin inhibitor.

Our results, together with those by Ihse and Lundquist suggest the possibility that the trophic effects of exogenous or endogenous PZ (and injections of caerulein) may serve the treatment of atrophic pancreas, chronic pancreatitis, partially resected pancreas and some cases of diabetes mellitus.

REFERENCES

1. Green GM, Lyman RL: Feedback regulation of pancreatic enzyme
   secretion as a mechanism for trypsin inhibitor-induced hypersecretion in rats. Proc. Soc. Exp. Biol. Med. 140, 6-12,1972

2. Mainz DL, Black O, Webster PD: Hormonal control of pancreatic
   growth. J. Clin. Invest. 52, 2300-2304, 1973

3. Fujita T, Yanatori Y, Murakami T: Insulo-acinar axis, its vascular basis and its functional and morphological changes
   caused by CCK-PZ and caerulein. In Endocrine Gut and Pancreas
   Edited by T Fujita, Amsterdam, Elsevier, 1976, p. 347-357

4. Yanatori Y, Fujita T: Hypertrophy and hyperplasia in the endocrine and exocrine pancreas of rats fed soybean trypsin inhibitor or repeatedly injected with pancreozymin. Arch.
   Histol. Jap. 39, 67-78, 1976

5. Solcia E, Vassallo G, Capella C: Selective staining of endocrine
   cells by basic dyes after acid hydrolysis. Stain. Technol.
   43, 257-263, 1968

6. Ihse I, Lundquist I, Arnesjö B: Oral trypsin-inhibitor-induced
   improvement of the exocrine and endocrine pancreatic function
   in alloxan diabet rats. Scand. J. Gastroent. 11, 363-368, 197

7. Ihse I, Arnesjö B, Lundquist I: Effects on exocrine and endocrine
   rat pancreas of long-term administration of CCK-PZ (Cholecystokinin-Pancreozymin) or synthetic octapeptide-CCK-PZ.
   Scand. J. Gastroent. 11, 529-535, 1976

8. Lundquist I, Ihse I, Arnesjö B: Carbohydrate metabolism in normal and diabetic rats following longterm oral trypsin inhibitor administration. Scand. J. Gastroent. 11, 369-375, 1976

ENTEROPANCREATIC AXIS

S. R. Bloom and J. M. Polak

Departments of Medicine and Histochemistry, Royal
Postgraduate Medical School, Hammersmith Hospital

London W12 OHS, England

INTRODUCTION

The pancreas houses insulin, the most powerful known metabolic
regulator. In addition to the β cell, Langerhan's pancreatic islets
also contain D cells producing somatostatin, α cells producing
glucagon and PP cells producing pancreatic polypeptide (Fig. 1).
Between them these cells have the power to stimulate or suppress
numerous physiological functions. A major and frequent challenge
to the constancy of the "milieu intérieur" is the daily ingestion
of food. The smooth assimilation of oral nutriments requires an
efficiently controlled digestive process and a precise adjustment
of metabolic regulators. To this end a complicated control system
exists. This is partly neural, thus allowing an early flow of information from the anticipation, smell and taste of food, and partly
hormonal, utilising the diffuse endocrine system of the gut to produce an integrated signal proportional to the amount and type of
food ingested. Only if the first two mechanisms fail to fully adjust
metabolic regulators, such as pancreatic insulin, would a significant
disturbance of circulating nutriments occur and stimulate the pancreatic islets directly. Thus it is to be noted that in the healthy
young adult insulin release after a meal is so adjusted that the
rise of plasma glucose is exceedingly small in spite of the ingestion
of a considerable quantity of carbohydrate. By comparison, if the
same amount of glucose were injected intravenously, thus bypassing
the early warning systems, glucose levels could easily rise more
than three-fold greater and yet evoke a smaller insulin release.
The malfunction of the oral mechanisms in human disease, for example
maturity onset diabetes, is of great importance. Our understanding

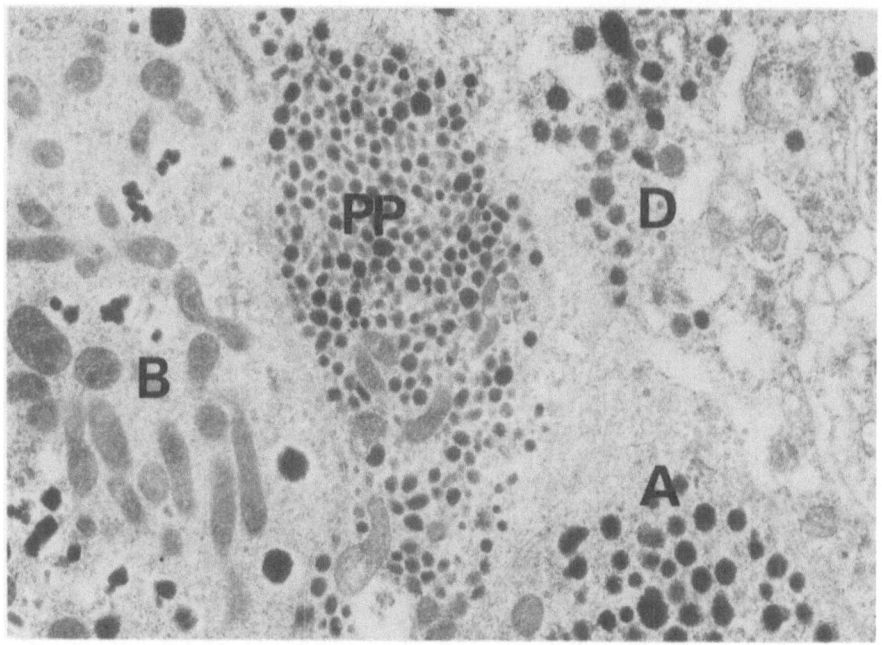

FIGURE 1.   Electron micrograph of islet of Langerhans showing β
cell (B), D cell (D), PP cell (PP), and α cell (A) (x 20,000).

of these control processes are still incomplete but some of what is
known concerning PP, glucagon and insulin release is described here.

PANCREATIC POLYPEPTIDE

     Pancreatic polypeptide (PP) is a 36 amino acid linear peptide
which is found almost entirely in the pancreas.[1] It is localised
to a particular endocrine cell (PP cell) which is found both in the
islets and between the acinar cells.[2] In the bird PP affects both
carbohydrate and lipid metabolism[3] but this has not so far been
demonstrated in man.[4] Infusions of bovine PP at a dose which gives
slightly supraphysiological blood levels shows in man that PP inhibits gall bladder contraction and inhibits pancreatic secretion of
both enzymes and bicarbonate.[4] No effect has been seen on the outpu
of either insulin or glucagon. After a meal plasma concentrations
of PP rise very rapidly, rivalling in magnitude those of insulin
(Fig. 2). However, volunteers receiving an intravenous infusion of
amino acids, lipids or glucose showed no significant change in their
PP values. Thus it would appear that an entero-PP axis exists.[5] As
a considerable release of PP occurs during insulin hypoglycaemia and
this release is blocked by vagotomy (Fig. 3), it is clear that a

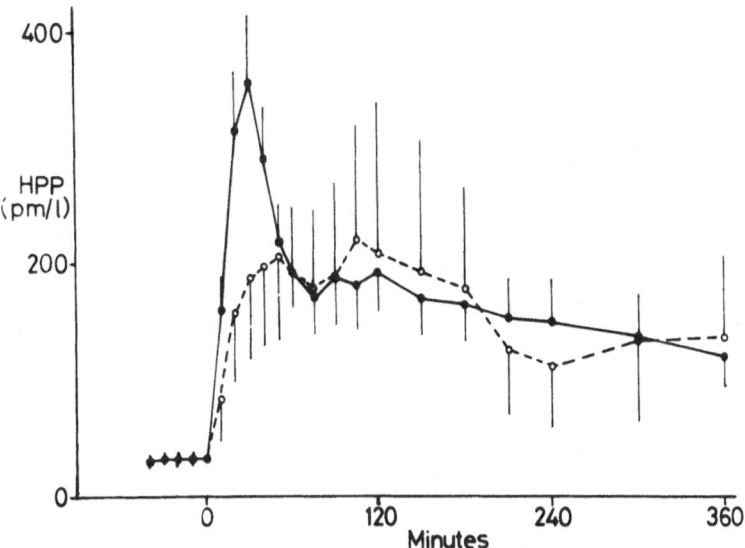

FIGURE 2.   Plasma PP in 10 normal subjects (●—●) and 6 patients with a complete truncal vagotomy (O---O) after eating lunch.[5]

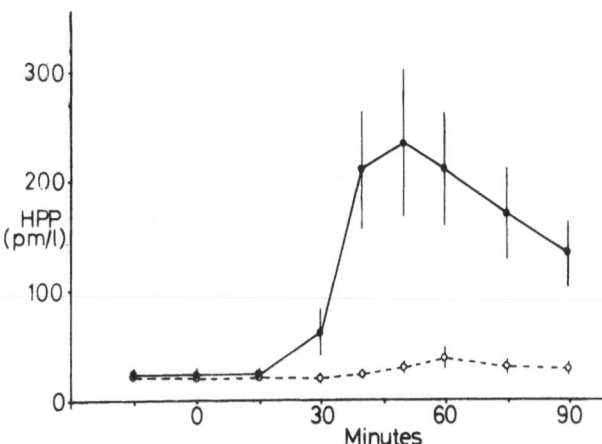

FIGURE 3.   Plasma PP following intravenous insulin (0.2 U/kg) in 8 normal subjects (●—●) and 17 patients after truncal vagotomy (O---O).   The degree of hypoglycaemia achieved was identical.[5]

vagal pathway of release exists. Post-vagotomy patients, however,
still have a highly significant release of PP after food, implying
the existence of other mechanisms of release (Fig. 2). An infusion
of either caerulein or Boot's secretin gives a large rise of PP,
thus demonstrating a hormonal component of the entero-PP axis. This
is further confirmed by work with the isolated perfused canine
pancreas. Here, addition of VIP, GIP and caerulein to the perfusate
produces a highly statistically significant release of PP[6] (Fig. 4).
It is interesting to note, however, that this release is greatly
enhanced when the stimulus follows acetyl choline (Fig. 5). This
implies that the entero-PP axis may involve a complex summation of
multifactorial signals.

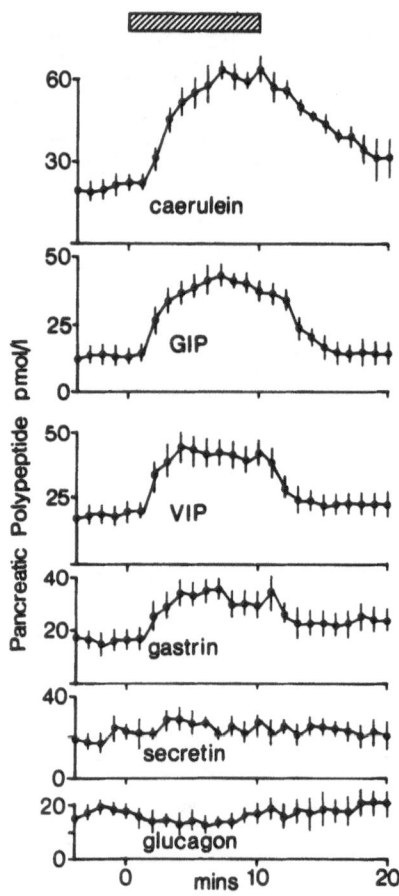

FIGURE 4. *The effect of a 10 minute perfusion of several gut*
hormones (1 nmol/l) on the efflux PP concentrations from the
isolated perfused canine pancreas.[6]

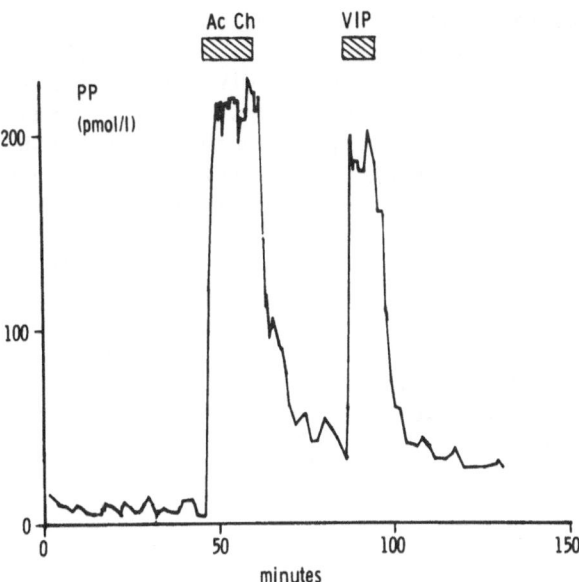

FIGURE 5.   *The enhanced release of PP evoked by 1 nmol/l VIP perfusion when following closely after a perfusion of 1 μmol/l acetyl choline in the isolated perfused canine pancreas.*[6]

PANCREATIC GLUCAGON

Pancreatic glucagon has been known for a long time to have powerful metabolic actions. However, its physiological role was difficult to determine until it was found that pharmacological infusions of somatostatin could suppress its release completely.[7] It was then observed that this caused a precipitous fall in glucose which occurred in spite of simultaneous suppression of basal insulin. The changes could be completely reversed by maintaining glucagon by means of an exogenous glucagon infusion.[8] Thus it became clear that pancreatic glucagon was a major regulator of the fasting hepatic glucose output and so controlled the fasting blood glucose concentration. Glucagon also appeared to have quite another role in addition to its homeostatic function. In situations of stress the glucagon level was found to be always very greatly increased even though there was concomitant hyperglycaemia.[9] In this situation glucagon acted as a stress hormone and was one of the factors maintaining a higher than normal plasma glucose, perhaps thereby acting to enhance the subject's readiness for instant action by maintaining a good supply of circulating metabolic fuel. From the latter function of glucagon it was clear that the original concept of control of glucagon output by the glucose concentration did not apply in all situations. Indeed further investigation showed that

the most powerful experimental stimulus to glucagon release was
electrical excitation of the sympathetic nerve to the pancreas.[10]
It soon also became apparent that the glucagon response to hypoglycaemia (Fig. 6), at least in the calf, was more dependent on signals
via the autonomic nervous system, presumably originating in a
cerebral gluco-regulatory centre, than the ambient glucose concentration acting directly on the α cells.[11] Thus, as with PP, nervous
control of glucagon release is of considerable importance. Unlike
the PP cell the isolated α cell does respond directly to changes in
nutriment concentration. Thus, while several gut hormones will
release glucagon at low glucose concentrations, this effect is very
greatly diminished or completely obtunded when glucose is elevated.[6]
A considerable release of glucagon is also produced by various amino
acids but this is difficult to demonstrate at concentrations near
physiological levels and may be of relatively little importance in
vivo.

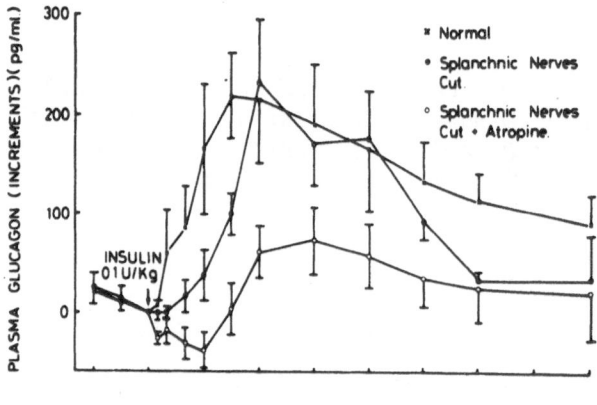

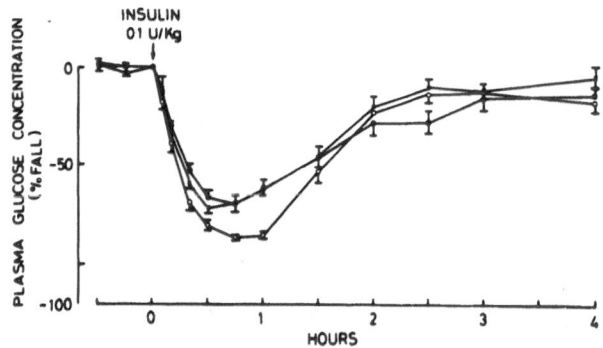

FIGURE 6.   Comparison of the changes in plasma glucagon and
glucose in conscious free-standing calves in response to
intravenous insulin.[11]

The changes of glucagon after a balanced meal are small, but
a high carbohydrate intake suppresses and high protein elevates
glucagon concentrations. These effects are of much greater magnitude than would be observed if the same stimuli were given intravenously. Thus an entero-α cell axis clearly exists. In addition
in some animal species the excitement of eating causes an early
rise in glucagon,[12] though this has not so far been demonstrated
in man. The elevated glucagon seen in the fasted state is rapidly
suppressed by food. One may conclude that while an entero-α cell
axis does exist and depends on nervous and hormonal influences as
well as the ambient nutriment level, its importance in human
pathophysiology appears small.

INSULIN

As mentioned in the introduction, insulin is probably the most
important metabolic regulatory hormone and the widespread occurrence
in civilised man of maturity onset diabetes makes the understanding
of its pathophysiology a major scientific goal. Of the three islet
hormones discussed, insulin release is most influenced by metabolite
concentration, particularly of glucose. Nonetheless the regulation
of insulin secretion by significant changes in metabolite concentration may be viewed as sub-optimum for the individual. It implies
that tight control of the "mileu intérieur" has already broken down.
It is much preferable that the β cell be primed in advance of
glucose changes so as to reduce them to an absolute minimum.

The search for an upper intestinal hormone which enhanced
insulin release has proceeded for many decades. With the discovery
of gastric inhibitory peptide (GIP)[13] a powerful insulin-releasing
hormone[14] was at last isolated. Indeed it has now been renamed
"Glucose-dependent Insulin-releasing Peptide," which term recognises
both its insulinotrophic action and the dependence on the ambient
glucose concentration, as it is quite ineffective when glucose is
low. GIP is released by glucose and fat and therefore rises greatly
after a meal. It has been shown that an intravenous infusion of
GIP to mimic postprandial plasma levels gives rise to a release of
insulin.[15] Thus, there seems little doubt that GIP is an important
component of the entero-insular axis. Several other gut hormones
have also been shown to release insulin, albeit at pharmacological
concentrations, and it is possible that they also play a part in
enhancing insulin release though their role may be relatively minor.
Further insight can be gained by the study of various gut diseases
where hormonal release is abnormal. For example we have recently
investigated coeliac disease, which is associated with a high
incidence of diabetes. The gluten-induced enteropathy in this
condition affects only the upper small intestine. Thus, we find a
quite normal release of PP and gastrin, but the release of secretin

and GIP is very greatly reduced (Fig. 7). In striking contrast,
two hormones located in the lower small intestine, neurotensin and
enteroglucagon (Fig. 8), are released in much increased amounts.[16]
Neurotensin has been shown to inhibit insulin release.[17] This and
the decreased postprandial rise of GIP may explain the decreased
glucose tolerance which was seen in the coeliac patients in our
study.

    GIP is confined to the upper small intestine (Fig. 9) and so
is not released until glucose is already being absorbed. Indeed,
after a meal, it is apparent that the rise in blood glucose occurs
slightly before that of GIP. Thus it seems that another mechanism
is responsible for the early insulin peak which has been shown to
be of considerable importance in maintaining normal glucose tolerance. The most likely mechanism is via the islet innervation.

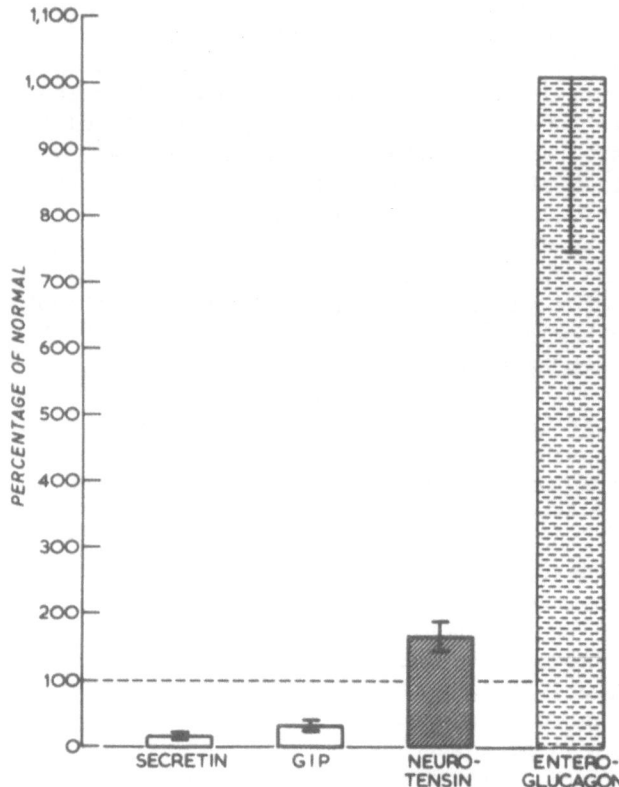

FIGURE 7. Total integrated 180 minute hormonal response to a test
breakfast in 11 patients with active coeliac disease expressed as
a percentage of the normal values obtained in 13 age and sex
matched normal controls.[16]

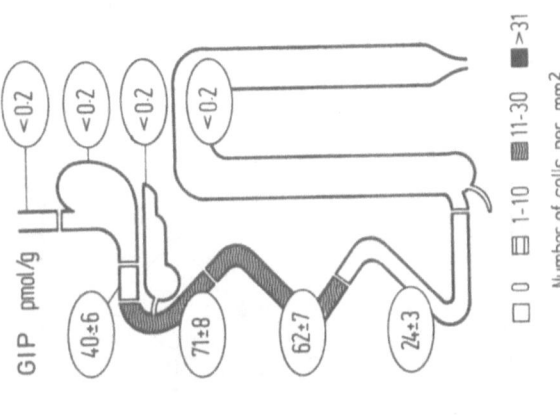

FIGURE 9. GIP concentrations in pmol/g wet weight whole bowel and GIP cells per mm² mucosa in fresh normal surgical specimens of human bowel (n > 4 for each area).

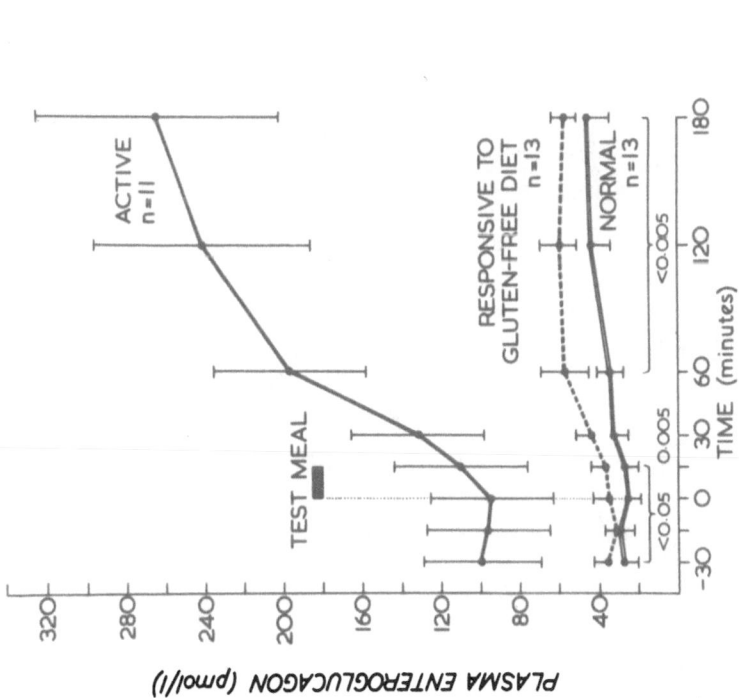

FIGURE 8. Plasma enteroglucagon concentrations following a test breakfast in patients with active coeliac disease, treated coeliac disease, and in healthy controls.16

It is well recognised that vagal stimuli can cause a rapid insulin release.[18] Similarly, sympathetic stimuli to the β cell acting through the α receptor can almost completely suppress insulin release.[19] Thus, reduced adrenergic tone and enhanced vagal tone could be important influences in the postprandial insulin rise. In the calf, autonomic blockade greatly retards insulin release postprandially so that instead of occurring pari passu with the rise in plasma glucose, it is significantly retarded (Fig. 10). The high

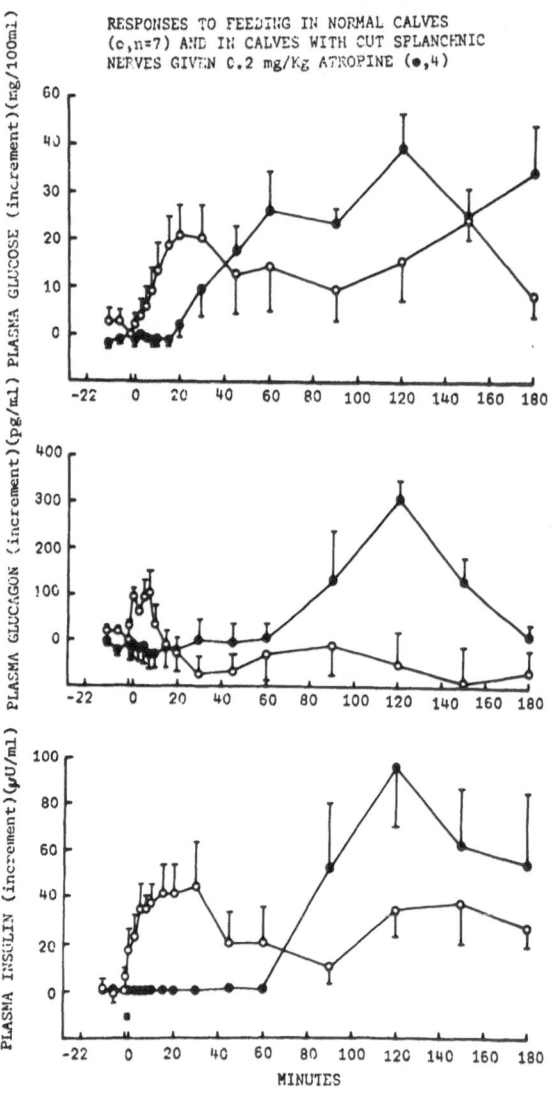

FIGURE 10. Comparison of the changes in arterial plasma glucose, glucagon and insulin concentrations in response to feeding in 3 to 5 week old calves given 1.7 l warm milk at the signal bar.[20]

rise of glucose and delayed insulin response are very reminiscent
of the pattern seen in maturity onset diabetes where, in spite of
the presence of considerable stores of insulin in the pancreas, a
failure of the normal release mechanism results in a delayed and
inadequate circulating insulin response.

CONCLUSIONS

The gut exerts a powerful influence on the release of the
three islet hormones--PP, glucagon and insulin--both via release
of circulating gut hormones and also by nervous reflexes. A number
of aspects are unclear at the present time. The role of somatostatin, for example, which is present in high concentrations in
the D cell of the islet[21] is unknown. It may act as a local hormone
(paracrine system) producing a tonic inhibition of insulin, glucagon
and PP release. Alternatively, it may itself be released directly
into the circulation and have as yet uncharacterised effects on the
liver and gut. The presence of a complex innervation of the islets
has been recognised since Langerhans drew attention to it in 1869.
We have recently recognised the major component of this innervation
is peptidergic (see chapter by Polak and Bloom) and that VIP nerves
form a major component of this. VIP has been shown to release all
three islet hormones[6] and its mediation may offer a solution for
previously problematic findings. For example, the release of glucagon by splanchnic nerve stimulation cannot be blocked by any of the
known adrenergic or cholinergic blocking agents.[22] This strongly
suggests that the mechanism is by the local release of a peptidergic
neurotransmitter. Further investigation on these mechanisms must
await the development of new tools such as specific blocking agents.

Ever since it was first shown that glucose released insulin
directly, the importance of other release mechanisms has tended to
be overlooked. This is clearly wrong as numerous experimental
situations demonstrate that direct nutriment control may well be
of minor importance in physiological situations. The recent advances
in our understanding of the nature of the neuroendocrine system now
offers the opportunity to fully quantitate the entero-pancreatic
axis and investigate its disturbance in disease.

REFERENCES

1. Adrian TE, Bloom SR, Bryant MG, Polak JM, Heitz Ph, Barnes AJ:
   Distribution and release of human pancreatic polypeptide.
   Gut 17: 940-944, 1976

2. Solcia E, Polak JM, Pearse AGE, Forssman WG, Larsson L-I,
    Sundler F, Lechago J, Grimelius L, Fujita T, Creutzfeldt W,
    Gepts W, Falkmer S, Lefranc G, Heitz Ph, Hage E, Buchan AMJ,
    Bloom SR, Grossman MI: Lausanne 1977 classification of
    gastroenteropancreatic endocrine cells. In: Gut Hormones,
    edited by SR Bloom, Churchill Livingstone, Edinburgh, 1978,
    p. 40-48

3. Kimmel JR, Hayden LJ, Pollock HG: Isolation and characterization
    of a new pancreatic polypeptide hormone. J. Biol. Chem.
    250: 9369-9376, 1975

4. Adrian TE, Greenberg GR, Besterman HS, McCloy RF, Chadwick VS,
    Barnes AJ, Mallinson CN, Baron JH, Alberti KGMM, Bloom SR:
    PP infusion in man - summary of initial investigation.
    In: Gut Hormones, edited by SR Bloom, Churchill Livingstone,
    Edinburgh, 1978, p. 265-267

5. Adrian TE, Bloom SR, Besterman HS, Barnes AJ, Cooke TJC,
    Russell RCG, Faber RG: Mechanism of pancreatic polypeptide
    release in man. Lancet I: 161-163, 1977

6. Adrian TE, Bloom SR, Hermansen K, Iversen J: Pancreatic polypep-
    tide, glucagon and insulin secretion from the isolated
    perfused canine pancreas. Diabetologia, in press

7. Mortimer CH, Carr D, Lind T, Bloom SR, Mallinson CS, Schally AV,
    Tunbridge WMG, Yeomans L, Coy DH, Kastin A, Besser GM:
    Growth hormone release inhibiting hormone: effects on circu-
    lating glucagon, insulin and growth hormone in normal,
    diabetic, acromegalic and hypopituitary patients. Lancet I:
    697-701, 1974

8. Nabarro JDN, Hall R, Besser GM, Coy DH, Kastin AJ, Schally AV:
    Glucagon control of fasting glucose in man. Lancet II:
    974, 1974

9. Bloom SR: Glucagon: a stress hormone. Postgraduate Medical
    Journal 49: 607-612, 1973

10. Bloom SR, Edwards AV, Vaughan NJA: The role of the sympathetic
    innervation in the control of plasma glucagon concentration
    in the calf. J. Physiol. 233: 457-466, 1973

11. Bloom SR, Edwards AV, Vaughan NJA: The role of the autonomic
    innervation in the control of glucagon release. J. Physiol.
    236: 611-624, 1974

12. Bloom SR, Edwards AV, Hardy RN, Malinowska K, Silver M:
     Cardiovascular and endocrine responses to feeding in the
     young calf. J. Physiol. 253: 135-155, 1975

13. Brown JC, Mutt V, Pederson RA: Further purification of a
     polypeptide demonstrating enterogastrone activity. J.
     Physiol. 209: 56-64, 1970

14. Pederson RA, Schubert HE, Brown JC: Gastric inhibitory polypep-
     tide. Diabetes 24: 1050-1056, 1975

15. Dupre J, Ross SA, Watson D, Brown JC: Stimulation of insulin
     secretion by gastric inhibitory peptide in man. J. Clin.
     Endocrinol. Metab. 37: 826-828, 1973

16. Besterman HS, Bloom SR, Sarson DL, Blackburn AM, Johnston DI,
     Patel HR, Stewart JS, Modigliani R, Guerin S, Mallinson CN:
     Characteristic gut hormone profile in coeliac disease.
     Lancet, in press

17. Brown M, Vale W: Effects of neurotensin and substance P on
     plasma insulin, glucagon and glucose levels. Endocrinology
     98: 819-825, 1976

18. Woods SC, Porte D: Neural control of the endocrine pancreas.
     Physiol. Rev. 54: 596-619, 1974

19. Bloom SR, Edwards AV: The release of pancreatic glucagon and
     inhibition of insulin in response to stimulation of the
     sympathetic innervation. J. Physiol. 253: 157-173, 1975

20. Bloom SR, Edwards AV, Hardy RN: The role of the autonomic
     nervous system in the control of pancreatic endocrine
     responses to the ingestion of milk in the conscious calf.
     J. Physiol., in press

21. Polak JM, Pearse AGE, Grimelius L, Bloom SR, Arimura A: Growth
     hormone release-inhibiting hormone in gastrointestinal and
     pancreatic D cells. Lancet I: 1220-1225, 1975

22. Bloom SR, Edwards AV: Effects of certain inhibitory or blocking
     agents on the release of pancreatic glucagon in response to
     stimulation of the splanchnic nerves. J. Physiol., in press

PANCREATIC POLYPEPTIDE (PP)

T. W. Schwartz

Institute of Medical Biochemistry

University of Aarhus, Denmark

Isolation

During purification of insulin a new hormone named pancreatic polypeptide (PP) was isolated independently by Kimmel[1] (avian PP) and Chance[2] (human, bovine, porcine, ovine and canine PP). PP is a 36 amino-acid straight chain polypeptide, with no apparent similarity in sequence with any known hormone. Only one or two amino-acids distinguish mammalian PP's, whereas they have about 16 amino-acids in common with the avian counterpart.

Localization

PP has been localized by L.-I. Larsson et al. using immunohistochemistry to secretory granules of a distinct endocrine cell type – different from the A, B, D and D_1 cells.[3,4] The PP cells are found both in the islets and in the acinar tissue and duct ephithelium. In many species PP cells are most numerous in the extra-insular tissue and concentrated towards the duodenal part of the pancreas; however, in the sheep insular PP cells nearly outnumber insulin cells.[4] In some species, e.g. the dog, a few PP cells are also in the stomach.[4]

Effect

The physiological role of mammalian PP has not yet been established although PP affects many gastrointestinal functions as demonstrated by Lin et al.[5,6] Pancreatic secretion - In low doses PP seems to augment secretin-induced HCO_3^- and water secretion and inhibits CCK-induced enzyme secretion; basal secretion is inhibited. Biliary motility - PP has the unique property of both relaxing the gallbladder and increasing choledocal pressure without affecting bile secretion. Gastric secretion - PP in high doses stimulates basal but inhibits pentagastrin-induced acid secretion. Intestinal

motility - In high doses PP stimulates gut motility and enhances
gastric emptying and intestinal transit. Glucose metabolism -
PP does not affect blood glucose levels.

Secretion

        During a protein rich meal PP plasma concentration increases[7,8,9]
in a biphasic response, composed of a rapid primary (<5-30 min.)
and a prolonged secondary phase (0.5- >5h).[8] The ingestion of protein is the main stimulating agent as demonstrated by Floyd et al.[7]
Since PP is almost exclusively found in the pancreas some entero-
pancreatic signal must mediate the PP response to food. Primary
response - PP increases in plasma within the first few minutes of
a meal before any other measurable G-I hormone, which rules out G-I
hormones as initiator of the PP response. However, we found the
rapid PP response was abolished by truncal vagotomy.[8] The importance of vagal, cholinergic influence on PP secretion is indicated
by the following results:[10] 1.) the large release of PP during
insulin hypoglycemia is nearly eliminated by atropine and totally
abolished by truncal vagotomy; 2.) electrical stimulation of the
vagal nerves increases portal PP concentration tenfold within 30
sec; this response is inhibited by atropine and abolished by
hexamethonium; 3.) acetylcholine in low doses (10^{-6}-10^{-7}M) is the
most potent stimulating agent in respect to maximal PP output from
the isolated, perfused, porcine pancreas. Furthermore we find that
PP is released in response to both "classical cephalic" stimulation
(sham feeding) and stimulation of a vagal, gastropancreatic reflex
(gastric distention)[11]. Probably the most important stimulus for
this vagal, gastro-pancreatic reflex is protein. The secondary prolonged response during a meal is reduced but not abolished by truncal vagotomy in man; thus, this response seems to be mediated by
synergistic action of vagal activity and humoral stimulation. Several groups have found that some G-I hormones of both the gastrin
and the secretin family can release PP[7,9,11] but it still remains
to be proven which G-I hormone(s) take part in this late "PP-entero
pancreatic axis" during physiological events. Circulating amino
acids also seem to affect PP secretion.[7]

Duodenal Ulcer

        As described above PP is very much under the influence of the
vagus. The finding that basal PP levels in some D.U. patients are
elevated preoperatively but not after a truncal vagotomy[8] indicates
that increased PP levels may reflect increased vagal actity in
these patients. This is supported by the finding that patients
with already elevated PP are relatively insensitive to cephalic
vagal stimulation, and furthermore atropine "normalizes" the elevated
PP[12].

CONCLUSIONS

PP is a hormone with known sequence, localized mainly in the pancreas. It may act as a local regulator of exocrine pancreatic secretion. PP is released in response to the ingestion of protein; the response is mediated principally by vagal stimulation. PP secretion may be used as an indicator of vagal activity in duodenal ulcer patients.

REFERENCES

1. Kimmel JR, Hayden LJ, Pollock HG: Isolation and characterization of a new pancreatic polypeptide hormone. J Biol Chem 250: 9369-9376, 1975

2. Chance RE, Johnson MG, Koffenberger Jr JE, et al: Isolation and amino acid sequence of a new pancreatic polypeptide. Unpublished

3. Larsson LI, Sundler F, Håkanson R: Immunohistochemical localization of human pancreatic polypeptide (HPP) to a population of islet cells. Cell Tiss Res 156: 167-171, 1975

4. Larsson LI, Sundler F, Håkanson R: Pancreatic polypeptide - a postulated new hormone: Identification of its cellular storage site by light and electron microscopic immunocytochemistry. Diabetologia 12: 211-226, 1976

5. Lin TM, Chance RE: Gastrointestinal actions of a new bovine pancreatic peptide (BPP). In Endocrinology of the gut. WY Chey and FP Brooks, editor. Thorofare, New Jersey. 1: 143-145, 1974

6. Lin TM, Evans DC, Chance RE, et al: Bovine pancreatic peptide: Action on gastric and pancreatic secretion in dogs. Am J Physiol 232 (3): E311-E315, 1977

7. Floyd JC, Fajans SS, Pek S, Chance RE: A newly recognized pancreatic polypeptide; plasma levels in health and disease. Rec Progr Horm Res, 1977

8. Schwartz TW, Rehfeld, JF, Stadil F, et al: Pancreatic-polypeptide response to food in duodenal-ulcer patients before and after vagotomy. Lancet I: 1102-1105, 1976

9.   Adrian TE, Bloom SR, Besterman HS, et al: Mechanism of pancreatic
     polypeptide release in man. Lancet I: 161-163, 1977

10. Schwartz TW, Holst JJ, Lindkaer Jensen S, et al: XIth Acta
     Endocrinologia Congress, abstract, 1977

11. Schwartz TW, Rehfeld JF: Mechanism of pancreatic-polypeptide
     release. Lancet I: 697-698, 1977

12. Schwartz TW, Stenquist S, Olbe L: Vagal influence on pancreatic
     polypeptide secretion. In "International Gut Hormone Sym-
     posium", SR Bloom, ed, 1977

PHYSIOLOGY AND PATHOPHYSIOLOGY OF GIP

*J.C. Brown, J. R. Dryburgh, J. L. Frost, S. C. Otte
and R. A. Pederson*

Department of Physiology, University of British Columbia

Vancouver, B.C., Canada V6T 1W5

Physiology and Pathophysiology of GIP

Two major physiological roles for GIP have been established. An enterogastrone-like action in dog, in which it has been shown that physiological doses of the hormone will inhibit gastric acid secretion stimulated by histamine, gastrin or insulin hypoglycaemia (1) and an insulinotropic action of the hormone in both dog and man (2,3). The important recent findings concerning the insulinotropic action of the hormone have led to the suggestion that the abbreviation GIP could also be interpreted as meaning glucose-dependent insulinotropic polypeptide (GIP) (4). These findings include the observations that glucose ingestion in man (5) and dog (2) will stimulate the release of immunoreactive GIP (IR-GIP) and that intravenous infusion of the hormone in man will potentiate glucose stimulated insulin release (6).

The insulinotropic action of GIP is dependent upon the prevailing glucose concentration in serum (man and dog) or perfusate (isolated rat pancreas) in that there is a threshold concentration of glucose, below which GIP is not insulinotropic (7). In both experimental models this is approximately 5.5 mM glucose and in man an increase over baseline glucose of 25 mg%. In the isolated perfused rat pancreas, the maximum immunoreactive insulin (IRI) response to glucose alone can be increased several fold by the addition of 5 ng/ml GIP into the perfusate. The relationship between the glucose concentration and the insulinotropic action of GIP requires that appropriate control of the glucose concentration be maintained in studies involving GIP. The glucose clamp technique has been utilized in such studies. This technique has allowed the effects of IR-GIP (endogenously released) (8) and porcine GIP (exogenously administered) on the release of insulin to be observed in a situation where the circulating glucose concentration is maintained constant. These studies have demonstrated that oral glucose results in a rapid and significant elevation in serum IR-GIP and IRI independent of any change in glucose concentration. The time course of the rise in plasma insulin after oral glucose in the clamped hyperglycaemic situation is almost identical with that for IR-GIP suggesting an interrelationship between these hormones.

In the situation where the serum glucose concentration is stabilized at the fasting level by simultaneous administration of insulin and intravenous glucose, oral glucose and oral fat will still release IR-GIP but there is no insulin release observed. Previous observations have also shown that fat ingestion will not provoke insulin release unless hyperglycaemia is present (9). These observations in man in which IR-GIP has been released endogenously by both oral glucose and oral fat administration have shown that there is a relationship between IRI release and IR-GIP release providing that blood glucose concentrations are elevated by at least 25 mg/dl above baseline. The insulinotropic action of exogenously administered porcine GIP has also been studied in the hyperglycaemic clamp situation. GIP administered in doses of 0.4 µg/kg-h significantly elevated IRI levels when the square wave of hyperglycaemia was maintained at basal plus 45 mg/dl. This glucose-dependent effect and the threshold effect seen in human studies are identical to those reported for the perfused rat pancreas.

Physiologically, GIP plays an important role in insulin secretion in that glucose is a secretagogue for GIP release and the circulating IR-GIP levels increase with increasing glucose load. Endogenously released GIP is not insulinotropic when the circulating glucose level is clamped at the basal level. The intravenous infusion of pure porcine GIP will mimic all the actions of endogenously released GIP.

Hyperinsulinaemia is associated with the clinical conditions of obesity and maturity onset diabetes. One possibility for this hyperinsulinaemia could be a disturbance of the entero-insular axis. GIP release has been studied in both these conditions. Two groups of obese individuals, one with a normal oral glucose tolerance (nOGT) and the other with a pathological glucose tolerance (pOGT) were administered a mixed meal (10) and serum glucose, IRI and IR-GIP were measured. The obese individuals all demonstrated a significantly greater release of IR-GIP over control subjects, the most exaggerated response was observed in the obese pOGT group, and the highest insulin response was also observed in this group.

Hypersecretion of IR-GIP has also been observed in patients with maturity onset diabetes (11,12), and as in the obese subjects there was concomitant hyperinsulinaemia.

It is tempting to speculate that the hyperinsulinaemia of obesity and maturity onset diabetes occurs as a result of hyper-secretion of GIP.

REFERENCES

1. Pederson RA and Brown JC, 1972. Gastroenterology 62: 393-400

2. Pederson RA, Schubert HE and Brown JC, 1975. Diabetes 24: 1050-1056

3. Dupre J, Ross SA, Watson D and Brown JC, 1973. J. Clin. Endocrinol. Metab. 37: 826-828

4. Brown JC and Pederson RA, 1976. Excerpta Medica (in Press)

5. Cataland S, Crockett SE, Brown JC and Mazzaferri EL, 1974. J. Clin. Endocrinol. Metabl. 39: 223-228

6. Brown JC, Dryburgh JR, Ross SA and Dupre J, 1975. Recent Progress in Hormone Research 31: 487-532

7. Pederson RA and Brown JC, 1976. Endocrinology 99: 780-785

8. Andersen DK, Brown JC, Elahi D, Tobin JD and Andres R, 1977. Submitted for publication

9. Dobbs R, Faloona GR and Unger RH, 1975. Metabolism 24: 69-75

10. Creutzfeldt W, Ebert R, Willms B, Frerichs H and Brown JC, 1977. Submitted for publication

11. Ross SA, Brown JC and Dupre J, 1977. Diabetes (in Press)

12. Creutzfeldt W and Ebert R, 1977. Excerpta Medica (in Press)

GASTRIC-GLUCAGON: PHYSIOLOGY AND PATHOLOGY

P. J. Lefebvre and A. S. Luyckx

Division of Diabetes, Institute of Medicine

University of Liège, B - 4020 Liège, Belgium

As recently reported by several groups of investigators[1-3], the plasma of totally depancreatized dogs contains normal, or even increased, quantities of a material immunometrically indistinguishable from pancreatic glucagon by radioimmunoassays regarded as highly specific for this hormone. As emphasized by Sasaki et al.[4], this post-pancreatectomy immunoreactivity cannot be attributed to a cross-reaction with high levels of "gut glucagon-like immunoreactivity", a group of immunometrically dissimilar polypeptides whose level is not elevated after pancreatectomy[5]. However, glucagon cannot be detected in the blood of dogs that have undergone complete abdominal evisceration[6]. Therefore, in the dog, the origin of extrapancreatic glucagon is likely to be an abdominal organ; other significant sources of glucagon, such as the salivary glands as reported in the rodents [7,8], are unlikely to exist.

Almost 30 years ago, Sutherland and De Duve[9] were the first to suggest that glucagon may be present in the gastro-intestinal tract. Since then, various groups[10-14] have reported that the oxyntic glandular mucosa of the canine stomach contains cells resembling pancreatic A cells. The joint efforts of Unger's group in Dallas and Orci and his coworkers in Geneva established the presence in the gastro-intestinal tract of a material biologically, physicochemically and immunometrically identical to pancreatic glucagon[4] and also established the existence of A cells in the dog gastric fundus[4,15].

Because of the difficulty involved in the selective investigation of gastric-glucagon secretion in pancreatectomized dogs, we designed an isolated perfused dog stomach system. The aim of the present study was to determine whether such an isolated system

173

responds to the intravascular infusion of arginine by a release of
gastric glucagon as does the pancreas, and whether such a response
is affected by glucose, insulin or somatostatin, which are all
factors previously demonstrated to modify profoundly the secretion
of pancreatic glucagon.

METHODS

The method used was reported in detail recently[16]. The perfusion system used was originally designed by Nizet and coworkers
[17-18] for the study of the isolated dog kidney. Glucagon was assayed using antiserum 30K provided by Dr. R.H. Unger (Dallas).

RESULTS

1. *Gastric-glucagon release under basal conditions*
    Basal glucagon release prior to arginine infusion averaged 1.4
$\pm$ 0.6 ng/100 gm stomach-min (n = 7; range 0.0 - 3.1). In a single
experiment, saline was infused at a rate of 1 ml/min into the artery
for 10 min; under these conditions basal glucagon release averaged
0.9 $\pm$ 0.1 ng/100 gm-min.

2. *Gastric-glucagon release in response to arginine infusion*
    Arginine was infused at a rate calculated to reach an arterial
plasma concentration of about 10mM. As shown in Fig. 1, arginine-
infusion resulted in rapid rise in glucagon output. Glucagon output began to increase 30 to 40 seconds after the beginning of the
intraarterial arginine infusion, reached a peak value at about 60
seconds and then declined progressively. A transient off-response
was sometimes observed immediately after cessation of the infusion.
The total amount of glucagon released during the arginine infusion
averaged 50.1 $\pm$ 10.4 ng/100 gm-5 min, that is, 6 to 10 times more
than the basal release.

3. *Effect of somatostatin on the arginine-induced gastric-glucagon*
    release
    Somatostatin (100 ng/ml) almost completely inhibited arginine-
induced gastric-glucagon release which, under these conditions,
averaged 8.8 $\pm$ 2.4 ng/100 gm-5 min a value similar to the release
obtained in basal conditions and which represents 17.6% of the release obtained with arginine alone.

4. *Effect of insulin on the arginine-induced gastric-glucagon*
    release
    When the blood perfusing the stomach was massively enriched with
exogenous insulin (final plasma concentration 10,000 μU/ml), arginine
elicited a glucagon release (43.7 $\pm$ 5.1 ng/100 gm-5 min) similar in

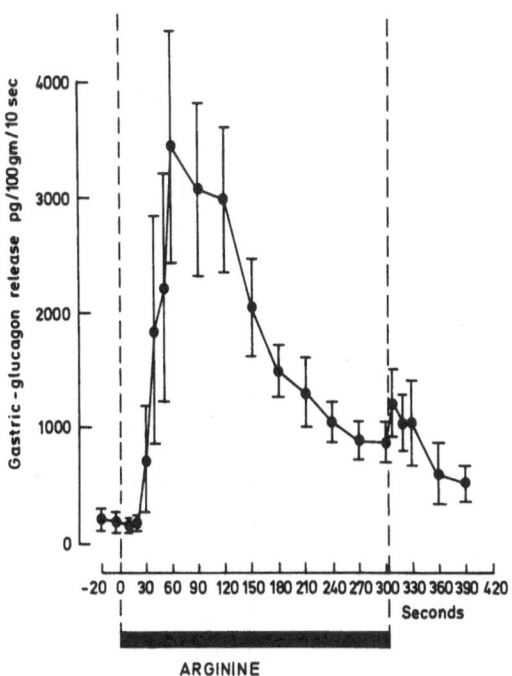

FIGURE 1.   Gastric-glucagon release in response to intra-arterial arginine infusion.   Results are expressed as mean ± SEM (n = 5). Data from Lefebvre and Luyckx[16].

magnitude to that obtained in standard conditions (50.1 + 10.4 ng/ 100 gm-5 min).

5.   *Effect of glucose alone or glucose-plus-insulin on arginine-induced gastric-glucagon release*

The arginine-induced gastric-glucagon release was unaffected by hyperglycemia (312 + 19 mg%) of the perfusing blood. When such hyperglycemia was concomitant with hyperinsulinemia of 115 ± 20 μU/ml, arginine-induced gastric-glucagon release was reduced by about 40%. As shown in Fig. 2, no major difference in the kinetics of the release was observed under either condition.

DISCUSSION

Under basal conditions, gastric glucagon release was absent or minimal, a finding which is in agreement with previous data obtained in the normal anesthetized dog[19]. The release of gastric glucagon increased markedly and rapidly after a short arginine infusion which reached a 10 mM concentration in the arterial plasma; this is indeed

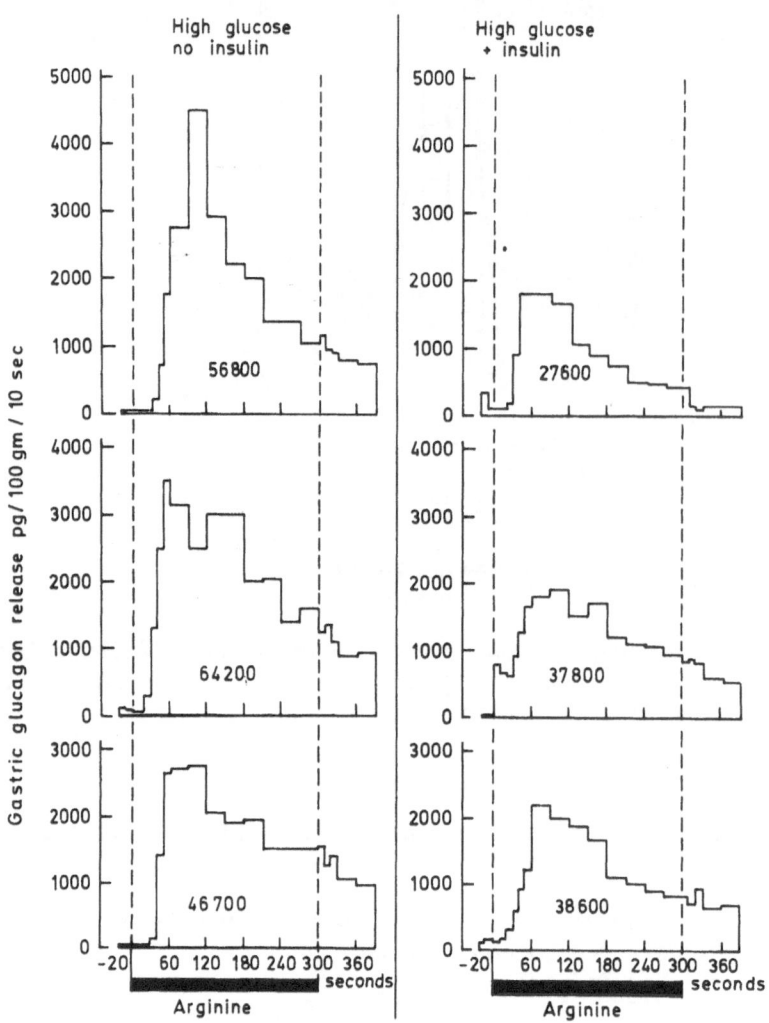

FIGURE 2.   Gastric-glucagon release in response to intra-arterial
arginine infusion.   On the left panel, the perfusing blood was
enriched with glucose alone (mean blood glucose concentration: 312
± 19 mg%); no exogenous insulin was added (mean plasma insulin concentration: 7 ± µU/ml).   On the right panel, the perfusing blood
was simultaneously enriched with glucose (319 ± 13 mg%) and insulin
(115 ± 20 µU/ml of plasma).   The figures in the histograms corresponded to the total amount of glucagon (in pg) released during the
5 min arginine infusion.   Data from Lefebvre and Luyckx[16].

a strong stimulus of pancreatic glucagon secretion both in vivo
(review in 20) and in vitro[21,22]. Therefore, the increased plasma
glucagon observed after arginine infusion in insulin-deprived depancreatized dogs[3,23] is likely to be, at least in part, of gastric
origin. Indeed, this was recently proven to be the case by Blazquez
et al.[24]. Somatostatin, which is a potent inhibitor of glucagon
secretion[25], almost completely blocked the arginine-induced gastric-
glucagon release. This finding is in agreement with the previously
reported decrease in plasma glucagon after somatostatin infusion in
the depancreatized dog[5,23]. Insulin alone, even at the massive concentration of 10,000 μU/ml, did not modify the arginine-induced release of gastric glucagon. This observation is in contradiction
with the concept proposed by Samols for the pancreatic A cell, that
insulin by itself might exert an inhibitory effect on glucagon secretion[26], a theory which was supported by the recent finding of
Weir et al.[27] that exogenous insulin (20,000 μU/ml) could reduce
arginine-induced glucagon secretion from the perfused pancreas of
streptozotocin-treated rats. Thus the gastric and pancreatic A cell
may differ with respect to their sensitivity to high concentrations
of surrounding insulin. Glucose has long been recognized to be an
inhibitor of glucagon release (review in 20). The mechanism by
which it exerts this effect is still largely unknown. Several observations suggest that insulin might play a major role in the
glucose-induced glucagon suppression[20] and it has been suggested
that the action of insulin in this respect would be to permit glucose to enter the A cell[28-30]. An experimental approach to this
question using the pancreas or isolated islets is particularly
difficult: even after streptozotocin administration, one is never
sure that some release of endogenous insulin does not occur[27],
causing appreciable intra-islet concentrations of insulin. The
isolated perfused dog stomach might provide a unique tool permitting
investigation of A cell function in the absence of endogenously released insulin: there is no evidence of the presence of B cells in
the stomach[31] and, as we have seen, the amounts of insulin in the
perfusion blood, obtained from overnight fasted dogs, were always
low. Furthermore, determinations of insulin levels in the effluent
blood, at the time of maximal arginine-induced glucagon release,
never showed any rise. In these conditions, it is extremely interesting that hyperglycemia alone did not reduce the arginine-induced
gastric-glucagon release, but did so only when it was accompanied
by a rise in plasma insulin within the physiological range. The
40% reduction in the arginine-induced release of glucagon in the
presence of glucose (300 mg%) and insulin (115 μU/ml) confirms the
important role of insulin in permitting glucose to inhibit glucagon
secretion. This observation supports the idea of an exquisite sensitivity of the gastric A cell to glucose[32] providing adequate concentration of insulin is present, and probably explains the rapid
fall in plasma glucagon after administration of insulin to the
diabetic pancreatectomized dog[5].

Even if no doubt persists concerning the presence of function-
ing A cells in the canine stomach, the presence of such cells in man
is still not certain[33]. Several authors reported the persistence
of circulating glucagon in pancreatectomized man[34-36] while others
did not[37,38]. With Vranic et al.[39], we suggest that the presence of
circulating insulin, even at relatively low concentration, might
favor the suppressive effect of glucose on extra-pancreatic A cells,
as demonstrated in the present study. Therefore, we think that the
concept of the "pancreatectomized man as a model of diabetes without glucagon"[37] must be accepted only if experimental data are provided of the total insulin lack of these patients at the time of
study. Prolonged insulin deprivation and appropriate assays of circulating insulin[40] showing that insulin lack is indeed complete are
prerequisites to the demonstration that there is no extrapancreatic
source of glucagon in man.

REFERENCES

1. Vranic M, Pek S, Kawamori R: Increased "glucagon immunoreactiv-
        ity" in plasma of totally depancreatized dogs. Diabetes 23,
        905-912, 1974

2. Matsuyama T, Foa PP: Plasma glucose, insulin, pancreatic, and
        enteroglucagon levels in normal and depancreatized dogs.
        Proc. Soc. Exp. Biol. Med. 147, 97-102, 1974

3. Mashiter K, Harding PE, Chou M, Mashiter GD, Stout J, Diamond D,
        Field JB: Persistent pancreatic glucagon but not insulin
        response to arginine in pancreatectomized dogs. Endocrinology
        96, 678-693, 1975

4. Sasaki H, Rubalcava B, Baetens D, Blazquez E, Srikant CB, Orci L,
        Unger RH: Identification of glucagon in the gastrointestinal
        tract. J. Clin. Invest. 56, 135-145, 1975

5. Dobbs R, Sakurai H, Sasaki H, Faloona G, Valverde I, Baetens D,
        Orci L, Unger R: Glucagon: role in the hyperglycemia of
        diabetes mellitus. Science 187, 544-547, 1975

6. Lefebvre PJ, Luyckx AS: Plasma glucagon after kidney exclusion:
        experiments in somatostatin-indused and in eviscerated dogs.
        Metab. Clin. Exp. 25, 761-768, 1976

7. Lawrence AM, Kirsteins L, Hojvat S, Rubin L, Mitton J, Pearce S,
        Kacherian R: Submaxillary gland hyperglycemic factor in man
        and animals; an extrapancreatic glucagon. Clin. Res. 26,
        364A, 1976 (Abstr.)

8. Dunbar JC, Silverman H, Kirman E, Foa PP: Salivary gland and
    kidney glucagon in the rat. Fed. Proc. 35, 218, 1976 (Abstr.)

9. Sutherland EW, De Duve C: Origin and distribution of the hyper-
    glycemic-glycogenolytic factor of the pancreas. J. Biol.
    Chem. 175, 663-674, 1948

10. Orci L, Forssmann WG, Forssmann W, Roullier C: Electron micro-
    scopy of the intestinal endocrine cells. Comparative study.
    In Electron Microscopy. 4th European Regional Conference.
    Boccharielli SD, editor. Tipografia Poliglotto Vaticana,
    Rome 2, 369-370, 1968

11. Cavallero C, Solcia E, Vassallo G, Capella C: Cellule endocrine
    della mucosa gastro-enterica ed ormoni gastro-intestinali.
    Rend. R.R. Gastroenterol. 1, 51-61, 1969

12. Cavallero C, Capella C, Solcia E, Vassallo G, Bussolati G:
    Cytology, cytochemistry and ultrastructure of glucagon-
    secreting cells. Acta Diabetol. Lat. 7, 542-556, 1970

13. Solcia E, Vassallo G, Capella C: Cytology and cytochemistry of
    hormone-producing cells of the upper gastro-intestinal tract.
    In Origin, Physiology and Pathophysiology of the Gastroin-
    testinal Hormones. Creutzfeldt W, editor. F.K. Schattauer-
    Verlag, Stuttgart, Germany, pp. 3-29, 1970

14. Kubes L, Jirasek K, Lomsky R: Endocrine cells of the dog gastro-
    intestinal mucosa. Cytologia (Tokyo) 39, 179-194, 1974

15. Baetens D, Rufener C, Srikant CB, Dobbs R, Unger R, Orci L:
    Identification of glucagon-producing cells (A cells) in
    dog gastric mucosa. J. Cell. Biol. 69, 455-464, 1976

16. Lefebvre PJ, Luyckx AS: Factors controlling gastric-glucagon
    release. J. Clin. Invest. 59, 716-722, 1977

17. Cuypers Y, Nizet A, Baerten A: Technique pour la perfusion de
    reins isolés de chien avec du sang hépariné. Arch. Int.
    Physiol. Biochim. 72, 245-255, 1964

18. Nizet A: The isolated perfused kidney: possibilities, limita-
    tions and results. Kidney Int. 7, 1-11, 1975

19. Munoz-Barragan L, Blazquez E, Patton GS, Dobbs RE, Unger RH:
    Gastric A-cell function in normal dogs. Am. J. Physiol.
    231, 1057-1061, 1976

20. Unger RH, Lefebvre PJ: Glucagon physiology. In Glucagon, Mole-
     cular Physiology, Clinical and Therapeutic Implications.
     Lefebvre PJ, Unger RH, editors. Pergamon Press Ltd., Oxford
     213-244, 1972

21. Assan R, Boillot J, Attali Jr, Soufflet E, Ballerio G: Diphasic
     glucagon release induced by arginine in the perfused rat
     pancreas. Nature (New Biol.) 239, 125-126, 1972

22. Pagliara AS, Stillings SN, Haymond MW, Hover BA, Matschinsky FM:
     Insulin and glucose as modulators of the amino acid-induced
     glucagon release in the isolated pancreas of alloxan and
     streptozotocin diabetic rats. J. Clin. Invest. 55: 244-255,
     1975

23. Vranic M, Gross G, Doi K, Lickley L: Effects of gastrointestinal
     glucagon on glucose turnover in depancreatized dog after
     withdrawal of insulin treatment. Diabetologia. 12, 425,
     1976 (Abstr.)

24. Blazquez E, Munoz-Barragan L, Patton GS, Orci L, Dobbs RE,
     Unger RH: Gastric A-cell function in insulin-deprived de-
     pancreatized dogs. Endocrinology, 99, 1182-1188, 1976

25. Koerker DJ, Ruch W, Chideckel E, Palmer J, Goodner CJ, Ensinck
     J, Gale CC: Somatostatin Hypothalamic inhibitor of the
     endocrine pancreas. Science 184, 482-484, 1974

26. Samols E, Tyler JM, Marks V: Glucagon-insulin interrelation-
     ships. In Glucagon, Molecular Physiology, Clinical and
     Therapeutic Implications. Lefebvre PJ, Unger RH, editors.
     Pergamon Press Ltd., Oxford. pp. 151-173, 1972

27. Weir GC, Knowlton SD, Atkins RF, Mc Kennan KX, Martin DB:
     Glucagon secretion from the perfused pancreas of strep-
     tozotocin-treated rats. Diabetes, 25, 275-282, 1976

28. Unger RH, Aguilar-Parada E, Muller WA, Eisentraut AM: Studies
     of pancreatic alpha-cell function in normal and diabetic
     subjects J. Clin. Invest. 49: 837-848, 1970

29. Massi-Benedetti F, Falorni A, Luyckx A, Lefebvre P: Inhibition
     of glucagon secretion in the human newborn by simultaneous
     administration of glucose and insulin. Horm. Metab. Res. 6:
     392-396, 1974

30. Östenson CG, Hellerström C: Effect of insulin on the glucose
     utilization of the pancreatic A_2-cell of the guinea-pig.
     Diabetologia 12, 413, 1976 (Abstr.)

31. Solcia E, Capella C, Vassallo G, Buffa R: Endocrine cells of
     the gastric mucosa. Int. Rev. Cytol. 42, 223-286, 1975

32. Unger RH: Diabetes and the alpha-cell. Diabetes 25, 126-151, 1976

33. Unger RH, editor. Metab. Clin. Exp. 25 (suppl. 1), 1481-1490,
     1976

34. Muller WA, Brennan MF, Tam Mh, Aoki TT: Studies of glucagon
     secretion in pancreatectomized patients. Diabetes 23, 512-
     516, 1974

35. Palmer JP, Werner PL, Benson JW, Ensinck JW: Plasma-glucagon
     after pancreatectomy. Lancet 1, 1290, 1976

36. Botha JL, Vinik AI: Plasma-glucagon after pancreatectomy. Lan-
     cet 1, 1290-1291, 1976

37. Barnes AJ, Bloom SR: Pancreatectomized man: a model for diabetes
     without glucagon. Lancet. 1, 119-221, 1976

38. Gerich JE, Karam JH, Lorenzi M: Diabetes without glucagon. Lan-
     cet 1: 855, 1976

39. Vranic M, Engerman R, Doi K, Morita S, Yip CC: Extrapancreatic
     glucagon in the dog. Metab. Clin. Exp. 25 (suppl. 1), 1469-
     1473, 1976

40. Heding LG: Determination of total serum insulin (IRI) in insulin-
     treated diabetic patients. Diabetologia 8, 260-266, 1972

THE GLUCAGONOMA SYNDROME

S. R. Bloom and J. M. Polak

Departments of Medicine and Histochemistry, The Royal
Postgraduate Medical School, Hammersmith Hospital

London W12 OHS, England

INTRODUCTION

Although there had been a number of previous reports of non-β
cell tumours of the pancreas, some of which might have been producing
glucagon, the first report with proper measurement of glucagon levels
did not appear until 1966.[1] The collection of clinical details from
9 separate glucagonoma patients in 1974[2] made it possible to deline-
ate the clinical features common to all cases and thus allowed us
to report the existence of a "glucagonoma syndrome." The main
features in this were a characteristic rash, necrolytic and migratory erythema, painful glossitis, angular stomatitis, normochromic
normocytic anaemia, severe weight loss, diabetes mellitus and the
symptoms of a pancreatic tumour. It is of interest that following
this description a considerable number of further case reports have
appeared reiterating these features.[3-11] Presumably such cases were
previously overlooked; this demonstrates the benefit of a proper
clinical description.

CLINICAL FEATURES

Diabetes. This is the first feature to be mentioned because
an abnormal glucose tolerance is almost invariable in the glucagonoma
syndrome. On the other hand overt diabetes is only seen in a proportion of the patients, perhaps just over half. The situation is thus
not very different from that seen in severe Cushing's disease or
acromegaly, where a considerable catabolic drive greatly increases
insulin requirements. The β cell is frequently unable to cope and
this output failure may or may not be irreversible. Thus we had

one patient who required 68 units of insulin a day to prevent
severe glycosuria but following resection of a benign pancreatic
glucagonoma the insulin could be immediately discontinued,[12] and
two years later her glucose tolerance curve was completely normal.
This provided the first proof that excessive glucagon alone could
provoke severe diabetes in man. The glucagonoma syndrome is one
of the few curable forms of diabetes.

In view of the current stress laid on the role of glucagon in
diabetes mellitus, even to the extent of diabetes being labelled a
bi-hormonal disease, it is of interest to note that the incidence
of ketoacidosis in glucagonomas is extremely low. Similarly there
is no evidence of an excessive incidence of arterial disease or
indeed the development of any of the diabetic complications, such
as retinopathy, neuropathy or nephropathy. As the glucagon levels
are many hundredfold higher than those seen in normal diabetics
the absence of these complications points away from glucagon as a
significant aetiological factor.

Skin Disease. The rash of the glucagonoma syndrome, a necrolytic migratory erythema, is sufficiently characteristic for many
cases to have been diagnosed by dermatologists before any other
information was available. Indeed, of patients referred to us by
a dermatologist as having the characteristic rash, just over half
had glucagon-producing tumours. On the other hand, there have
been many reports of glucagonomas without any rash and it is clear
that other factors are involved. It seems possible that the rash
only develops in patients with longstanding tumours and the finding
of a somewhat higher incidence in the elderly suggests the possibility of a slowly developing deficiency syndrome.

A more general familiarity with the skin appearance should
enhance the detection rate considerably as, unfortunately, many
cases are still found at autopsy. Figures 1-3 show some aspects

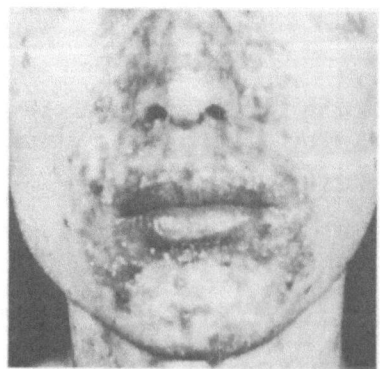

FIGURE 1.   *Close up of facial rash in glucagonoma syndrome.*[3]

of the rash but fuller descriptions and some colour photographs
may be obtained from the literature.[7,13-15] There is destruction
of the superficial epidermis, a marked erythema and a tendency for
the lesion to migrate. Secondary infection, however, can confuse
the picture. Perhaps the earliest abnormality is scattered pink
macules and papules. Later slightly raised reddened oedematous
patches can appear which tend to spread. The condition can be made

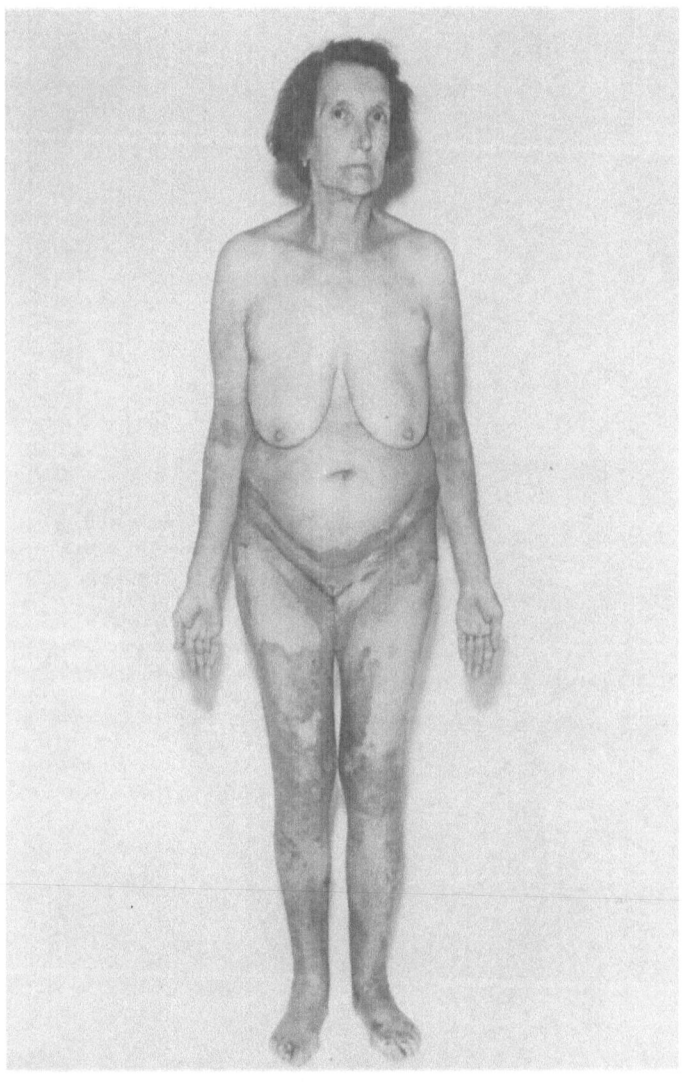

FIGURE 2.   Necrolytic migratory erythema in female with glucagonoma
            syndrome.[2]

worse by trauma which may remove the superficial layers of epidermis leaving a red raw surface. In more severe cases there is a superficial blistering followed by crusting. As the edge of the lesions spread so the older central portions tend to heal leaving a brown pigmented area. There is a tendency to exacerbations and remissions and these may be complete. The rash is most frequently localised

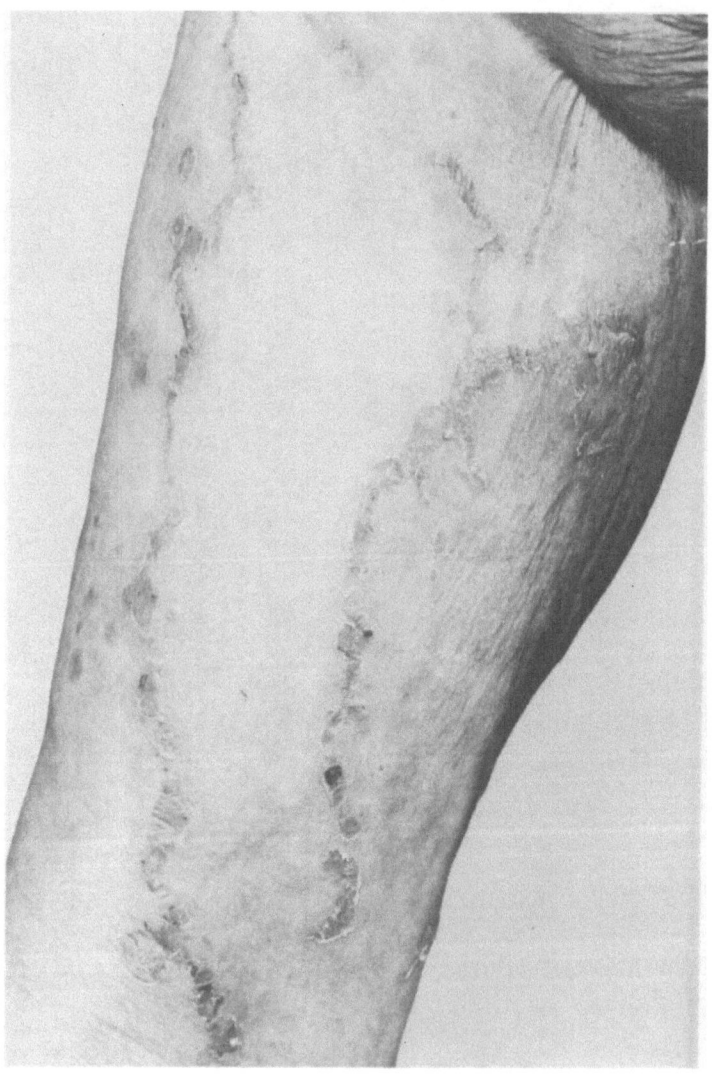

FIGURE 3.   *Close up of Figure 2.*

to skin folds and dependent parts as well as areas which can be traumatised. The commonest sites are the lower abdomen, groins, perineum, thighs and legs. The other commonly associated features are painful red glossitis, which can cause considerable discomfort to the patient, and angular stomatitis with circumoral crusting of unpleasant appearance. Histology shows necrosis and lysis in the upper half of the epidermis with the dead surface layers separating with the formation of superficial vesicles or bullae. On occasion there is a pronounced infiltrate of leucocytes and histiocytes with pustule formation. The acantholysis, typical of pemphigus vulgaris, is absent and anti-epithelial antibodies are undetectable.

Other Features. The presence of normochromic normocytic anaemia is common but no specific abnormalities are detected on examination of the bone marrow. Similarly, weight loss, often of marked degree, is a relatively constant finding. These features are seen in all catabolic states and is perhaps a reflection that glucagon is an extremely powerful stimulus to catabolism. Perhaps because of this the glucagonoma syndrome patients are peculiarly likely to develop secondary infections, such as pneumonia. In addition, they have a tendency for venous thrombosis and overall nearly a third of the patients have developed this complication at some time or another and many had pulmonary embolic episodes which were sometimes fatal. It is uncertain whether this thrombotic tendency is the result of a specific abnormality in platelet function or clotting mechanisms, or whether it is merely a nonspecific consequence of debilitation. No such tendency is observed, however, with non-endocrine pancreatic carcinoma and the glucagonoma patients do not have the condition of thrombophlebitis migrans. Many of the patients also develop depression which is often out of proportion to their physical condition. Various abdominal symptoms have been described including constipation and intermittent short episodes of diarrhoea but it is difficult to prove that either condition is related to the glucagonoma syndrome itself. Biopsies of the intestinal mucosa have usually proved to be normal but a few cases have villous hypertrophy[6] and in this respect are reminiscent of the changes reported in the only described enteroglucagon-producing tumour.[16]

BIOCHEMICAL INVESTIGATIONS

Whereas the fasting plasma glucagon concentration in a healthy young adult is usually below 30 pmol/l, patients with the glucagonoma syndrome have levels in excess of 300 pmol/l, sometimes grossly so. Thus a single fasting plasma glucagon estimation is usually enough to confirm the diagnosis. Care must be taken that the sample is properly collected to prevent glucagon degradation with addition of Trasylol, rapid separation, and storage of the plasma at $-20^{\circ}C$ within approximately 15 minutes of the initial venepuncture. Some problem may be encountered in seriously ill patients, perhaps with metastatic carcinoma, whose glucagon levels are often significantly elevated. These cases very rarely exceed 300 pmol/l, however, and would certainly not be repeatedly above that concentration. Analysis of the molecular forms of glucagon in the circulation frequently will show predominance of a large molecular weight glucagon.[9,10] This material migrates on gel columns at the 9,000 molecular weight position and is also somewhat raised in some other conditions such as renal failure. The proportion of big glucagon in normal plasma is quite low. The tumour output varies from 100% production of material of apparently normal size to 100% production of the big glucagon (Fig. 4). The large glucagon may well be of diminished biological activity but so far no close correlation between the type of material being produced by the tumour and the clinical picture has been shown.

Clearly as awareness of the glucagonoma syndrome increases less severe cases will be investigated and the need for a dynamic diagnostic test may arise. Glucose inhibits the normal output of

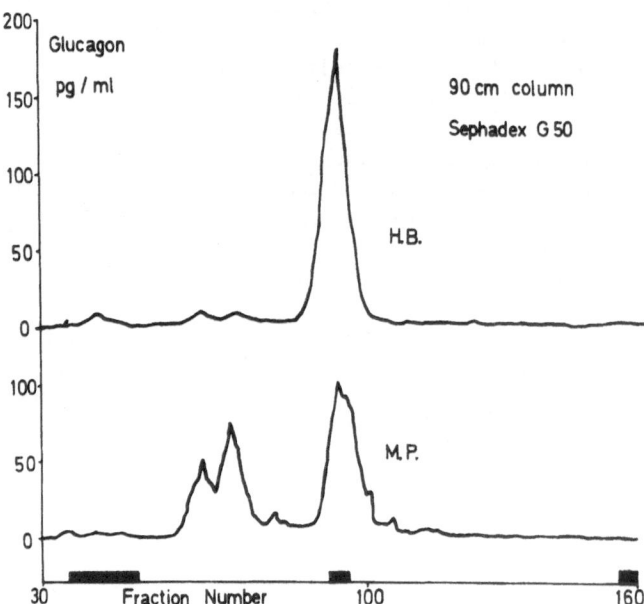

FIGURE 4. Gel chromatograms of plasmas from two patients with the glucagonoma syndromes. HB plasma shows glucagon immunoreactivity eluting in position of pure porcine pancreatic glucagon (central black marker bar) while MP plasma has several peaks of higher molecular weight material.

glucagon while arginine stimulates it and either substance might
form the basis of a glucagonoma test. Unfortunately, tumour
behaviour is wildly variable. In some cases glucose stimulates
the output of glucagon, while in others it suppresses it as in
normals. Arginine may produce a massive release of glucagon but
sometimes produces no change at all. It is noteworthy that in one
case we investigated the most impressive stimulus to glucagon
release was the sulphonylurea tolbutamide.[4] This may indicate that
caution should be observed when treating the diabetes of the
glucagonoma syndrome with sulphonylureas.[5]

     The biological activity of tumour glucagon has not been care-
fully investigated. Nonetheless, when glucagon output from the
glucagonoma has been suppressed by an acute infusion of somatostatin,
hypoglycaemia rapidly followed.[17] Similarly, the glucagonoma
patients have almost universally been found to have markedly depres-
sed plasma amino acid levels.[2] Following tumour resection these
returned to normal or even above normal. The insulin levels in
these patients tend to be high but in spite of this glucose tolerance is reduced. Thus in the cases studied so far there seems
little doubt that at least a proportion of the glucagon secreted
is biologically effective.

     The old concept that each type of pancreatic endocrine tumour
was quite separate has now been greatly undermined by the finding
that many tumours produce several different hormones simultaneously.
Indeed it might be nearer the truth to call all the pancreatic
endocrine tumours "islet tumours." Thus in more than half of the
glucagonoma patients we have studied, the plasma PP levels were
markedly elevated,[18] being continuously higher than the peak level
of PP seen after a meal in normal subjects. Indeed in some individuals PP levels as high as 20,000 pmol/l have been observed. So far
no clinical correlate of these elevated PP concentrations has been
discovered. The suggestion that they might be associated with, for
example, watery diarrhoea is not sustained. Infusions of PP in man
result in inhibition of gall bladder contraction and a reduction of
pancreatic enzyme and juice production.[19] It may well be, however,
that escape of tissue response occurs from the effects of sustained
elevations of PP. Elevated levels of insulin and gastrin have also
been described with glucagonomas, but so far there have been no
reports of raised plasma somatostatin concentrations. Glucagonomas
can form part of the multiple endocrine adenomatosis syndrome and
be associated with distinct lesions in the pituitary or parathyroids.
The indicence of glucagonomas in the MEA syndrome is, however,
quite small.

TUMOURS

     The tumours can be shown to contain pancreatic glucagon both
by extraction and immunocytochemistry. The total content, however,

varies widely. Some tumours appear to secrete almost all the
glucagon they synthesise while others maintain a reasonably intact
storage system. In the latter cases electron microscopy demonstrates
typical α cell type granules (Fig. 5). It is quite common to find
significant numbers of other types of endocrine cells in between
the tumorous α cells. The most common cell type produces PP and
accounts for the high incidence of PP secretion by glucagonomas.
Other cells may contain insulin, gastrin or somatostatin, though
the number of somatostatin cells is usually quite small. Extraction
of primary or secondary glucagonoma tissue always yields some
somatostatin immunoreactivity.[20] This can be shown by chromatography
to behave in the same manner as synthetic somatostatin. It is thus
quite likely to be biologically active and, as the somatostatin-
producing D cell is a member of the locally active paracrine system,
it may well exert a local influence on hormone production. Thus
there is some evidence that the tumours containing the highest
concentration of somatostatin are in the least active state for
glucagon secretion.

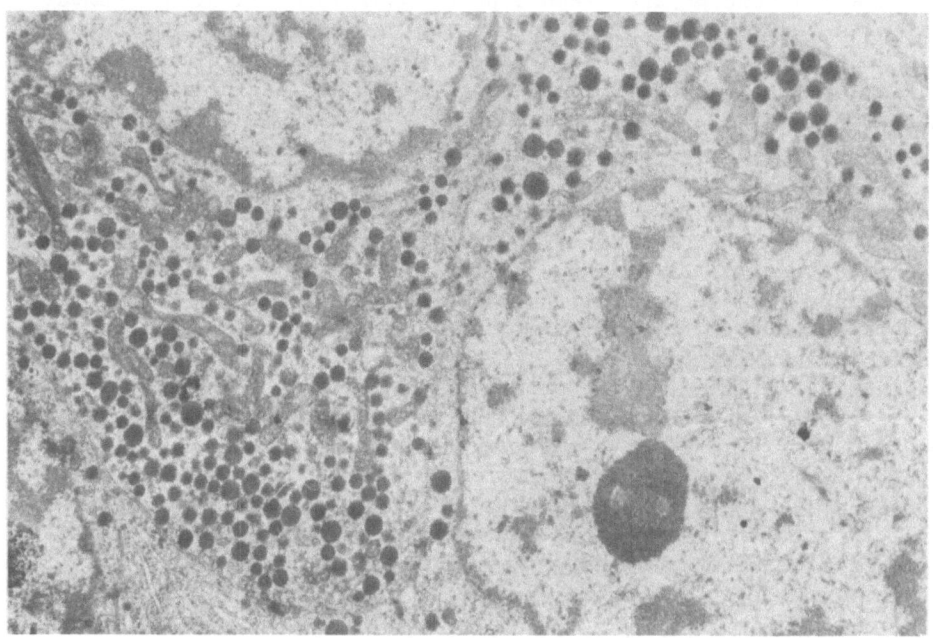

FIGURE 5. Ultramicrograph of a glucagonoma showing numerous
electron dense secretory granules with a dense core, typical of
pancreatic A cells (x 12,000).

TREATMENT

The treatment of choice is total surgical resection. This is
possible in less than half the cases. Resection of a benign adenoma
can be expected to give complete cure and it is possible that resection of a malignant tumour prior to local invasion or the development
of significant metastases may greatly improve the prognosis. Glucagonomas are frequently quite slow growing and if an active primary
can be resected it may be many years before the secondaries develop
sufficiently to be a nuisance. Once the diagnosis has been confirmed
by glucagon measurement, it is helpful to accurately localise the
tumour in the pancreas. If angiography fails to do this, suggesting
the tumour is small, percutaneous transhepatic portal venography,
with selective venous sampling,[21] is usually successful. It is
useful during this investigation to take simultaneous hepatic vein
samples so that any spontaneous fluctuation in tumour hormone output
can be detected, thus avoiding false localisation.

If surgery is not possible there are no very satisfactory
alternative treatments. The cytotoxic drug streptozotocin may be
effective in about a third of cases. One unpublished patient that
we have been following for several years now has a completely normal
glucagon level and an undetectable tumour mass following multiple
courses of streptozotocin in spite of extremely high hormone levels
and numerous hepatic secondaries at the time of his original laparotomy. Because of the slow growing nature of the tumour and the fact
that the morbidity is often related to hormone levels, tumour debulk-
ing is usually helpful. We have recently treated a patient whose
problem was the result of bulky liver secondaries by blocking the
hepatic artery by injection of numerous small pieces of gelatin
foam directly through an hepatic artery catheter.[22] This procedure
caused very little discomfort to the patient and, perhaps surprisingly,
hardly affected hepatic function. Glucagon levels fell dramatically
and the patient's well-being greatly improved. The very unpleasant
skin condition has been refractory to most commonly used dermatolo-
gical treatments. Mallinson has recently been able to show, however,
that zinc administration causes a complete remission.[23] There is
some evidence to suggest that the glucagonoma patients are zinc
deficient and the rash has features in common with that seen in zinc
deficiency due to other causes. A short-term treatment that may
prove of benefit in preparing these rather cachexic patients for
surgery is the use of long-acting somatostatin analogues. The
Des[1,2,4,5,12,13], D Try[8], D Cys[14] analogue is effective for 12 hours
when given subcutaneously.[24] Thus a twice-daily injection regime
is a practical proposition, though the suppression of insulin
release will necessitate adjustment of the diabetic treatment.

CONCLUSIONS

The frequency of the glucagonoma syndrome, as a curable cause
of diabetes and a most unpleasant skin rash, is still not fully
known. In a recent series of diabetic autopsies, 1% of the patients
were found to have an α cell tumour of the pancreas.[25] This
syndrome may thus be commoner than we think. There is no doubt
that a greater clinical awareness should lead to earlier diagnosis
and an improved prognosis. The scientific value of these natural
experiments in understanding the role of glucagon in man should
not be underestimated. Glucagon can be seen to be a highly catabolic
hormone with little evidence of escape from long continued elevated
concentrations. On the other hand, no evidence can be adduced in
favour of the theory that elevated glucagon may be responsible for
any of the complications of diabetes mellitus.

REFERENCES

1. McGavran MH, Unger RH, Recant L, Polk HC, Kilo C, Levin M: A
       glucagon-secreting alpha-cell carcinoma of the pancreas.
       N. Engl. J. Med. 274: 1408-1413, 1966

2. Mallinson CN, Bloom SR, Warin AP, Salmon PR, Cox B: A
       glucagonoma syndrome. Lancet II: 1-4, 1974

3. Scully RE, McNeely BU: Case of glucagonoma in case records of
       the Massachusetts General Hospital. N. Engl. J. Med.
       273: 1117-1123, 1975

4. Soler NG, Dates GD, Malins JM, Cassar J, Bloom SR: Glucagonoma
       syndrome in a young man. Proc. Roy. Soc. Med. 69: 429-431,
       1976

5. Cho KJ, Wilcox CW, Reuter SR: Glucagon-producing islet cell
       tumor of the pancreas. Am. J. Roentgenol. Radium Ther.
       Nucl. Med. 129: 159-161, 1977

6. Stevens FM, Flannagan RWJ, Buchanan KD, O'Gorman DJ: The
       first Irish glucagonoma. Ir. J. Med. Sci. 146: 350, 1977

7. Kahan RS, Perez-Figaredo RA, Neimanis A: Necrolytic migratory
       erythema. Arch. Dermatol. 113: 792-797, 1977

8. Recant L, Perrino PV, Bhathena SJ, Danforth Jr DN, Lavine RL:
       Plasma immunoreactive glucagon fractions in four cases of
       glucagonoma: increased "large glucagon-immunoreactivity."
       Diabetologia 12: 319-326, 1976

9.   Weir GC, Horton ES, Aoki TT, Slovik D, Jaspan J, Rubenstein AH:
     Secretion by glucagonomas of a possible glucagon precursor.
     J. Clin. Invest. 59: 325-330, 1977

10. Danforth Jr DN, Triche T, Doppman JL, Beazley RM, Perrino PV,
     Recant L: Elevated plasma proglucagon-like component with
     a glucagon-secreting tumor. N. Engl. J. Med. 295: 242-245,
     1976

11. Holst JJ, Pedersen NB: Glucagon-producing tumors of the
     pancreas. Acta Endocrinol. [Suppl.] 199: 379, 1975

12. Lightman SL, Bloom SR: Cure of insulin-dependent diabetes
     mellitus by removal of a glucagonoma. Br, Med, J. 1: 367-
     368, 1974

13. Sweet RD: A dermatosis specifically associated with a tumour
     of pancreatic alpha cells. Br. J. Dermatol, 90: 301-308,
     1974

14. Pedersen NB, Jonsson L, Holst JJ: Necrolytic migratory erythema
     and glucagon cell tumor of the pancreas: the glucagonoma
     syndrome. Acta Dermatol. (Stockholm) 56: 108, 1976

15. Wilkinson DS: Necrolytic migratory erythema with carcinoma of
     the pancreas. Transactions of St. John's Hospital Dermato-
     logical Society, 59: 244-246, 1973

16. Bloom SR: An enteroglucagon tumour. Gut 13: 520-523, 1972

17. Mortimer CH, Carr D, Lind T, Bloom SR, Mallinson CN, Schally AV,
     Tunbridge WMG, Yeomans L, Coy DH, Kastin A, Besser GM:
     Growth Hormone release inhibiting hormone: effects on circu-
     lating glucagon, insulin and growth hormone in normal,
     diabetic, acromegalic and hypopituitary patients. Lancet
     I: 697-701, 1974

18. Polak JM, Bloom SR, Adrian TE, Heights P, Bryant MG, Pearse AGE:
     Pancreatic polypeptide in insulinomas, gastrinomas, VIPomas
     and glucagonomas. Lancet I: 328-330, 1976

19. Adrian TE, Greenberg GR, Besterman HS, McCloy RF, Chadwick VS,
     Barnes AJ, Mallinson CN, Baron JH, Alberti KGMM, Bloom SR:
     PP infusion in man - summary of initial investigations.
     In: Gut Hormones, ed. SR Bloom, Churchill Livingstone,
     Edinburgh, 1978, p 265-267

20. Bloom SR, Polak JM, West AM: Somatostatin content of pancreatic
     endocrine tumours. In: Proceedings of the 1st International
     Symposium on somatostatin. Metabolism, in press

21.   Ingemansson S, Lunderquist A, Holst J: Selective catheteriza-
         tion of the pancreatic vein for radioimmunoassay in
         glucagon-secreting carcinoma of the pancreas. Radiology
         119: 555-556, 1976

22.   Allison DJ, Modlin IM, Jenkins WJ: Treatment of carcinoid
         liver metastases by hepatic-artery embolisation. Lancet
         II: 1323-1325, 1977

23.   Mallinson CN, Adrian TE, Hanley J, Bryant M, Bloom SR:
         Metabolic and clinical responses in patients with pancreatic
         glucagonomas. Ir. J. Med. Sci. 146: 37, 1977

24.   Bloom SR, Adrian TE, Barnes AJ, Long RG, Hanley J, Mallinson CN,
         Rivier JE, Brown MR: New specific long-acting somatostatin
         analogues in the treatment of pancreatic endocrine tumours.
         Gut, in press

25.   Lomsky R, Langer F, Vortel V: Demonstration of glucagon in
         islet cell adenomas of the pancreas by immunofluorescent
         technique. Am. J. Clin. Path. 51: 245-250, 1967

VIP: THE CAUSE OF THE WATERY DIARRHOEA SYNDROME

I. M. Modlin, S. R. Bloom and S. Mitchell

Departments of Surgery and Medicine, Royal Postgraduate
Medical School

Du Cane Road, London W12 OHS

INTRODUCTION

Verner and Morrison in 1958 described a syndrome of refractory
watery diarrhoea and hypokalaemia associated with non-insulin secre-
ting tumours of the pancreatic islets. In the following years many
such cases were reported and suggestions made as to the causative
agent[1]. In fact at one time or another almost all the characterised
GI hormones have been put forward as the cause of the watery diarrhoea
syndrome. In 1973 it was reported that the responsible tumour contained large quantities of the newly discovered polypeptide VIP and
that plasma VIP levels were extremely elevated. The relationship
of VIP to the syndrome seemed likely because the biological actions
of VIP closely fitted the described symptoms of the patients. Thus,
VIP is a potent stimulant of small intestinal juice production and a
powerful inhibitor of gastric acid secretion. It also increases the
output of hepatic glucose, stimulates pancreatic bicarbonate secretion, relaxes the gall bladder and causes vasodilation[2]. These may
explain the frequent clinical findings of hypochlorhydria, diabetes,
flushing attacks, high resting pancreatic juice production and a
flaccid distended gall bladder. It was proposed that VIP was the
cause of the Verner-Morrison syndrome and therefore that demonstration of a raised plasma VIP level would be helpful in diagnosis[3].
This proposal, however, has not been generally accepted, especially
since until now no direct evidence exists that VIP produced the
watery diarrhoea syndrome[4].

We therefore sought to correlate information from both clinical
and experimental situations to demonstrate that VIP-producing tumours
(VIPomas) are almost certainly the most frequent cause of the watery
diarrhoea syndrome.

PLASMA VIP MEASUREMENT

Considerable care is necessary both during specimen collection and in assay since subtle errors in technique may introduce error. VIP is rapidly destroyed by proteolytic enzymes as it possesses two separate double basic amino acid sequences which are particularly liable to degradation by trypsin-like enzymes. In addition it contains a methionine residue which renders it liable to oxidative damage and two asparagine residues which are easily deamidated. As VIP is a highly basic molecule it is rapidly absorbed onto active surfaces, particularly those which are negatively charged. Blood should therefore be taken into a proteolytic enzyme inhibitor (aprotinin 1,000 KIU/ml blood) and then rapidly centrifuged and the plasma frozen within 15 minutes of venipuncture. Steady loss of immunoreactive VIP occurs at -20 C storage whilst thawing and refreezing accentuates this process. Lyophilising plasma is now considered the best technique for storage and transport without significant degradation.

Using an assay system sensitive to 1 pmol/l plasma, we have found the normal plasma VIP concentration to be less than 10 pmol/l and the absolute upper limit in healthy subjects to be 20 pmol/l[5]. Other reports have suggested that completely healthy individuals may have grossly elevated plasma VIP. We have not had this experience. The antibody used in this assay system was carefully chosen so as to be insensitive to the non-specific effects of plasma. It also combines most avidly with whole VIP but relatively little with VIP fragments. There was no cross reaction with other named hormones and assay specificity was further confirmed by showing no cross reaction with other peptides in a crude extract of human ileum. Thus on gel chromatography only a single peak of immunoreactivity was detected eluting in exactly the same position as both pure porcine and human tumour-produced VIP. This indicates that there are no other materials in the gut which are detected by this assay and that there is therefore no cross reaction even with as yet undiscovered ileal hormones[8].

CLINICAL RESULTS

A. *VIP-producing Tumours*

We have assayed samples of nearly one thousand patients suffering from unexplained severe diarrhoea. In 39 of these patients the plasma VIP level was grossly elevated and in each case a tumour was located. The majority of these lesions were located in the pancreas, but in 7 individuals (4 children) the tumour was a ganglioneuroma or ganglioneuroblastoma. Analysis of the tumour in each case always demonstrated high values of VIP. In more than half of these 39 cases there was no evidence of metastases and removal of the primary tumour resulted in plasma VIP returning to normal levels with abrupt cessation of diarrhoea.

B. *Treatment of Metastatic VIPomas in Cases with Metastases*

Although the course of the diarrhoea is fluctuant, death may occur in weeks or months. The reason for this is often that the diagnosis is made only in extremis by which time severe metabolic sequelae are already present. Clinical awareness of the syndrome and thus early diagnosis by plasma VIP measurement is therefore of great importance. Streptozotocin, a nitrosourea antibiotic that powerfully inhibits DNA synthesis, has proved very effective in treating disseminated pancreatic VIPomas. Remission periods of several years have been reported. We have demonstrated that lowering of the plasma VIP level by streptozotocin therapy closely parallels the reduction in diarrhoea and clinical remission[6]. Monitoring the plasma VIP level is a good early indication of the imminence of a relapse and recurrence of the diarrhoea and suggests the need for further preventive cytotoxic therapy. In some cases high dosages of steroids (50-100 mg of prednisolone daily) are effective in temporarily controlling the diarrhoea but only lead to a slight fall in plasma VIP concentrations. In this situation the effect of the steroids is probably directly on the bowel mucosa.

C. *Other Diarrhoeal Conditions*

In infective diarrhoea and inflammatory bowel disease (Crohn's and ulcerative colitis) plasma VIP levels are not raised[7]. Analysis of samples from patients with severe diarrhoea suffering from medullary carcinoma of the thyroid, purgative addiction, carcinoma of the lung, carcinoid syndrome and villous adenoma of the rectum revealed no elevation of plasma VIP levels. A rise in plasma VIP levels is found in bowel ischaemia, both clinically and experimentally[8]. This situation often produces diarrhoea but the mechanism may not be the same, as in patients with a VIPoma the levels are persistently raised. The role of high plasma VIP levels in the pathogenesis of bowel ischaemia requires further study. In three patients with slightly elevated VIP levels diarrhoea fluctuated in parallel with the plasma VIP concentration but so far no pancreatic tumour has been demonstrated in these subjects in spite of extensive investigation. It is possible that these individuals may have another mechanism for abnormal VIP release.

This situation differs slightly from the 11 patients we have examined with the classical clinical picture of the watery diarrhoea syndrome, complete with severe hypokalaemia and in most cases hypochlorhydria but no evidence of pancreatic tumour at laparotomy. These cases all had normal plasma VIP levels. Occasionally pancreatic islet cell hyperplasia has been found in such patients, who are usually treated by partial pancreatic resection, though in some cases only a subsequent total pancreatectomy cured the diarrhoea. The term pseudo-Verner-Morrison syndrome has been applied to these cases and it almost certainly represents a different aetiology[3].

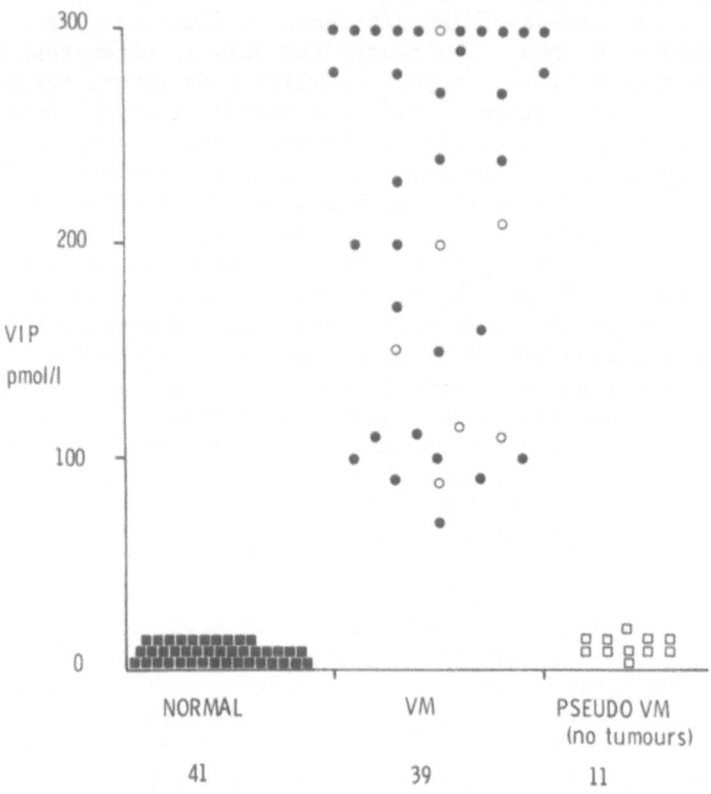

Figure 1. *The dark circles indicate the plasma VIP levels of patients*
with pancreatic VIPomas. The open circles indicate the plasma VIP
levels for patients with ganglioneuromas. The absolute upper limit
of normal is 20 pmol/l. In the patients with pseudo Verner-Morrison
syndrome plasma VIP levels are within normal limits.

        In seventy percent of the VIP-producing tumours plasma PP
levels are grossly elevated and it is conceivable that pancreatic
polypeptide may play a part in the development of the diarrhoeal
syndrome. PP, however, has never been demonstrated physiologically
or pharmacologically to produce diarrhoea and in fact in massive
single doses produces only defaecation, without diarrhoea. Furthermore, many patients with gastrinomas, glucagonomas and insulinomas
have very elevated PP levels and no diarrhoea. It is now almost
certain that there are some tumours in which other agents (for example,prostaglandins)are responsible for the diarrhoea.

EXPERIMENTAL VIP-INDUCED DIARRHOEA

Previous data on the production of watery diarrhoea by VIP has been rejected on two bases. Firstly, the dosage necessary to elevate mucosal cyclic AMP or produce secretion in bowel loops was far in excess of the levels obtained in patients with VIP-producing tumours[9]. Secondly, that while VIP appeared to act by increasing cyclic AMP, patients with the Verner-Morrison syndrome did not have raised mucosal cyclic AMP levels[10]. These findings may, however, merely indicate that VIP can act by other mechanisms.

Since the only VIP isolated to date is porcine VIP and no information is currently available as to the possibility of species difference, only porcine experimental data can be really valid. We therefore infused pure porcine VIP into eight healthy ambulant and unsedated pigs 72 hours after two vascular cannulae had been inserted. The infusion rate was increased stepwise from 2 pmol/kg-min to 10 pmol/kg-min over a 12-hour period. Initially severe facial flushing and a marked tachycardia were noted. Diarrhoea began in three pigs after four hours and was marked by six hours. By eight hours six of the eight pigs had developed severe watery diarrhoea (20-30 ml/kg). This effect was seen at plasma VIP levels of 70-90 pmol/1. In addition, serum potassium levels fell significantly. Two of the animals did not develop diarrhoea. In one pig the plasma VIP levels never rose above 60 pmol/1. In the other animal where the plasma VIP was 100 pmol/1 no obvious special factor could be found. It is evident that experimental administration of VIP to produce constant elevated plasma levels above 60 pmol/1 can mimic the watery diarrhoea syndrome. Some patients' diarrhoea is cyclical, however, despite continually high plasma VIP levels and this variable responsiveness might explain the absence of diarrhoea in the one pig with a plasma level of 100 pmol/1 of VIP.

CONCLUSIONS

In patients suffering from the watery diarrhoea syndrome who have pancreatic and neural tumours , plasma VIP is elevated. The VIP concentrations in the tumour (immunocytochemistry and radioimmunoassay) are increased and selective venous sampling of the lesion also demonstrates an extremely high VIP production locally. The pharmacological actions of VIP include a potent stimulation of small intestinal juice secretion. Removal of the tumour or ablation of the metastases with streptozotocin produces reduction of plasma VIP levels and a parallel simultaneous cessation of the diarrhoea.

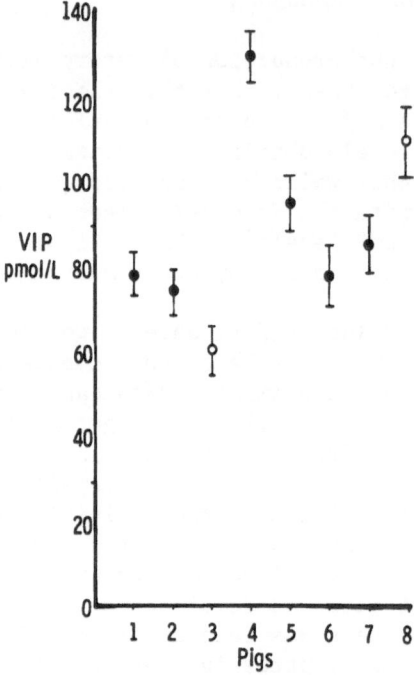

Figure 2. The dark circles indicate the plasma VIP level of those pigs who developed diarrhoea after VIP infusion. The two open circles express the plasma values of VIP in the two pigs who did not develop diarrhoea. Persistent plasma VIP levels of greater than 60 pmol/l for more than six hours produced copious watery diarrhoea.

     Since porcine VIP is apparently immunochemically and chromato-graphically identical to human VIP the production of diarrhoea in pigs infused with pure porcine VIP would strongly suggest a casual role of VIP in the pathogenesis of the human watery diarrhoea syndrome. The fact that this is achieved at plasma levels closely comparable to those found in humans adds further weight to the suggestion

     Since VIP has not been demonstrated to be elevated in any other diarrhoeal situation, plasma measurement is of exceptional diagnostic value. The mortality rate in the watery diarrhoea syndrome is directly related to the length of delay in detection and diagnosis of the tumour. Early detection of a raised plasma VIP level may thus be life saving.

REFERENCES

1. Bloom SR, Polak JM, Pearse AGE: Vasoactive intestinal peptide
    and watery diarrhoea syndrome. Lancet, II, 14-16, 1973

2. Bloom SR, Polak JM: The role of VIP in pancreatic cholera.
    Gastrointestinal Hormones, (ed) JC Thompson, Austin, Univer-
    sity of Texas Press, 635-642, 1975

3. Bloom SR, Polak JM: VIP measurement in distinguishing Verner-
    Morrison syndrome and Pseudo Verner-Morrison syndrome.
    Clin Endocr. 5, 223s-228s, 1976

4. Modlin IM, Bloom SR, Mitchell SJ: The role of VIP in diarrhoea.
    Gut, 18, A418, 1977

5. Mitchell SJ, Bloom SR, Modlin IM: VIP assay and intestinal
    ischaemia. (In Press)

6. Gagel RF, Costanza ME, Delellis RA, Norton RA, Bloom SR, Miller
    HH, Ucci A, Nathanson L: Streptozotocin-treated Verner-
    Morrison syndrome: plasma vasoactive intestinal peptide and
    tumour responses. Arch intern Med, 136, 1429-1435, 1976

7. Fairclough PD, Bloom SR, Mitchell SJ: Plasma-VIP in ulcerative
    colitis. Lancet, I, 964, 1976

8. Modlin IM, Bloom SR, Mitchell S: VIP release in intestinal
    ischaemia. Gut, (In Press) 1977

9. Schwartz CJ, Kimberg DV, Sheerinhe V, Field M, Said SI: Vaso-
    active intestinal peptide stimulation of adenolate cyclase
    and active electrolyte secretion in intestinal mucosa.
    J Clin'Invest, 54, 536-544, 1974

10. Klaeveman HL, Conlon TP, Levy AG, Gardner JD: Effect of gastro-
    intestinal hormones on adenylate cyclose activity in human
    jejunal mucosa. Gastroenterology, 68, 667-675, 1975

SECRETIN RELEASE IN MAN: CURRENT STATUS

O. B. Schaffalitzky De Muckadell and J. Fahrenkrug

Department of Clinical Chemistry, Bispebjerg Hospital

Copenhagen, Denmark

All radioimmunoassays for secretin are developed against porcine secretin and the structure and biological activity of human secretin are for the moment unknown. So, in the following, secretin concentrations are expressed as pmol porcine secretin per liter plasma.

However, with our antibodies only one molecular form of secretin is found in human plasma and the elution position after gel filtration is identical to that of porcine secretin. Furthermore, dilution curves of extracts of human small intestine or plasma are superimposable on standard curves.

The assay used can detect 0.6 pmol secretin per liter[1,2,3] plasma and in the fasting state secretin concentrations in 70 normal subjects ranged from zero to 5 pmol per liter with a median of 0.9 pmol liter^{-1}. In the fasting state higher values are found in the portal vein and the highest values are found in the pancreaticoduodenal vein. Thus secretin is secreted but the problem is whether and how this release changes during physiological conditions.

The effect of HCl, amino acids, ethanol, fat, isotonic glucose, hypertonic glucose, or isotonic saline given intraduodenally was studied in seven normal persons[4]. All the solutions except hydrochloric acid had a pH of 7.

Only hydrochloric acid increased the secretin levels significantly. Median peak concentration was 13.2 pmol liter^{-1}.

The effect of endogenous acid was studied during insulin induced hypoglycemia. Each person was studied twice, in one experiment gastri acid was aspirated and in the other one no aspiration was performed.

Median peak secretin concentration was 8.6 pmol liter^{-1} when acid was allowed to enter the duodenum and no significant increase in secretin concentration was found when gastric acid was aspirated. Thus gastric acid does drive secretin secretion.

The pH threshold to secretin release has been determined by perfusion of the duodenum with buffers of various pH[5]. A high flow rate of 50 ml per minute was used in order to ensure a stable, well defined and reproducible duodenal acidity. Infused buffer was aspirated from the third part of duodenum.

Secretin levels increased at pH levels between 2 and 3 and so did the bicarbonate secretion estimated from the disappearance rate of acid. This threshold pH might seem surprisingly low. Thus, it was of importance to measure the intraduodenal pH during more physiological conditions, e.g., in the fasting state and during a meal

Intraduodenal pH was measured in situ proximally in the second part of the duodenum, before and after ingestion of a meal consisting of 200 g beef steak, vegetables and a beer. In this part of the duodenum the pH at rare intervals reached the critical values of pH 2-3, but lower pH values are found more proximally in the duodenum. These rapid falls in duodenal pH simply indicate that some acid had reached the glass-electrode and tell nothing about the pH at the surface of the secretin cell.

In different subjects the pH changes occur at different times before and after the meal; therefore it might be difficult to detect an effect on secretin levels in a group of subjects. Thus, each subject should be evaluated separately.

It was found that increments in secretin concentrations coincided with rapid falls in intraduodenal pH.

Furthermore, the figures indicated that secretin is released at short intervals and blood samples should be drawn frequently.

As a kind of control experiment the meal-test was repeated after suppression of gastric acid secretion by 600 mg of cimetidine.

Both changes in secretin concentration and duodenal pH are diminished after cimetidine. Separate experiments showed that cimetidine had no effect on the secretin response to exogenous hydrochloric acid.

It should be noticed that also in the fasting state small increments in secretin concentrations seem to be preceded by rapid falls in duodenal pH. So it was concluded that secretin is released during physiological conditions in man and that secretin levels are fluctuating and depend on acid in the duodenum.

The major role of secretin is believed to be the stimulation of pancreatic bicarbonate secretion. So it was of interest to see if such low concentrations of secretin as measured during a meal exerted any effect on the secretion of bicarbonate into the duodenum. For that reason porcine (GIH) secretin was given intravenously in doses of 0.004 to 0.36 clinical units per kg per hour. A non-absorbable marker was infused into the duodenum and the concentration of marker and bicarbonate was measured in duodenal aspirates. Each of 10 normal subjects received 4 different doses of secretin.

Physiological levels were obtained by doses lower than 0.1 CU/kg-h. Both secretin levels and bicarbonate secretion increased significantly after a dose of 0.01 CU/kg-h.

The relation between the concentration of porcine secretin in plasma and the bicarbonate output after infusion of porcine secretin in graded doses was compared to the relation between the concentration of endogenous secretin and the bicarbonate output induced by acid. At the same secretin level there was a tendency to higher bicarbonate secretion after acid than after exogenous secretin.

Preliminary results from measurements of plasma secretin and pancreatic bicarbonate output in patients with diverted pancreatic juice have further substantiated the relation between plasma secretin and pancreatic bicarbonate secretion. In these patients fasting secretin levels were elevated and the relation between plasma secretin and bicarbonate output was close to that found after exogenous secretin.

The bicarbonate secretion induced by secretin is small. However, during physiological conditions, at least during a meal, secretin is not acting alone but together with CCK, gastrin, VIP, and vagal impulses.

It is concluded that:

1) Secretin is released by acid in the duodenum.
2) Secretin is secreted during physiological conditions in man.
3) The physiological levels of secretin might by themselves influence pancreatic bicarbonate secretion.

REFERENCES:

1. Schaffalitzky De Muckadell OB, Fahrenkrug J: Radioimmunoassay
    of secretin in plasma. Scand. J. Clin. Lab. Invest. 37,
    155-167, 1977

2. Fahrenkrug J, Schaffalitzky De Muckadell OB, Rehfeld J: Produc-
    tion and evaluation of antibodies for radioimmunoassay of
    secretin. Scand. J. Clin. Lab. Invest. 36, 281-287, 1976

3. Schaffalitzky De Muckadell OB, Fahrenkrug J: Preparation of[125]
    I-labelled synthetic porcine secretin for radioimmunoassay.
    Scand. J. Clin. Lab. Invest. 36, 661-668, 1976

4. Fahrenkrug J, Schaffalitzky De Muckadell OB: Plasma secretin
    concentration in man: Effect of intraduodenal glucose, fat,
    aminoacids, ethanol, HCl, or ingestion of a meal. Eur. J.
    Clin. Invest. 7 (3), 201-203, 1977

5. Fahrenkrug J, Schaffalitzky De Muckadell OB, Rune SJ: pH thres-
    hold for release of secretin in normal subjects and in
    patients with duodenal ulcer and patients with chronic pan-
    creatitis. Scand. J. Gastroent., in press

SECRETIN, GASTRIN AND PANCREATIC BICARBONATE RESPONSES TO MEALS VARYING IN pH LEVELS

O. L. Llanos, S. J. Konturek, P. L. Rayford, J. C. Thompson

University of Texas Medical Branch, Department of Surgery
Galveston, Texas

Release of endogenous secretin by duodenal acidification has been shown to occur in man and other species. Ingestion of food has, however, resulted in no consistently detectable increase in plasma secretin levels measured by radioimmunoassay. This study was designed to test the effect of the acidification of a meal on secretin and gastrin levels and on pancreatic bicarbonate and gastric acid secretion.

METHODS

The studies were performed in five fasting dogs prepared with chronic gastric and pancreatic fistulas. Solutions of liver extract (LE) at different pH levels ranging from 7 to 2 were used as test meals; 400 ml of each solution were introduced into the stomach and the preselected pH was kept constant by intragastric titration for 45 minutes. Gastric acid secretion was recorded automatically by autoburet.

In another study, 400 ml of LE at pH 7 were introduced into the stomach of the dogs and allowed to remain without exogenous acidification. The pH of the gastric contents was measured periodically.

In different tests, an LE meal at pH 7 was administered in the stomach and the pH kept constant by intragastric titration and graded doses of secretin (0.25 to 4.0 U/kg-hr) were infused intravenously. In control experiments, secretin was infused without an LE meal.

In all studies, pancreatic secretion was collected from the pancreatic cannula for measurement of volume and bicarbonate. Blood samples were collected periodically for gastrin and secretin determinations by specific radioimmunoassay.

RESULTS

Administration of LE meal at pH 7 evoked release of gastrin which rose from a basal of 120 pg/ml to values nearly twice as high (Fig. 1). Acid output was stimulated by LE meal at pH7 to 16 mEq/30 minutes. Acidification of the meal provoked a pH-dependent reduction of acid and gastrin secretion. Plasma secretin levels did not change with the meal at pH from 7 to 5. On acidification of the meal, secretin was significantly elevated at pH 3 and 2 (Fig. 2). Pancreatic secretion showed a sharp increase with acidification of the LE meal, reaching an output of 5.2 mEq/30 min at pH 3.

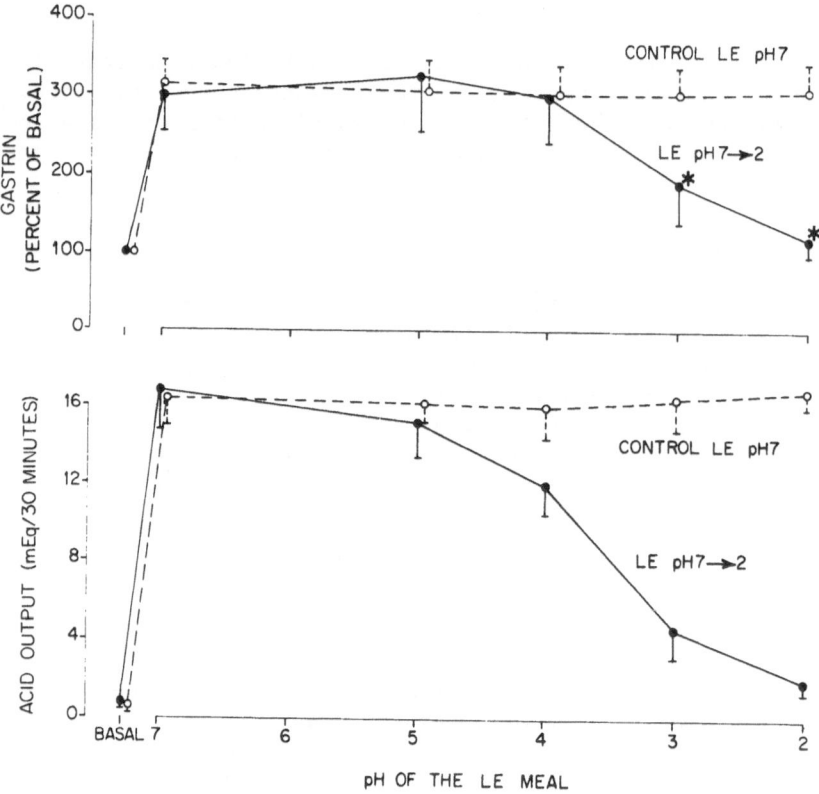

FIGURE 1. *Serum gastrin and gastric acid responses to liver extract (LE) meal at different pH levels. Basal gastrin = 100%.* * = *suppressed below peak (p < 0.05).*

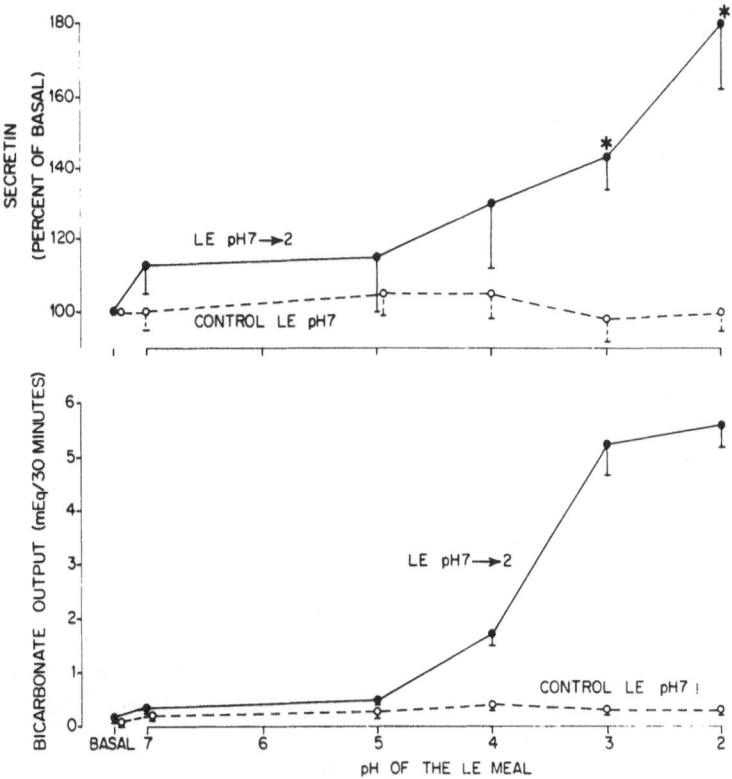

FIGURE 2. *Plasma secretin and pancreatic bicarbonate responses to liver extract (LE) meal at different pH levels. Basal secretin = 100%. * = significant increase above basal (p < 0.05).*

When a LE meal at pH 7 was introduced into the stomach, without intragastric titration (and allowed to undergo endogenous acidification by gastric acid secretion), there was a gradual fall in the pH of the gastric contents from 7 (the initial pH of the meal) to pH 2, at about 90 minutes after administration of the meal. In this experiment, gastrin rose initially to 130% of basal (80 pg/ml) and then gradually decreased with endogenous acidification of the gastric content (Fig. 3). Secretin levels rose below pH 4 and significant release of secretin was observed when the gastric content reached pH 2. Pancreatic bicarbonate secretion increased sharply when the pH of the gastric content descended below 4.

Infusion of graded doses of exogenous secretin produced a maximal output of 9 mEq/30 minutes when the infusion was combined with administration of LE meal at pH 7. This maximal output was about 30% higher than the maximal output of bicarbonate secretion achieved with the infusion of secretion alone (7 mEq/30 minutes).

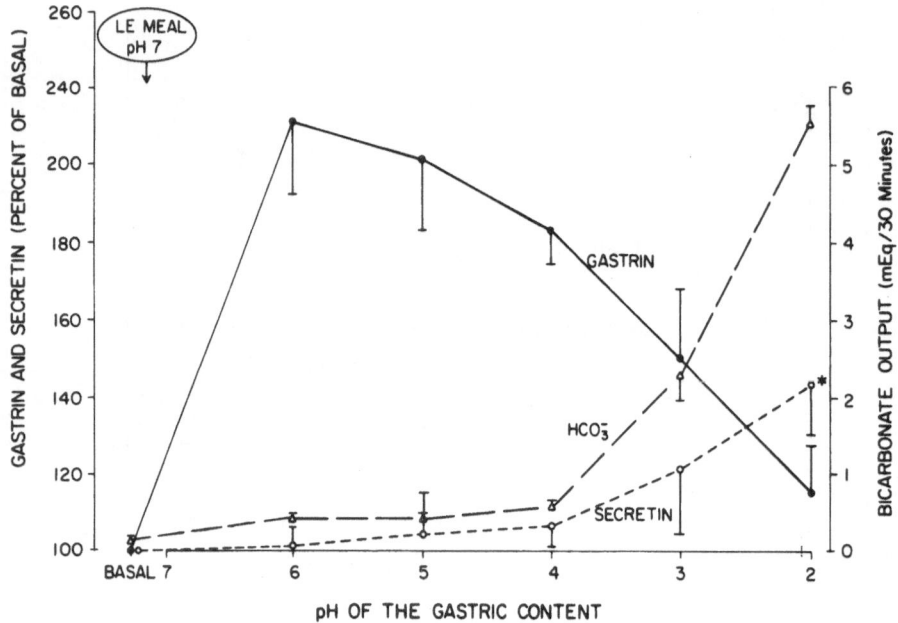

*FIGURE 3. Serum gastrin and secretin and pancreatic bicarbonate responses to a liver extract meal, pH 7 left to normal digestion in the stomach. Basal gastrin and secretin = 100%. * = significant increase of secretin above basal (p < 0.05).*

DISCUSSION

Our results demonstrate that acidification of a meal produces a pH-dependent inhibition of gastric acid and gastrin secretion. On the other hand, acidification of the meal provokes a pH-dependent stimulation of pancreatic bicarbonate and release of endogenous secretin at pH below 4. The studies in which the LE meal was allowed to be acidified by normal gastric secretion show that in dogs with pancreatic fistulas, secretin in immunoassayable amounts was released by the meal. The enhancement of pancreatic bicarbonate secretion in response to exogenous secretin, observed when secretin infusion was accompanied by a LE meal at pH 7, shows that the secretin effect on bicarbonate secretion is potentiated by other stimuli released by food. We conclude that postprandial duodenal acidification releases immunoassayable quantities of secretin. The normal pancreatic bicarbonate response to food may depend partially upon potentiation of the secretin effect by other neurohumoral stimuli.

ROLES OF THE VAGUS IN ENDOGENOUS RELEASE OF SECRETIN AND EXOCRINE PANCREATIC SECRETION IN DOG

K. Y. Lee, W. Y. Chey and H. H. Tai (with technical assistance of Y. Lee)

The Isaac Gordon Center for Digestive Diseases and the Department of Medicine, Genesee Hospital and University of Rochester School of Medicine and Dentistry

Rochester, New York, U.S.A.

It has been well established that secretin[1] is released from upper small intestinal mucosa in response to hydrogen ion delivered in the duodenum in man[2] and dog.[3] As pH of the proximal duodenum decreased below 4.0 during the postprandial period, plasma secretin concentration increased significantly in man.[4] We investigated a possible role of the vagus nerve in release of endogenous secretin in conscious dogs as well as anesthetized dogs.

Studies on conscious dogs

In fasting dogs prepared with gastric fistula and a modified Herrera's pancreatic fistula,[3] studies were carried out to determine whether or not the intravenous administration of atropine or 2-deoxyglucose (2-DG) affects release of endogenous secretin in response to HCl. Release of endogenous secretin was produced by constant intraduodenal infusion of 0.05 N HCl and serial plasma secretin concentrations were measured by radioimmunoassay. As shown in figure 1, plasma secretin concentrations and output of bicarbonate from the pancreas were determined while 0.05 N HCl was infused in the duodenum through the intestinal limb of the pancreatic cannula at a rate of 1.1 ml/min for 45 minutes. The acid infusion resulted in a prompt increase in the plasma secretin concentrations as well as pancreatic secretion of bicarbonate. When atropine sulfate was infused intravenously at a rate of 100 μg/kg-hr, during the duodenal infusion of HCl, both pancreatic secretion of water and bicarbonate were markedly suppressed. The increase in the plasma

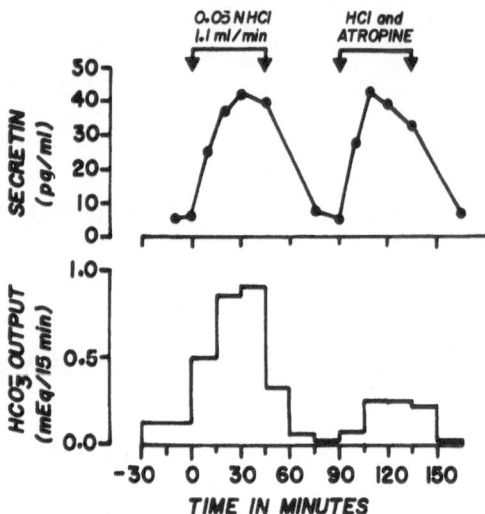

Figure 1. Plasma secretin concentrations and pancreatic secretion of bicarbonate in a conscious dog with pancreatic fistula and gastric fistula in response to intraduodenal infusion of 0.05 N HCl, 1.1 ml/min, alone and with simultaneous intravenous administration of atropine sulfate, 100 μg/kg-hr.

secretin concentrations during the same period, however, had not changed. When 2-DG, 100 mg/kg-hr, was administered intravenously during the acid infusion, the bicarbonate output was much greater than that produced by the same amount of acid infusion alone (Fig. 2) The magnitude of increase in the plasma secretin concentrations during the acid infusion was not changed by 2-DG. Nor have we found any increase in the secretin concentrations when the same amount of 2-DG was administered without the acid infusion although 2-DG produced a significant increase in pancreatic secretion of water and bicarbonate (Fig. 3). Similar trends of pancreatic secretory response to 2-DG were observed in 3 other dogs.

Effects of bilateral vagotomy in anesthetized dogs
     Under chloralose anesthesia, both cervical vagi were identified in 5 dogs. The stomach was exposed and the pyloro-duodenal junction was ligated in order to prevent leakage of gastric juice into the duodenum. A polyethylene tube with inner diameter of 3 mm was placed in the first portion of the duodenum through the pylorus in order to infuse saline or HCl solution. As shown in figure 4, plasma secretin concentration increased markedly during intraduodenal infusion of 0.05 N HCl at rates of 1.1 ml/min and 2.2 ml/min. The peak concen-

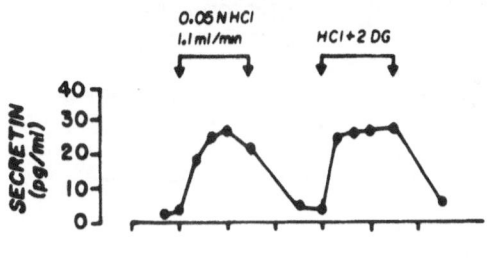

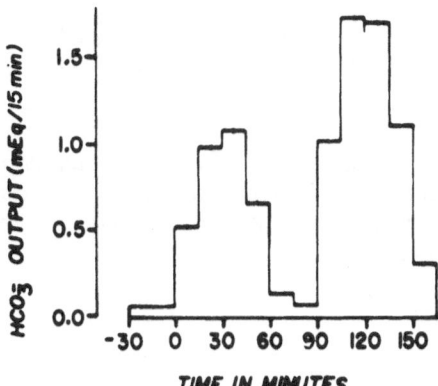

Figure 2. Plasma secretin concentrations and pancreatic secretion of bicarbonate in a conscious dog with pancreatic fistula and gastric fistula in response to intraduodenal infusion of 0.05 N HCl, 1.1 ml/min, alone and with simultaneous intravenous administration of 2 deoxyglucose (2-DG), 100 mg/kg-hr, and intraduodenal infusion of 0.05 N HCl.

tration of plasma secretin during HCl infusion at 2.2 ml/min was greater than that observed during infusion 1.1 ml/min. Following bilateral cervical vagotomy, in the identical experiment, the plasma secretin concentration increased again as 0.05 N HCl solution was infused. Similar trends of the plasma secretin response were observed in the remaining four dogs.

Comment:

In the present study we studied the influence of cholinergic nerves on the endogenous release of secretin as well as pancreatic secretion of bicarbonate in dogs. Stimulation of the vagus by 2-DG or inhibition of the vagus by atropine had little influence on plasma secretin concentration in response to intraduodenal HCl infusion. As the vagus was stimulated by 2-DG, however, the output of bicar-

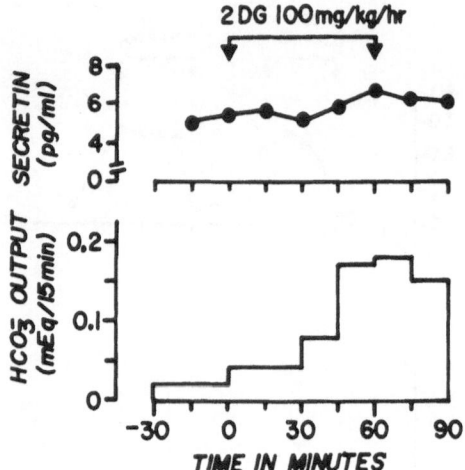

Figure 3. Plasma secretin concentrations and pancreatic secretion of bicarbonate in response to intravenous administration of 2 deoxyglucose (2-DG), 100 mg/kg-hr, in a conscious dog with pancreatic fistula and gastric fistula.

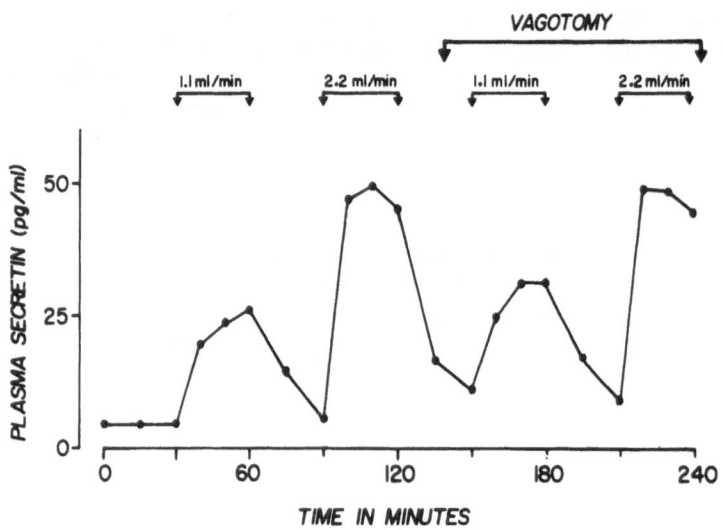

Figure 4. Plasma secretin concentrations of an anesthetized dog in response to intraduodenal infusion of 0.05 N HC1 at rates of 1.1 ml/min, before and after bilateral cervical vagotomy.

bonate during intraduodenal infusion of acid was much greater than that during the acid infusion alone. It is possible that greater stimulation of pancreatic bicarbonate secretion by 2-DG during duodenal acid infusion in the present study could be due to a mechanism which may not be vagally mediated. Atropine, on the other hand, suppressed bicarbonate secretion significantly. However, neither atropine in conscious dogs, nor bilateral cervical vagotomy in anesthetized dogs suppressed the increase in plasma secretin levels in response to intraduodenal acid infusion. Thus, endogenous release of secretin by duodenal acidification is not affected by the vagus in dog. This observation strongly suggests that the vagus plays a significant role in modulating the action of endogenous secretin on the pancreatic secretion of water and bicarbonate without affecting release of endogenous secretin.

SUMMARY

1. A possible role of the vagus nerves in endogenous release of secretin was investigated in dogs with gastric fistula and pancreatic fistula on anesthetized dogs.

2. Plasma secretin concentrations during intraduodenal infusion of 0.05 N HCl, 1.1 ml/min, were not altered by 2-DG or atropine, although 2-DG produced pancreatic secretion of bicarbonate much greater than that during the acid infusion alone and atropine inhibited bicarbonate secretion markedly.

3. In anesthetized dogs, the magnitude of increase in plasma secretin levels during duodenal acidification did not change following bilateral vagotomy.

4. The vagus may play a role in pancreatic secretion of bicarbonate.

REFERENCES

1. Bayliss WM, Starling EH: The mechanism of pancreatic secretion. J Physiol 28: 325-353, 1902

2. Rhodes RA, Tai HH, Chey WY: Observations on plasma secretin levels by radioimmunoassay in response to duodenal acidification and to a meat meal in humans. Amer J Digest Dis 21: 873-879, 1976

3. Lee KY, Tai HH, Chey WY: Plasma secretin and gastrin responses
    to a meat meal and duodenal acidification in dogs. Am J
    Physiol 230: 784-789, 1976

4. Chey WY, Hendricks J, Tai HH: Plasma secretin in fasting and
    postprandial state in man. Gastroenterology 72: 1156, 1977

5. Herrera F, Kemp DR, Tsukamoto M, Woodward ER, Dragstedt LR: A
    new cannula for study of pancreatic function. J Appl Physiol
    25: 207-209, 1968

6. Tai HH, Chey WY: Simultaneous immunoassay of secretin and gas-
    trin. Anal Biochem 74: 12-24, 1976

7. Debas HT, Konturek SJ, Grossman MI: Effect of extragastric and
    truncal vagotomy on pancreatic secretion in the dog. Am J
    Physiol 228: 1173-1177, 1975

8. Konturek SJ, Tasler J and Obtulowiczw: Effect of atropine on
    pancreatic response to endogenous and exogenous secretin.
    Am J Digest Dis 16: 385-394, 1971

THE EFFECT OF ATROPINE ON SECRETIN RELEASE AND PANCREATIC

BICARBONATE SECRETION AFTER DUODENAL ACIDIFICATION IN MAN

L. E. Hanssen

Research Laboratory of Gastroenterology, Dept. IX
Internal Medicine, Ullevål Hospital

Oslo, Norway

INTRODUCTION

The aim of the present study was to investigate whether the release of secretin was under cholinergic control in man. This was postulated by O'Connor and co-workers[3] but Ward and Bloom[5] found vagotomy to be without effect on acid stimulated secretin release. Both groups employed radioimmunoassays for secretin determinations, but these investigations were carried out as unpaired comparisons. The present report gives the results of a randomized, paired investigation of immunoreactive secretin (IRS) release from the intestine after duodenal acidification with or without atropine infusion, together with the effect on duodenal aspirates.

MATERIALS AND METHODS

Eight healthy young volunteers of both sexes were investigated on two different days with an interval of at least one week. Following an overnight fast and a basal period of 30 min, a constant infusion of isotonic saline 37 ml/h with or without 0.75 mg/h atropine sulphate (+ bolus injection 0.5 mg) was given into an arm vein throughout the rest of the experiment. 30 min later, 40 ml 100 mmol/l HCl containing trace amounts of ^{57}Co were infused over 5 min into the midpart of the duodenum (37°C).

Blood was collected for analysis of IRS every 15 min before the acid infusion, then every minute for 10 min, and less frequently thereafter (Fig. 1). IRS was determined by radioimmunoassay as described by us[2], with a detection limit in plasma of 2.5 pmol/l,

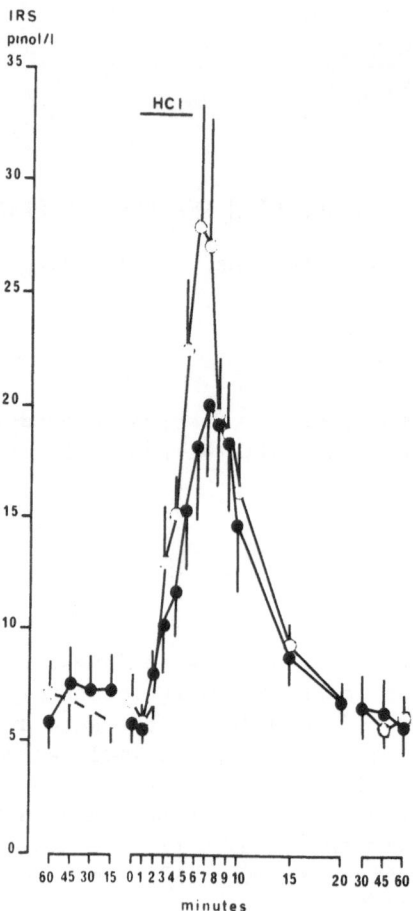

Figure 1.   Concentration of IRS (immunoreactive secretin) in plasma before and after duodenal acidification with 40 ml 100 mmol/l HCl over 5 min (indicated).   Open symbols, without atropine; closed symbols, with atropine infusion. Mean ± S.E.M., n = 8.

a within assay precision of 10 per cent and a between assay precision of 15 per cent at 17 pmol/l.

Duodenal juice was collected in 15-min portions, except during duodenal acidification, when collection was interrupted for 10 min, and then a 5-min collection was made.   In duodenal aspirates volume, bicarbonate and ^{57}Co were determined.   Gastric juice was collected in 15-min portions, and volume and ^{57}Co were determined.   The volume and bicarbonate content of duodenal aspirates were corrected for the volume and alkaline-neutralizing capacity of the infused acid still in the duodenum as judged by the amount of ^{57}Co in the aspirated duodenal juice[1].   All results are given as mean ± S.E.M., n = 8.

The Wilcoxon test for paired comparison was used in all statistics, and p-values less than or equal to 0.05 were considered significant. IRS levels were compared to the mean of the basal determinations, and the integrated IRS response to acid was compared with and without atropine infusion. Values for pancreatic flow rate and bicarbonate output found during 20 min after duodenal acidification were compared to mean values of the 30 min basal period.

RESULTS

IRS increased in all subjects after duodenal acidification whether atropine was given or not (Fig. 1), $p < 0.005$. The integrated IRS response to acid was slightly decreased by atropine infusion, $p < 0.05$. However, the pancreatic response to acidification was significantly reduced by atropinization; both flow rate and bicarbonate output were inhibited, $p < 0.005$ (Fig. 2).

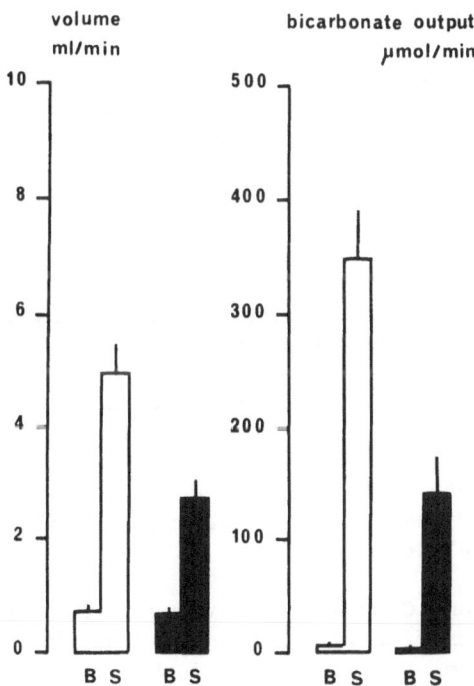

Figure 2. Pancreatic flow rate (left scale and curves) and bicarbonate output (right scale and curves). Open bars, without atropine; closed bars, with atropine infusion. Basal values (B) and stimulated values (S) after duodenal acidification with 40 ml 100 mmol/l HCl over 5 min. Mean ± S.E.M., n = 8.

DISCUSSION

This study shows a small decrease in IRS release by atropine after duodenal acidification in man. However, a much more significant effect was seen on the pancreatic response.

Earlier work has suffered from the disadvantage of not having a method for accurate determinations of secretin levels in the blood. Only O'Connor and co-workers and Ward and Bloom have hitherto investigated the effects of anticholinergics or vagotomy on secretin release in man. The effect of atropine may be manifold. Not only the effect of the vagus can be inhibited, but also local cholinergic reflexes, a direct effect on the duct cells, and possibly central effects must be considered.

Our work shows only a small inhibition of integrated IRS release, and this effect might be secondary to other effects of atropine than just the effect on the secretin cell (e.g.,inhibition of motility in the intestine) so that fewer secretin cells might come into contact with the acid.

Recently Tai and co-workers[4] have shown atropine to be without effect on acid stimulated secretin release in the dog.

REFERENCES

1. Hanssen LE, Hanssen KF and Myren J. Inhibition of secretin release and pancreatic bicarbonate secretion by somatostatin infusion in man. Scand. J. Gastroent. 1977, 12, 391-394

2. Hanssen LE and Torjesen P. Radioimmunoassay of secretin in human plasma. Scand. J. Gastroent. 1977, 12, 481-488

3. O'Connor FA, Buchanan KD, Connon JJ and Shahidullah M. Secretin and insulin: Response to intraduodenal acid. Diabetologia 1976, 12, 145-148

4. Tai HH, Lee KY and C ley WY. Does vagus effect endogenous release of secretin? International symposium on gastrointestinal hormones and pathology of the digestive system, Rome, 1977

5. Ward AS and Bloom SR. Effect of vagotomy on secretin release in man. Gut 1975, 16, 951-956

IMMUNOREACTIVE SECRETIN RELEASE AND PURE PANCREATIC JUICE AFTER DUODENAL INFUSION IN BILE IN MAN

L. E. Hanssen, M. Osnes, O. Flaten and J. Myren

Research Laboratory of Gastroenterology, Dept. IX
Internal Medicine, Ullevål Hospital

Oslo, Norway

INTRODUCTION

The improved technique of endoscopic cannulation of the main pancreatic duct has made a collection of pure pancreatic juice possible in man[2,6,7,8]. We have described[7] an endoscopic model for the study of the exocrine pancreatic secretion after intraduodenal stimulation with test solutions. Using a side-viewing Olympus duodenoscope (model JF B2) the main pancreatic duct was cannulated with the standard Olympus catheter used for endoscopic cannulation. Prior to the examination another similar catheter was attached to the outside of the duodenoscope for instillation of test solutions.

Previously, it has been shown that bile in the duodenum augments the stimulating effect of exogenous secretin on pancreatic enzyme secretion[1,9]. The effect of bile on endogenously stimulated enzyme secretion[4] and the effect on the exogenously stimulated bicarbonate secretion[1,9] is uncertain.

The purpose of the present investigation was to study the effect of intraduodenal bile infusion on immunoreactive secretin (IRS) release, and its effect on basal pancreatic secretion obtained from the main pancreatic duct after endoscopic cannulation[6].

MATERIALS AND METHODS

Six patients with a normal endoscopic pancreatogram were investigated in the fasting state as described above. Anticholinergics

were not given, but one individual received a small dose of diaze-
pam. After successful cannulation of the main pancreatic duct,
juice was collected in 5-min samples. After 20 min 6 g cattle bile
in 60 ml distilled water (iso-osmolar, pH 6-7, H^+ 7 mmol/1, 37°C)
was infused into the duodenum over 5 min. Juice was collected for
four 5-min periods, and the another 6 g cattle bile in 40 ml distilled water (hyper-osmolar) was infused as described. Again juice
was collected for four 5-min periods; blood was frequently drawn
from an indwelling catheter in a peripheral vein.

Plasma immunoreactive secretin (IRS) was determined by radioimmunoassay as described by us[3]. In pancreatic juice volume, bicarbonate and α-amylase (E.C. 3.2.1.1.) were determined in 5-min
portions. The Wilcoxon test for paired comparison was used and
results are given as mean ± S.E.M., n = 6.

RESULTS

IRS increased in all studies; peak levels were found 10 min
after the start of the infusions ($p < 0.02$), (Fig. 1). The flow
rate of pancreatic juice reached its peak value in the third 5-min
period after instillation of the bile solutions ($p < 0.02$),(Fig. 2).
Peak bicarbonate output was reached in the same periods ($p < 0.02$),
(Fig. 3). Peak amylase output was found in the second period after
the first bile infusion ($p < 0.02$), (Fig. 4).

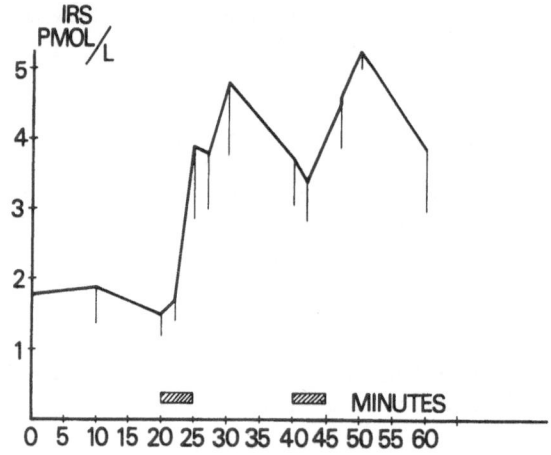

*Figure 1. Immunoreactive secretin (IRS) after intraduodenal bile
infusions (indicated). Mean ± S.E.M., n = 6.*

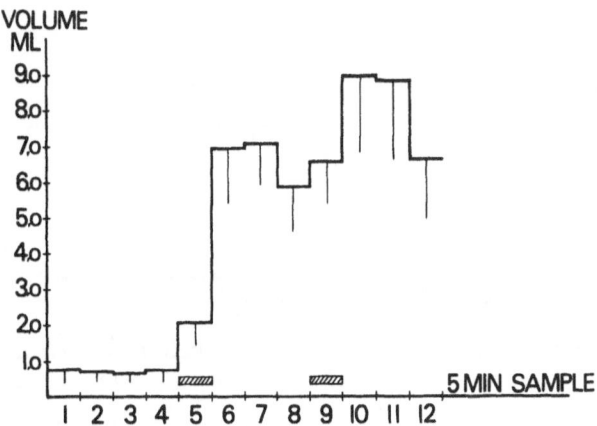

Figure 2.   Volume of each 5-min portion of pancreatic juice before and after infusion of bile solutions (indicated).   Mean ± S.E.M., n = 6.

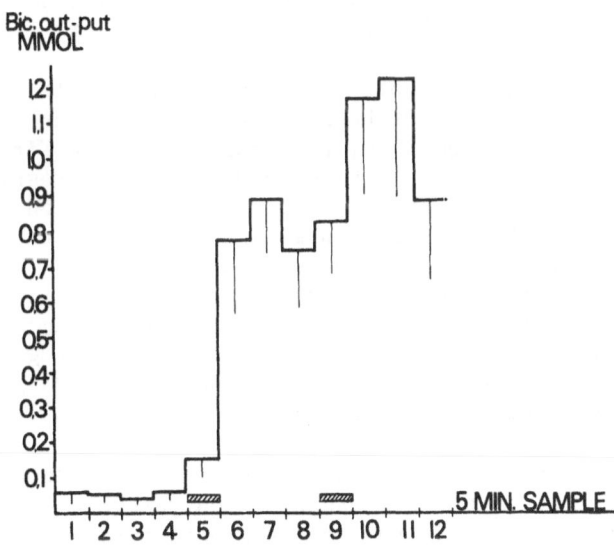

Figure 3.   Bicarbonate output in pancreatic juice before and after infusion of bile solutions (indicated).   Mean ± S.E.M., n = 6.

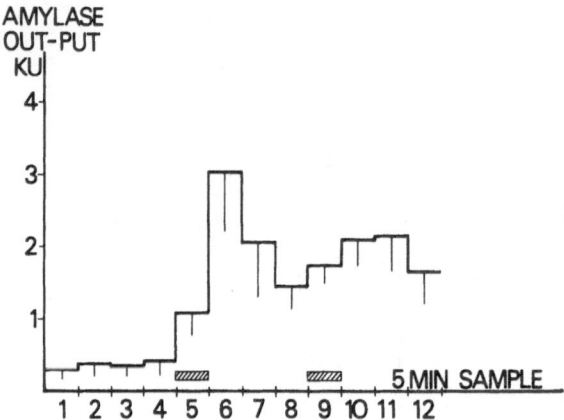

Figure 4. α-*Amylase output in pancreatic juice before and after infusion of bile solutions into the duodenum. Mean ± S.E.M., n = 6*

DISCUSSION

The present study shows that bile is able to cause the release of IRS. This was followed by a significant stimulation of basal pancreatic secretion of water, bicarbonate and α-amylase, obtained by endoscopic cannulation of the main pancreatic duct.

The mechanisms by which bile or bile salts stimulate the pancreatic secretion have not yet been defined. Our study confirms the hypothesis of Melanby[5] from 1926, by showing, for the first time, that secretin is released after bile infusions. Interestingly, the infused bile was nearly neutral in pH, but the diversion of bile and pancreatic juice in these experiments might have made the mucosa more susceptible to the small amounts of H^+ present in the infused bile. The release of secretin, as seen in this study, is probable not the sole explanation for the increase in pancreatic secretion found. Other mechanisms, such as the release of other gastrointestinal hormones, local nervous reflexes and circulatory alterations also influence pancreatic secretion.

The improved technique for duodenoscopy with cannulation of the papilla of Vater has made the present method possible. By this procedure juice is obtained from the main pancreatic duct after intr duodenal instillation of solutions which may stimulate the pancreati secretion. It is also an advantage that the secretion of enzymes, bicarbonate and other components may be studied simultanously with the release of gastrointestinal hormones.

REFERENCES

1. Forell MM, Otte M, Kohl HJ, Lehnert P and Stahlheber HP: The
    influence of bile and pure bile salts on pancreatic secre-
    tion in man. Scand. J. Gastroent. 6, 261-266, 1971

2. Hanssen LE, Osnes M and Myren J. The exocrine pancreatic
    secretion obtained by endoscopic cannulation of the main
    pancreatic duct and secretin release after duodenal acidi-
    fication in man. Scand. J. Gastroent., 1977, in press

3. Hanssen LE and Torjesen P. Radioimmunoassay of secretin in
    human plasma. Scand. J. Gastroent. 12, 481-488, 1977

4. Malagelada JR, DiMagno EP, Summerskill WHJ and Go VLM. Regula-
    tion of pancreatic and gallbladder function by intraluminal
    fatty acids and bile salts in man. J. Clin. Invest. 58,
    493-499, 1976

5. Melanby J. The secretion of pancreatic juice. J. Physiol.
    (Lond.) 61, 419-435, 1926

6. Osnes M, Hanssen LE, Flaten O and Myren J. The exocrine pancre-
    atic secretion and immunoreactive secretin (IRS) release
    after intraduodenal instillation of bile in man. Gut, 1977,
    in press

7. Osnes M, Hanssen LE and Myren J. An endoscopic method for the
    study of the exocrine pancreatic secretion in man. Endoscopy,
    8, 124-126, 1976

8. Robberrecht P, Cremer M, Vandermeers A, Vandermeers-Piret MC,
    Cotton P, De Neef P and Christophe J. Pancreatic secretion
    of total protein and of three hydrolases collected in healthy
    subjects via duodenoscopic cannulation. Effects of secretin,
    pancreozymin, and caerulein. Gastroenterology 69, 374-379,
    1975

9. Wormsley KG. Stimulation of pancreatic secretion by intraduodenal
    infusion of bile salts. Lancet 2, 586-588, 1970

SOMATOSTATIN AND GASTROINTESTINAL SECRETION AND MOTILITY

S. J. Konturek

Institute of Physiology, Medical Academy

Krakow, Poland

Somatostatin, a hypothalamic growth hormone-release inhibiting hormone (GH-RIH), has been shown to suppress the secretion of many hypophyseal and extrahypophyseal hormones including gastrointestinal hormones such as gastrin, secretin, cholecystokinin, motilin, and vasoactive intestinal peptide[1]. It was also found in laboratory animals and in man to be a potent inhibitor of gastrointestinal secretions induced both by exogenous and endogenous stimuli[2-5]. Recently, many analogs of somatostatin have been prepared in an attempt to increase biological activity and some of the analogs were found to preferentially suppress the secretion of some hormones[6].

Somatostatin, originally isolated from the hypothalamus, was subsequently found in other regions of the brain and localized by immunocytochemistry in nerve cell bodies and in nerve terminals indicating that it may play a role of a neural transmitter in a special neuronal system of the brain[7]. Immunoreactive somatostatin has also been detected in high concentrations in the distinctive paracrine-endocrine cells in the gastrointestinal mucosa and in the pancreas[8], where it might play a role as a local factor affecting digestive functions. In view of the rapid clearance from the circulation and relatively high doses required to produce inhibitory effects, somatostatin probably plays the role of a local hormone, a member of Feyrter's paracrine system[9]. Furthermore, deficiency of somatostatin in the gastric mucosa has been found in duodenal ulcer patients and suggested to be responsible for gastric hyperacidity in these patients[10].

This presentation will attempt to describe the effects of somatostatin and its analogs on gastric and pancreatic secretions, formation of experimental peptic ulcers in animals, gastric secretion and gastrin release in duodenal ulcer patients and motility pattern of the small bowel.

SOMATOSTATIN AND GASTRIC SECRETION

The results obtained from animals (dogs, cats) indicate that natural somatostatin is a potent inhibitor of gastric acid secretion induced by a variety of secretagogues and that it is a relative more potent inhibitor of pepsin than acid secretion[3]. Meal-induced secretion appears to be the most sensitive to somatostatin inhibition; it is accompanied by a suppression of the release of gastrin and a reduction in the mucosal blood flow[3].

Somatostatin was shown previously to inhibit pentagastrin and betazole-induced gastric secretion in healthy subjects[2] and to diminish basal and postprandial serum gastrin level in healthy subjects and in patients with pernicious anemia and the Zollinger-Ellison syndrome[11]. We have compared the effects of graded doses of somatostatin on pentagastrin and meal-induced gastric acid secretion in healthy subjects and in duodenal ulcer patients.

Gastric secretion in response to pentagastrin and a liver extract meal was induced by a method described previously[12]. Pentagastrin was infused intravenously in a dose (2μg/kg-hr) producing near maximal acid output and gastric juice was collected by the standard aspiration technique. The meal-induced secretion was provoked by a modified method of intragastric titration using 10% liver extract (Reheis Chemical Co, Chicago, Ill.) as a stimulus.

Somatostatin given intravenously in healthy human subjects caused a potent and dose-related suppression of gastric acid and pepsin responses to pentagastrin, being a more potent inhibitor of pepsin than acid secretion. Gastric acid and pepsin responses to pentagastrin in duodenal ulcer patients were significantly higher and somatostatin administered in graded doses caused less inhibition of gastric acid secretion than in healthy subjects. The calculated $ID_{50\%}$ (dose producing 50% inhibition) for duodenal ulcer patients was about three times higher than for normals. Duodenal ulcer patients also showed less pronounced suppression of pepsin secretion by somatostatin but the difference from normals was less marked than with regard to acid secretion (Fig.1).

Meal-induced acid secretion in control tests (without somatostatin) in both groups of subjects was well sustained throughout the study and similar to that achieved with pentagastrin. It was accompanied by a significant rise in serum gastrin level which was higher and more pronounced in duodenal ulcer patients than in healthy controls. Somatostatin given in graded doses during the

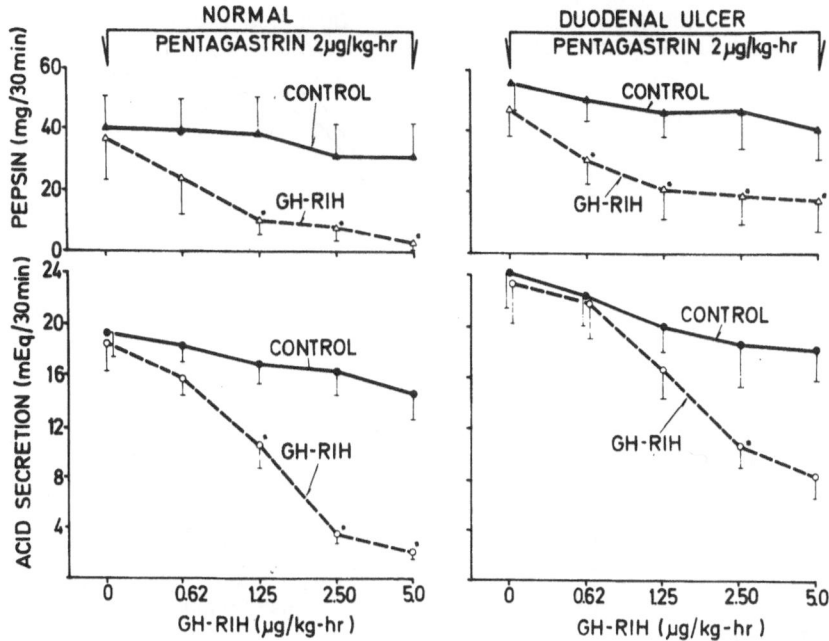

Figure 1. Effect of graded doses of somatostatin given intravenous-ly on gastric acid and pepsin responses to pentagastrin in healthy subjects and duodenal ulcer patients. Mean ± SEM of 6 tests on 6 subjects. Asterisks denote significant difference (p < 0.05) from control.

meal test resulted in a dose-dependent inhibition of postprandial acid response in both groups of subjects.

A dose of 2.5 µg/kg-hr somatostatin almost completely inhibited acid output in healthy persons, whereas in duodenal ulcer patients, it produced only about 50% inhibition (Fig. 2). It is of interest that serum gastrin level in duodenal ulcer patients was less affect-ed by somatostatin than in normals. This study suggests that the resistance to the inhibitory effect of somatostatin may be regarded as one of many defects underlying gastric acid hypersecretion in duodenal ulcer disease.

Somatostatin has been suggested in the treatment of certain forms of peptic ulcer in man. We examined this concept in an exper-imental model of cats with peptic ulcers produced by prolonged in-fusion of gastric secretagogues such as pentagastrin or histamine[13]. Somatostatin given in a dose producing about 50% inhibition of pen-tagastrin-stimulated gastric acid secretion caused a marked reduc-tion in both ulcer incidence and ulcer area. The major factor responsible for peptic ulcers, which develop in this model only in

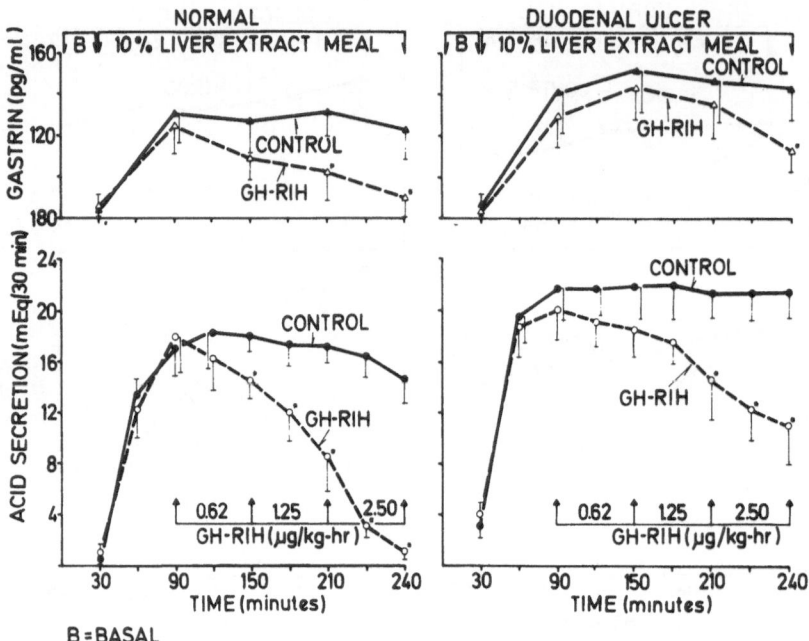

*Figure 2. Effect of graded doses of somatostatin given intravenous-
ly on gastric acid and serum gastrin responses to liver extract meal
in healthy subjects and duodenal ulcer patients. Mean±SEM of 6 test
on 6 subjects. Asterisks denote significant difference (p < 0.05)
from control.*

the duodenum, is an inadequate neutralization of highly acid gastric
juice in the duodenum. Somatostatin inhibits pancreatic bicarbonate
secretion but it causes relatively stronger inhibition of gastric
acid and pepsin secretion and this is probably the major mechanism
by which somatostatin prevents the formation of duodenal ulcers
(Fig. 3).

Somatostatin analogs, in which certain amino acid residues
were replaced by their D-isomers (D-Trp[8]-somatostatin and D-Cys[14]-
somatostatin) were reported to have dissociated actions on the release of growth hormone, insulin and glucagon. These analogs compared with regard to their action on gastric secretion were found
to cause similar competitive inhibition of pentagastrin-induced
gastric secretion and to be essentially equipotent to natural somatostatin on gastric secretion[14].

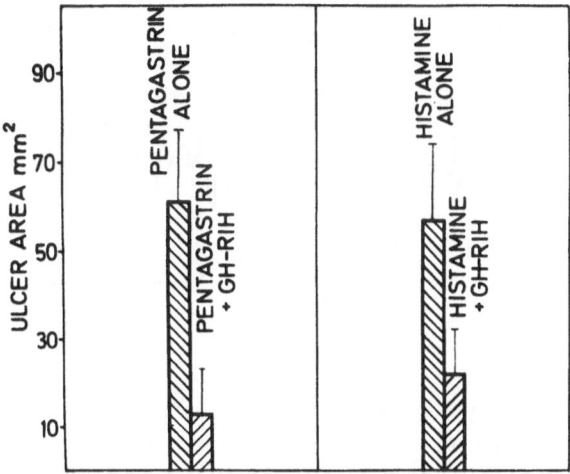

Figure 3.   Effect of somatostatin (2.5 µg/kg-hr) on histamine (160 µg/kg-hr)-induced duodenal ulcerations in cats.   Mean ± SEM of 10 observations in 10 cats.

SOMATOSTATIN AND PANCREATIC SECRETION

The influence of somatostatin on exocrine pancreatic secretion has been studied by several authors[4,15], who reported that this peptide inhibits pancreatic fluid, bicarbonate and enzyme secretion induced both by duodenal acidification and exogenous secretin. It was also found that somatostatin suppresses pancreatic response to a meal and to cholecystokinin released endogenously by duodenal perfusion with amino acid mixture or fat. The kinetic analysis showed that the interaction between somatostatin and secretin affecting pancreatic bicarbonate secretion possesses the characteristics of competitive inhibition[4].

Somatostatin analogs (D-Trp[8]-somatostatin and D-Cys[14]-somatostatin) cause similar competitive inhibition of secretin-induced pancreatic bicarbonate secretion indicating that these analogs are essentially equipotent to somatostatin itself on pancreatic secretion[14].

Somatostatin was also found to inhibit pancreatic response to feeding and to suppress the release of cholecystokinin by endogenous stimuli such as amino acid mixture, peptone meal or fat[4]. The effect of somatostatin on pancreatic response to exogenous cholecystokinin is more controversial. In humans, somatostatin inhibited the effect

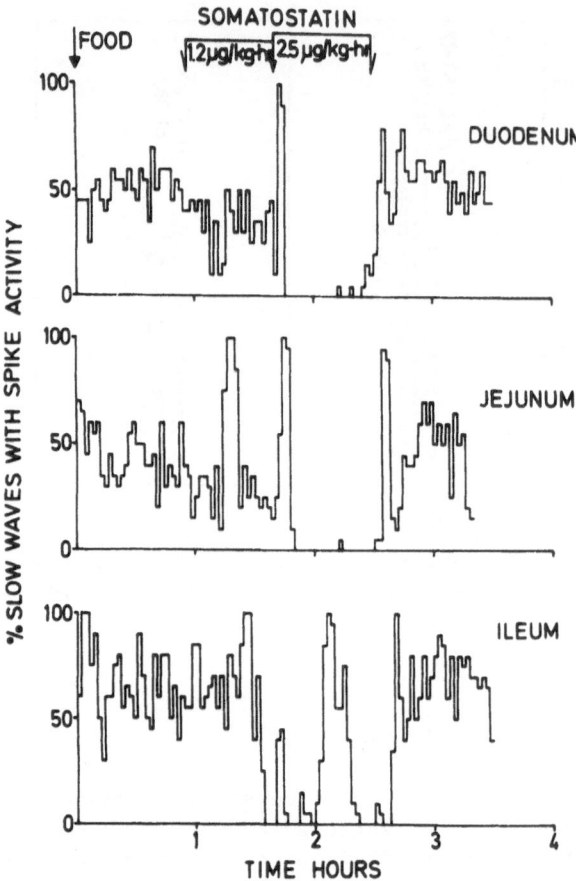

Figure 4. Temporal distribution of slow waves with spike potentials in fed dog infused with somatostatin. Tracings from three electrodes in one dog.

of cholecystokinin on pancreatic enzyme secretion and on gallbladder contraction[15] whereas in dogs it was found to have no effect upon cholecystokinin-induced pancreatic protein secretion. Although this might represent a species difference, it is also likely that the inhibitory effect of somatostatin depends upon the dose and the nature of the stimulant used and that it may be overcome by increasing the doses of the stimulant (competitive mechanism).

In view of the marked inhibitory influence of somatostatin on gastric and pancreatic secretions and its ability to reduce requirements for exogenous insulin, it may have potential value in the treatment of acute pancreatitis.

EFFECT OF SOMATOSTATIN ON INTESTINAL MOTILITY

There are only a few studies devoted to study of the action of somatostatin on gastrointestinal motility. The preliminary results indicate that this hormone delays gastric emptying probably due to the suppression of motilin release[16] and inhibits the contractions of gallbladder, presumably because of the suppression of the release and action of cholecystokinin[14]. The influence of somatostatin on intestinal motility has not been studied previously.

The motility pattern of the small intestine can be determined by studying its myoelectric activity, which consists of two basic types: slow waves and spike potentials. We have studied the influence of somatostatin on this motility pattern in conscious dogs prepared with electrodes spaced 25 cm apart along the entire small bowel. Recordings of myoelectric activity were made with a Beckman type R-611 Dynagraph. Somatostatin infused intravenously in fasting dogs increased the frequency of the interdigestive complexes while in fed dogs or those infused with gastrin or cholecystokinin, it caused a dose-dependent reduction in spike potentials and introduced patterns like those seen in fasting animals (Fig. 4). These results indicate that somatostatin may play a promoting role in the maintenance of the interdigestive myoelectric complexes and inhibit the spike potentials evoked by feeding because of the suppression of the postprandial release and action of gastroduodenal hormones implicated in the regulation of the myoelectric activity of the gut[17].

REFERENCES

1. Konturek SJ: Somatostatin and the gastrointestinal secretions. Scand J Gastroent 11: 1-4, 1976

2. Arnold R, Creutzfeldt W: Hemmung der pentagastrin-induzierten Sauresekretion des Magens beim Menschen durch Somatostatin. Dtsch Med Wschr 100: 1014-1016, 1975

3. Konturek SJ, Tasler J, Cieszkowski M: Effect of growth hormone release-inhibiting hormone on gastric secretion, mucosal blood flow and serum gastrin. Gastroeneterology 70: 737-741, 1976

4. Konturek SJ, Tasler J, Obtulowicz W: Effect of growth hormone release inhibiting hormone on hormones stimulating exocrine pancreatic secretion. J Clin Invest 58: 1-6, 1976

5. Konturek SJ, Swierczek J, Kwiecień N: Effect of somatostatin on meal-induced gastric secretion in duodenal ulcer patients. Gastroenterology 72: 818, 1976

6. Meyers C, Arimura A, Gordin A et al: Somatostatin analogs which inhibit glucagon and growth hormone more than insulin release. Biochem Biophys Res Commun 74: 630-635, 1977

7. Hokfelt T, Efendic S, Hellerstrom C et al: Cellular localization of somatostatin in endocrine like cells and neurons of the rat with special reference to the A_1 cells of the pancreatic islets and to the hypothalamus. Acta Endocrinolog (suppl) 80: 5-40, 1975

8. Arimura A, Sato H, Dupont A et al: Somatostatin abundance of immunoreactive hormone in rat stomach and pancreas. Science 189: 1007-1009, 1975

9. Feyrter F: Über die peripheren endocrinen (parakrinen) Drüsen des Menschen. Wien-Düsseldorf, Verlag W. Maudrich 1953, p.27

10. Polak JM, Bloom SR, McCrossman MV, et al: Studies on gastric D cell pathology. Gut 17: 400, 1976

11. Bloom SR, Mortimer CH, Thorner MG, et al: Inhibition of gastrin and gastric acid secretion by growth hormone release inhibiting hormone. Lancet 2: 1106-1109, 1974

12. Konturek SJ, Biernat J, Kwiecień N, et al: Effect of glucagon on meal-induced gastric secretion in man. Gastroenterology 68: 448-454, 1975

13. Konturek SJ, Radecki T, Pucher A: Effect of somatostatin on gastrointestinal secretions and peptic ulcer production in cats. Scand J Gastroent (in press)

14. Konturek SJ, Tasler J, Król R et al: Effect of somatostatin analogs on gastric and pancreatic secretion. Proc Soc Exp Biol Med (in press)

15. Creutzfeldt W, Lankisch PG, Folsch UR: Hemmung der Sekretin- und Cholecystokinin-Pancreozymin-induzierten Saft und Enzymesekretion des Pankreas und der Gallenblasenkontraktion beim Menschen durch Somatostatin. Dtsch Med 100: 1135-1138, 1975

16. Bloom Sr, Ralphs DN, Besser GM, et al: Effect of somatostatin on motilin levels and gastric emptying. Gut 16: 834, 1975

17. Weisbrodt NW, Copeland EW, Thor PJ, et al: The myoelectric activity of the small intestine of the dog during total parental nutrition. Proc Soc Exp Biol Med 153: 121-124, 1976

THE INHIBITORY ACTION OF SOMATOSTATIN ON THE STOMACH

E. Schrumpf

Research Laboratory of Gastroenterology, Dept. 9

Ulleval University Hospital, Oslo, Norway

INTRODUCTION

The finding of somatostatin-like immunoreactivity in the D-cells of the pancreas and in the lamina propria in the gastric wall[1] has given evidence for a role of this peptide in the regulation of pancreatic and gastric functions. The recent discovery by Uvnäs-Wallensten et al. of somatostatin-like material in the gastric secretion[2] and the description by Larsson et al. of a patient with a somatostatinoma[3] give further support to the idea of a regulatory role of somatostatin in this area. In this presentation, the effects of somatostatin on the stomach will be reviewed and discussed.

GASTRIC EFFECTS OF SOMATOSTATIN

If somatostatin had effects on the stomach, these might be: 1. effect on the secretion of gastrin; 2. effect on the secretion of acid, pepsin and intrinsic factor; 3. effect on gastric motility.

Secretion of gastrin

Bloom et al.[4] were first to demonstrate a fall in basal serum gastrin concentration after large amounts of somatostatin in patients. Our group,[5] however, failed to obtain any effect on the basal concentration of gastrin in healthy students after 50 μg/h and 500 μg/h of somatostatin. The largest dose is similar to that one used by Bloom et al. Schlegel and coworkers[6] have found an inhibitory effect of somatostatin on gastrin release after a meal with a dose close to 50 μg/h. This dose corresponds to the smallest dose capable of inhibiting growth hormone release. In conclusion, small amounts of somatostatin inhibit meal stimulated gastrin release, whereas it is uncertain whether somatostatin affects the basal gastrin secretion.

Gastric secretion

It is well established that somatostatin inhibits gastric
secretion and that this effect may be gastrin independent. It has
been suggested that this antisecretory effect of somatostatin is
entirely secondary to the strong inhibitory effect of somatostatin
on the splanchnic blood flow. This has, however, been ruled out in
a study of Konturek et al. [7]. Figs. 1-4 demonstrate the effect of
a small dose of cyclic somatostatin (50 μg/h) on the gastric secretion after stepwise increasing doses of pentagastrin. The experiments have been performed in our laboratory in healthy volunteers.

During the smallest dose of pentagastrin the acid output was
reduced to 18 per cent by somatostatin. At the largest dose step
it was reduced only to 79 per cent. This finding indicates competitive kinetics, which is also found in the case of pepsin and intrinsic factor secretion. This finding stresses the importance of doing kinetic studies when testing inhibitory agents. If the effect
of somatostatin had been tested against the largest dose of pentagastrin only, the inhibitory effect would hardly have been seen.

Motility

Figure 5 demonstrates the effect of somatostatin on gastric
motility.

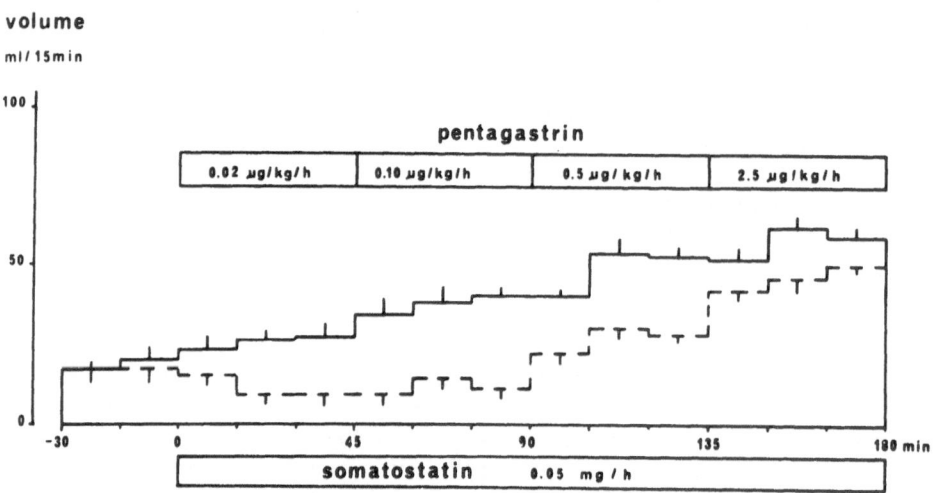

*Figure 1. Mean volume output in 6 healthy volunteers before and
after stepwise increasing doses of pentagastrin alone (unbroken
line) or in combination with a continuous infusion of somatostatin
(broken line). Vertical bars indicate SEM.*

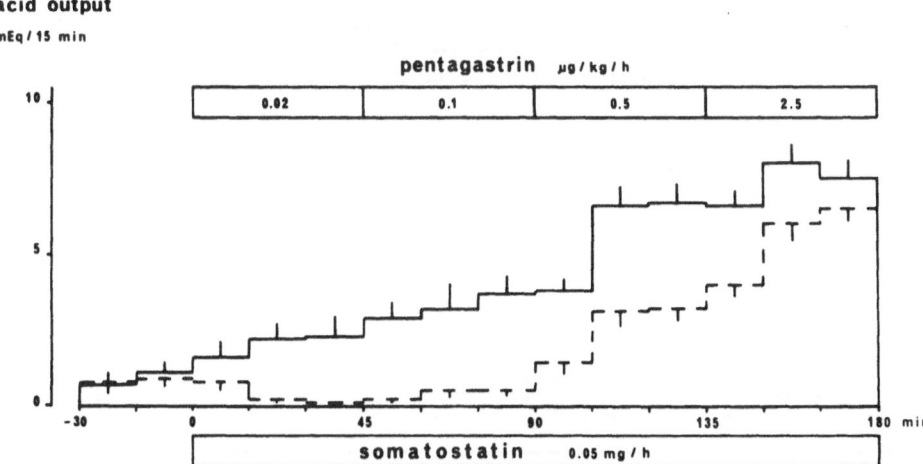

Figure 2. Mean acid output during the same experiments as explained in Figure 1.

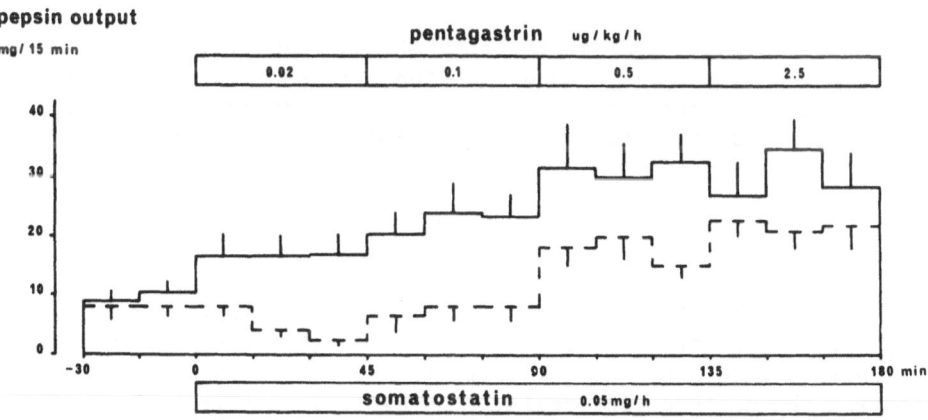

Figure 3. Mean pepsin output during the same experiments as explained in Figure 1.

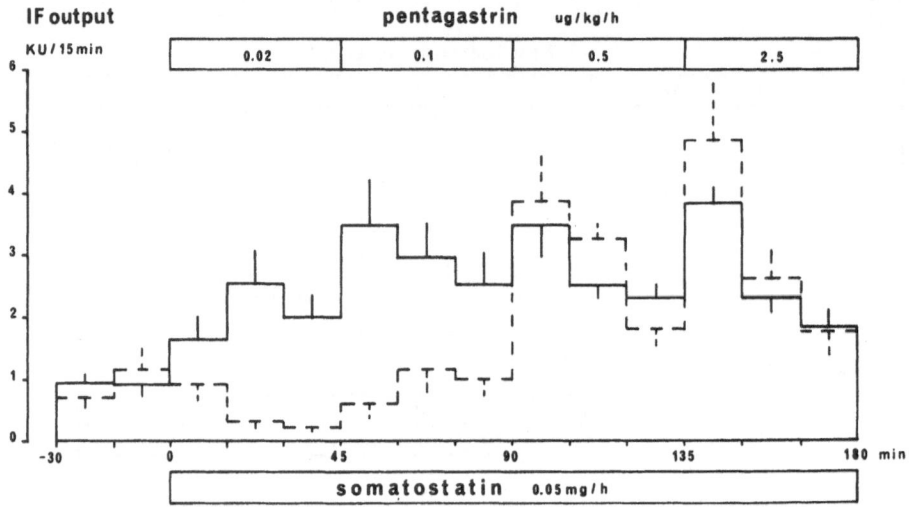

Figure 4. Mean IF output during the same experiments as explained in Figure 1.

A balloon was placed in the stomach and then filled with water in a stepwise manner. By connecting the balloon with a transducer we were able to record pressure changes in the stomach. After a small dose of somatostatin (50 μg/h) the rhythmic pressure variations were strongly reduced ($p < 0.05$). These results are reproducible[8].

DISCUSSION

Whether the effects of somatostatin on the gastric functions are physiological or not is an open question, and it may prove difficult to find out. As somatostatin may be a paracrine substance affecting the neighbouring cells, a physiological effect should be obtained after amounts giving concentrations around the secretory cell similar to those seen after a physiological stimulus for somatostatin release. The somatostatin concentration on the outside of the cell may, however, be impossible to measure. For the time being it is, therefore, impossible to conclude that any effects obtained with somatostatin are physiological ones. It is, however, more likely that effects obtained with small amounts of somatostatin are physiological than those obtained with large doses.

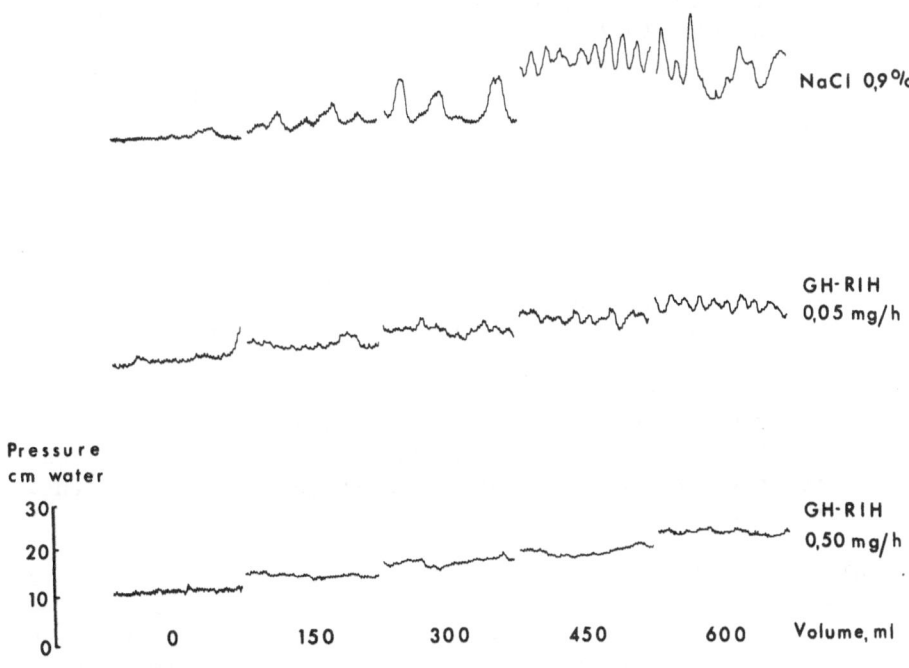

Figure 5. Typical individual pressure recordings in response to distention alone (upper curve) and during somatostatin infusions (lower curves).

Conclusion

Somatostatin may play a physiological role in the control of the gastric function. In small doses somatostatin is able to inhibit both the release of gastrin, the secretion of acid, pepsin and intrinsic factor, and to inhibit gastric motility.

REFERENCES

1. Hokfeldt T, Efendic S, Hellstrom C, Johansson O, Luft R, Arimura A: Cellular localization of somatostatin in endocrine-like cells and neurons of the rat with special reference to the A_1-cells of the pancreatic islets and to the hypothalamus area. Acta Endocrin. (Kbh). Suppl. 200, 5-41, 1975

2. Uvnäs-Wallensten K, Efendic S, Luft R: Vagal release of soma-
    tostatin into the antral lumen of cats. Acta Physiol. Scand.
    99, 126-128, 1977

3. Larsson LI, Hirsch MA, Holst JJ, Ingemansson S, Kühl C, Linkäer
    Jensen S, Lundquist G, Rehfeld JF, Schwarz TW: Pancreatic
    somatostatinoma. Clinical features and physiological im-
    plications. Lancet I, 666-668, 1977

4. Bloom SR, Mortimer CH, Thorner MO, Besser GM, Hall R, Gomez-Pan
    A, Roy VM, Russel RGG, Coy DH, Kastin AJ, Schally AV: In-
    hibition of gastrin and gastric acid secretion by growth
    hormone release-inhibiting hormone. Lancet II, 1106-1109,
    1974

5. Vatn MH, Schrumpf E, Hanssen KF, Myren J: The effect of soma-
    tostatin on the pentagastrin-stimulated gastric secretion
    and on the release of gastrin in man. Scand. J. Gastroent.
    In press

6. Schlegel W, Raptis S, Dollinger HC, Pfeiffer EF: Inhibition of
    secretin-pancreozymin and gastrin release and their biologi-
    cal activities by somatostatin. pp. 361-377 in Bonfils et
    al. (eds). First International Symposium on Hormonal Re-
    ceptors in Digestive Tract Physiology. Elsevier/North-
    Holland Biomedical Press, 1977

7. Konturek SJ, Tasler J, Ciesckowski M, Coy DH, Schally AV: Effect
    of growth hormone release-inhibiting hormone on gastric secre
    tion, mucosal blood flow, and serum gastrin. Gastroenterolog
    70, 737-741, 1976

8. Schrumpf E, Stadaas J, Hanssen KF: Somatostatin inhibits gastric
    motility in response to distention. Scand. J. Gastroent.
    11, suppl. 38, 111, 1976

RECENT ADVANCES IN MOTILIN RESEARCH: ITS PHYSIOLOGICAL AND CLINICAL SIGNIFICANCE

Z. Itoh, S. Takeuchi*, I. Aizawa*, R. Takayanagi*,
K. Mori**, T. Taminato**, Y. Seino**, H. Imura**
and N. Yanaihara****

*G. I. Laboratories, Department of Surgery, Gunma
University School of Medicine, Maebashi, Japan
**Third Division, Department of Internal Medicine, Kobe
University School of Medicine, Kobe, Japan
***Laboratory of Bio-organic Chemistry, Shizuoka College
of Pharmaceutical Science, Shizuoka, Japan

Since 1975, when motilin was synthesized in Japan[1,2], we concentrated our efforts on various aspects of motilin in man and dog. We have found that: 1.) motilin induces the cyclic recurring episodes of caudad-moving bands of strong contractions that move from the lower esophageal sphincter (LES) to the terminal ileum, 2.) motilin has no significant influence upon the gastrointestinal contractile activity during the digestive state, and 3.) the naturally-occurring and motilin-induced gastric contractions are completely abolished by the ingestion of food or the i.v. infusion of pentagastrin[3,4]. These actions of motilin are all related to the interdigestive function of the gut and, therefore, we have proposed the name "interdigestive hormone" for motilin[5]. Recently, we have developed a RIA for motilin and this then enabled us to measure changes in plasma motilin concentration. Moreover, we synthesized fragments and an analogue of motilin, the biological activity of which was also investigated.

METHODS AND MATERIALS

Twenty five healthy mongrel dogs of both sexes, weighing 9 to 15 kg, were used in the present study. A silastic tube was introduced into the external jugular vein through a branch vein so that the tip of the tube was placed into the superior vena cava. The other end of the tube was drawn through a subcutaneous tunnel and brought

out through a stab wound near the right scapula on the right lateral
neck[6]. This tube was used for postoperative fluid transfusion or
withdrawal of blood specimens. In the next step, extraluminal force
transducers were sutured on the serosal surface of the gastrointestinal tract from the LES to the terminal ileum. Lead wires were
drawn out of the abdominal cavity through a stab wound made on the
left abdominal wall and passed through a subcutaneous tunnel made
on the left costal flank and pulled out through a skin incision
made between the scapulas. Two weeks after the surgery, measurements of the changes in gut motor activity were started and were
continuously made with a pen-writing recorder; the details of this
procedure have already been reported elsewhere[7]. The dogs were fed
regularly with a dry type of dog food (Gaines meal, Ajinomoto-General
Foods Corporation, Tokyo, Japan). In five dogs, the splenic vein
was cannulated with a silastic tube (Dow Corning) which was used
for intraportal vein infusion of motilin. Five additional dogs were
prepared with chronic pancreatic fistulas, and two force transducers
were sutured on the serosal surface of the gastric body and antrum.
In these dogs, pancreatic juice secreted into the duodenal pouch was
first exteriorized through an implanted tube and its volume measured
by means of a drop counter and then returned into the duodenum by
means of a peristalic infusion pump. Gastrointestinal hormones used
in the present study were pentagastrin (ICI), secretin (synthesized
by N. Yanaihara), vasoactive intestinal polypeptide (VIP) (synthesized by N. Yanaihara), gastric inhibitory polypeptide (GIP) (synthesized by N. Yanaihara), glucagon (provided by Dr. T.M. Lin, Eli
Lilly), and motilin (synthesized by N. Yanaihara). These hormonal
preparations were dissolved in 0.9% saline to a desired concentration
and infused into the superior vena cava or the portal vein by a
Harvard infusion pump through chronically implanted tubes.

For radioimmunoassay for motilin[8], blood was withdrawn into
heparinized disposable syringes from the chronically implanted jugular tube. Two ml samples of blood for the determination of motilin
were placed promptly into chilled tubes containing 2,000 KU of
Trasylol in a volume of 0.2 ml. The mixture was immediately centrifuged at 4°C and plasma was separated and frozen at -20°C until assay
Plasma motilin was measured by dextran-coated charcoal radioimmunoassay using antisera raised in a guinea-pig against synthetic motilin prepared by N. Yanaihara. Synthetic motilin was used as standard
and also for labeling. Labeled hormone was prepared by the chlora-
mine-T method. The minimal detectable quantity by this method was
35 pg/ml. Intra- and inter-assay variation were 4.7% and 9.6%,
respectively. The addition of synthetic motilin to normal plasma
gave a mean $\pm$ SE recovery rate of 98 + 12%. In our radioimmunoassay system, substance P (synthesized by N. Yanaihara), secretin,
VIP, human gastrin-I (ICI), somatostatin (synthesized by H. Yajima,
Kyoto University, Kyoto, Japan), GIP, porcine monocomponent
glucagon (Novo, Denmark), and porcine insulin (Novo, Denmark) did
not cross-react with motilin antiserum. Serial dilution of plasma

obtained from normal fasted and fed dogs gave curves parallel to that of standard motilin. Three synthetic preparations of motilin (MTL), fragments MTL_{1-6}, MTL_{7-22}, MTL_{12-22}, were prepared by N. Yanaihara. Moreover, an analogue to motilin, 15-Gln-motilin, was also synthesized by N. Yanaihara. These synthetic preparations were assayed for their gastric motor stimulating activity using conscious dogs in our system. All data obtained in the present study were statistically analysed by the Student's t-test; p values less than 0.05 were considered to show a significant difference between paired data.

In the clinical study, gastric motor activity was measured by means of a balloon method on 10 healthy volunteer medical students. An argyl stomach tube with an attached thin rubber balloon was introduced into the stomach through the nose and placed so that the balloon was positioned in the gastric body by aid of fluoroscopy. The stomach tube was connected to a pressure tranducer to measure changes in gastric motor activity.

RESULTS

1. Chemical structure and biological activity relationship
     Three synthetic preparations of motilin fragments, MTL_{1-6}, MTL_{7-22} and MTL_{12-22} were assayed for their gastric stimulating activity in conscious dogs during the interdigestive state using MTL_{1-22} as a reference standard. It was found that MTL_{1-6} and MTL_{12-22} had no activity to stimulate or inhibit gastric contractions in a dose up to 100 µg/kg-hr; however, MTL_{7-22} in a dose of 200 to 300 µg/kg-hr showed a gastric stimulating activity as shown in Figure 1. MTL_{7-22} was examined for its inhibitory activity upon the naturally-occurring interdigestive contractions in the stomach; however, it was found that regular occurrence of the interdigestive contractions in the stomach was retained even after 1 hr. i.v. infusion. 15-Gln-motilin had $154 \pm 18.3\%$ activity of MTL_{1-22} in the same dose level. However, if the first residue of the N-terminus was replaced, the compound lost activity to less than 1/300 of MTL_{1-22}.

Therefore, it is concluded that the whole molecule is essential for biological activity of motilin (Table 1).

2. Metabolism and release
     Gastric motor stimulating activity of MTL_{1-22} was compared when it was infused into the systemic circulation and the portal vein in conscious dogs. It was observed that gastric motor stimulating activity of MTL_{1-22} in doses of 0.3 and 0.9 µg/kg-hr was only slightly inactivated by a single passage through the liver (Table 2). Major sites of motilin metabolism other than the liver are suggested. On the other hand, if pancreatic juice was diverted and replaced by

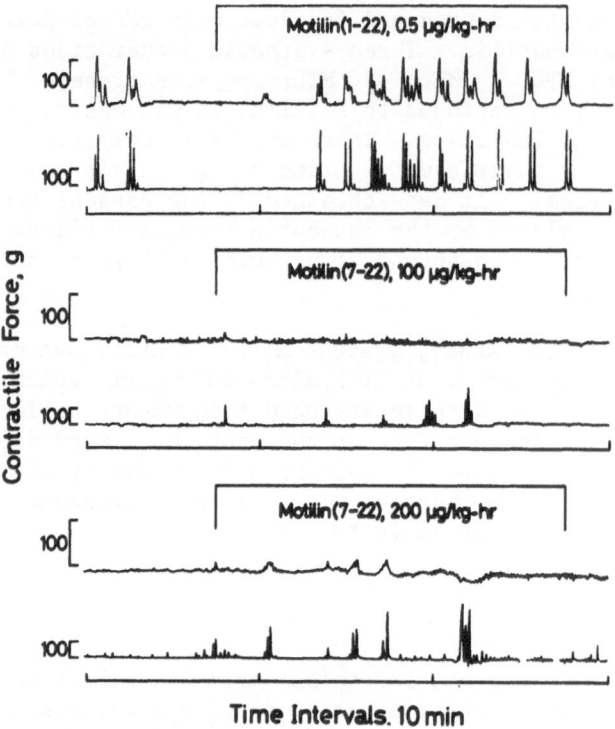

Figure 1. Comparison of gastric motor stimulating activity between MTL_{1-22} and MTL_{7-22}.

TABLE 1 - Structure-activity relationship of motilin

Fragments	Gastric Motor	
	Stimulating Activity	Inhibiting Activity
MTL_{1-6}	0	0
MTL_{7-22}	1/200–1/300[*]	?
MTL_{12-22}	0	0

[*]MTL_{1-22} =1

TABLE 2 - *Metabolism of motilin*

Synth. Motilin (µg/kg–hr)	Administration Route	
	Systemic	Portal
0.3	35.1 ± 0.5[*]	33.8 ± 1.1
0.9	45.2 ± 1.3	46.6 ± 1.5

[*] Motor Index

the same amount of saline for 24 hrs, it was always observed that
regular occurrence of the interdigestive contractions in the stomach
was completely inhibited. However, when pancreatic juice was diverted only during the latter 12 hours of a 24 hr fast, regular occurrenc
of the interdigestive contractions was not disturbed. These findings
suggest that pancreatic juice is important for regular occurrence of
the interdigestive contractions in the dog (Table 3).

3. *Action of motilin on gut motor activity and pepsin secretion*
     As reported previously, it was found that the 24-hr gastrointes-
tinal motor activity consisted of the two different major patterns:
the digestive and interdigestive patterns as shown in Figure 2.
The interdigestive motor activity was characterized by cyclic recurring, caudad-moving bands of strong contractions interrupted by
long-lasting motor quiescence. When one band of strong contractions
reached the distal ileum, another developed in the LES, stomach,
and duodenum again and propagated in a caudad direction. Such re-
cycling episodes repeatedly occurred until the next meal. After
ingestion of food, gastrointestinal motor activity was continuous
and such characteristic interdigestive patterns were not observed.
Synthetic motilin in a dose from 0.3 to 2.7 µg/kg-hr was assayed
for its motor stimulating activity in both states.   In the digestive state, an i.v. infusion of motilin had no influence upon the
motor activity even if the dose was increased up to 9.0 µg/kg-hr.
On the other hand, when motilin was infused during the interdigestive state, it induced a pattern precisely like the naturally-
occurring interdigestive contractions as shown in Figure 3. Figure
4 represents action of motilin on the LES, which is one of the

TABLE 3 - Effect of exteriozation of pancreatic juice

Replacement Amounts by Saline	Occurrence of Interdigestive Contraction
Total, 24 hr	not occurred
First half, 12 hr	greatly disturved
Latter half, 12 hr	not disturved

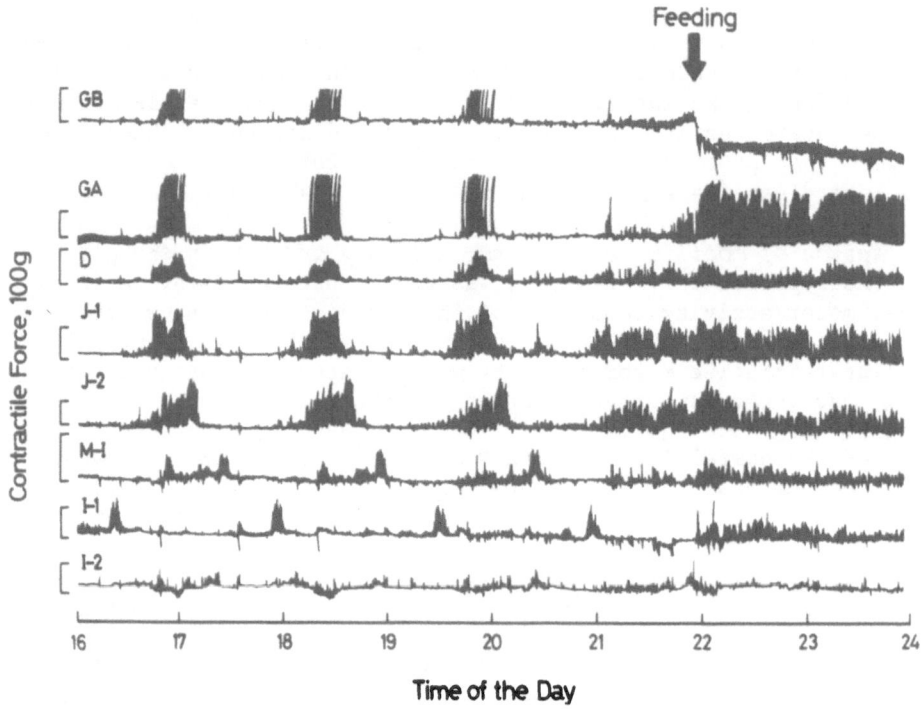

Figure 2. Eight-hr changes in gastrointestinal motor activity from the stomach to the terminal ileum before and after feeding. Recording speed: 1 mm/min. (Reprinted from: Scand J Gastroent 11 (Suppl 39) 93-110, 1976)

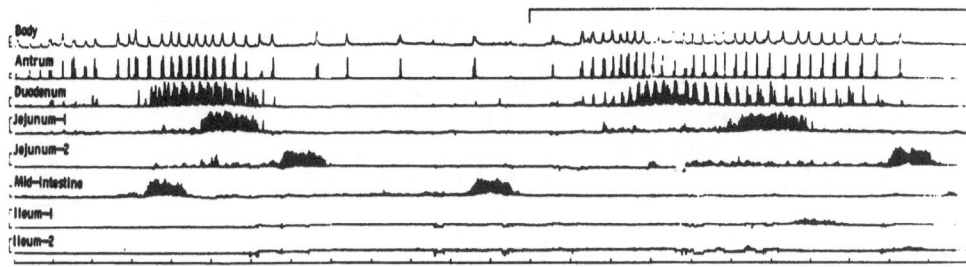

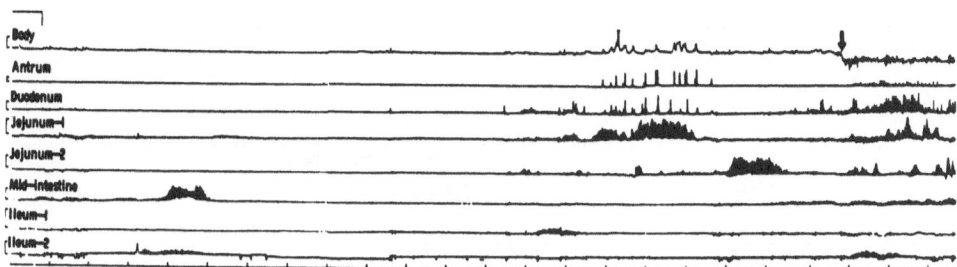

Figure 3. Effect of i.v. infusion of motilin in a dose of 0.3 µg/kg-hr during the interdigestive state. The upper and lower tracings are continuous. Recording speed: 10 mm/min. (Reprinted from: Scand J Gastroent 11 (Suppl 39): 93-110, 1976)

most important findings in relation to the clinical aspects of GE reflux. Motilin is the only substance known of the GI peptides that induces a contractile pattern in the LES similar to that observed during the interdigestive state. Motilin is also known to stimulate pepsin secretion and we also confirmed this activity in Heidenhain pouch dogs. However, according to our electronmicroscopic study, there is no evidence to indicate that motilin stimulates pepsin secretion in the interdigestive state. We, therefore, consider that pepsin might be squeezed by the strong contractions induced by motilin.

4. Interaction between motilin and other GI peptides

Interaction between motilin and other GI peptides was investigated on the interdigestive gastric motor activity in conscious dogs. It was found that motilin (0.3 µg/kg-hr)-induced gastric contractions were instantly inhibited by an i.v. infusion of pentagastrin (0.2-1.9 µg/kg-hr) as shown in Figure 5. However, secretin, VIP, GIP and glucagon had no effect on the motilin-induced gastric contractions in a dose of 5-10 µg/kg-hr. Similar findings were also obtained for the spontaneously-occurring interdigestive contractions. These results indicate that the gastrin family peptides control the inter-

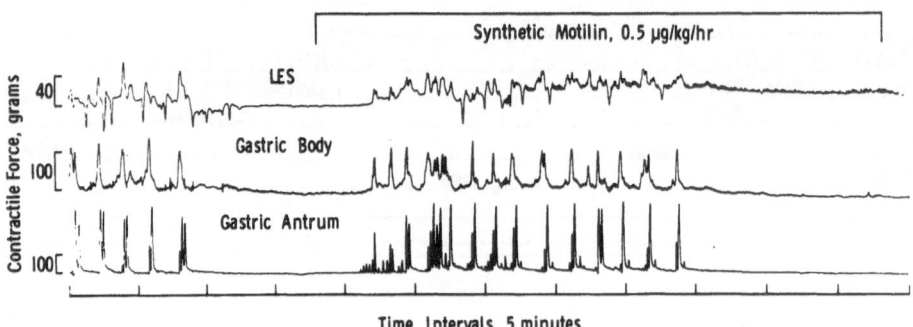

Figure 4. Induction of coordinated motor increase in the LES and the stomach by i.v. infusion of motilin in a dose of 0.5 µg/kg-hr in the interdigestive state.

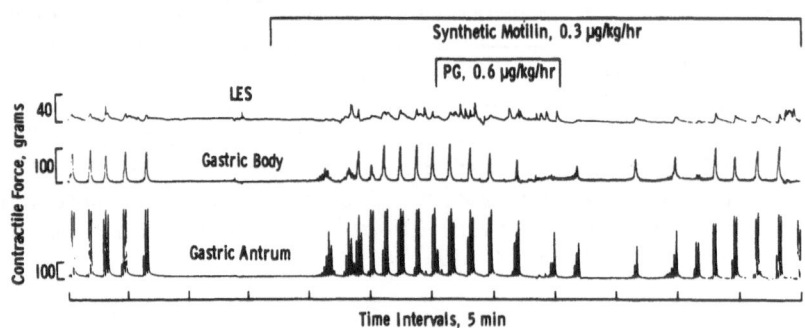

Figure 5. Inhibitory effect of pentagastrin on motilin-induced contractions in the stomach.

digestive motor activity of the gut. However, the secretin family of peptides seems to have no influence upon the interdigestive motor activity of the gastrointestinal tract (Table 4). Feeding also strongly inhibited the naturally-occurring and motilin-induced (Figure 6) contractions in the LES and stomach, which were described in detail in our chapter on the LES.

5. *Changes in plasma motilin concentration*

Simultaneous measurements of plasma motilin concentration and gastric contractile activity were made in conscious dogs. In the fasted state, when the stomach showed interdigestive contractions, it was found that plasma motilin concentration was elevated in all dogs. This high plasma motilin level was lowered by ingestion of food and its concentration remained low as long as the gastric motor activity remained in the digestive pattern. Changes in mean plasma motilin level as well as that of gastrin before and after ingestion of food are shown in Figure 7, in which it will be seen that plasma

TABLE 4 - *Effect of GI peptide on motilin-induced contractions*

G I Peptides	Actions
pentagastrin(ICI)	inhibit(0.2 μg/kg-hr)
CCK(Karolinska)	inhibit or disturved(3 IDU/kg-hr)
secretin(synth.by Yanaihara)	
GIP(")	
VIP(")	not inhibit(9 μg/kg-hr)
glucagon(Eli Lilly)	

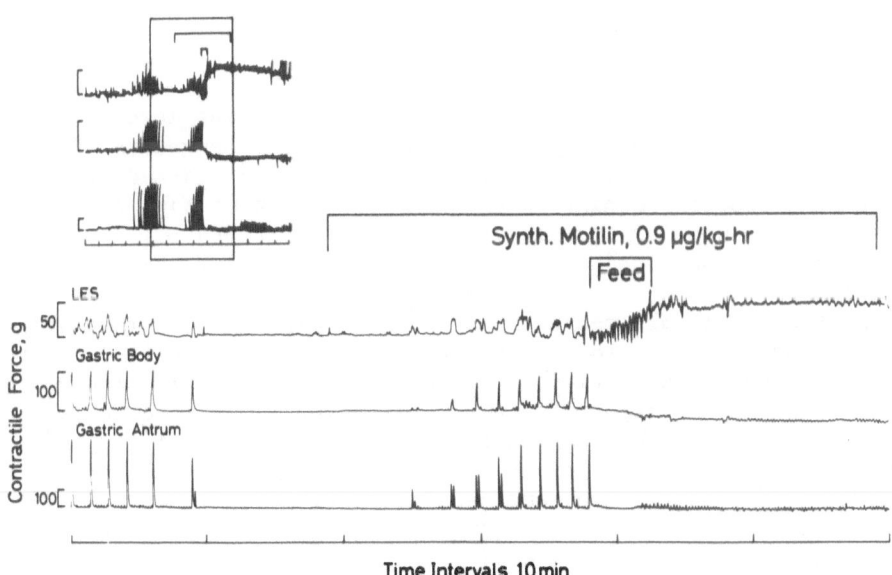

Figure 6. *Inhibitory effect of feeding upon motilin-induced gastric contractions*

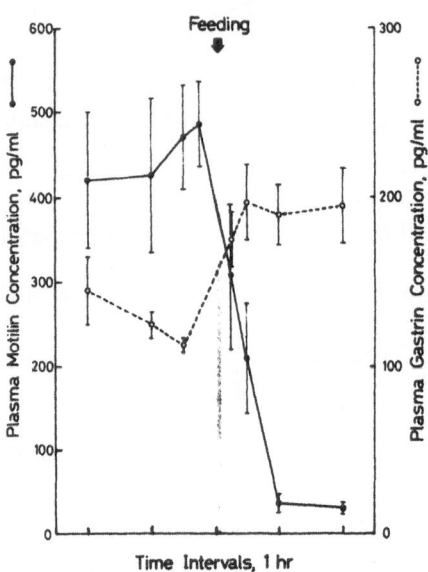

Figure 7. Changes in mean plasma motilin and gastrin concentration 2-hr before and 2-hr after feeding measured twice in each of 4 dogs. Rapid decrease in plasma motilin concentration was observed after ingesting food.

motilin concentration at 1-hr postprandial period was lowered in all cases. However, when gastric motor activity changed to the intermediate pattern, plasma motilin concentration started to increase and returned to high levels again by the time gastric motor activity was in its full interdigestive pattern. These changes are demonstrated in Figure 8. These parallel changes in plasma motilin concentration and gastric motor pattern were always observed in all measurements made in all dogs examined. Figure 9 demonstrates interdigestive changes in plasma motilin concentration and gut motor activity. As demonstrated in this Figure, it was found that plasma motilin concentration fluctuated in complete association with occurrence of gastric and jejunal contractions during the interdigestive state. When the stomach was in motor quiescence, plasma motilin remained low. However, it seems likely that the strong interdigestive contractions in the lower part of the small bowel are not directly mediated by plasma motilin concentration. In human studies, similar findings were obtained: ingestion of food always lowered the plasma motilin concentration and 2-hr postprandial concentration was always below the minimum detectable concentration. Plasma motilin concentration during prolonged fasting also fluctuated at approximately 100 min intervals which just paralleled the occurrence of the interdigestive gastric contractions. In the present study, we demonstrated the existence of interdigestive contractions in human subjects, which are also under control of plasma motilin concentration. Gastrin may play an important role in its mechanism.

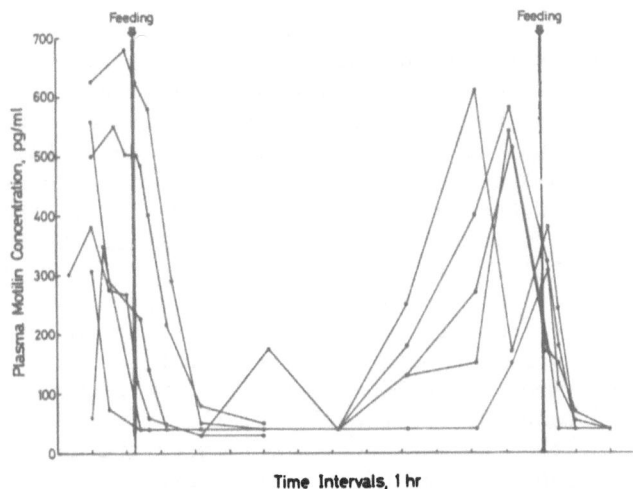

Figure 8. Fourteen-hr changes in plasma motilin concentration in 7 dogs. In this case, the dogs were fed 100 g dog food at 10.00 and 22:00. The digestive pattern lasted for about 7.5 hr. Contractile pattern of the stomach 10 hr after feeding was a typical interdigestive pattern.

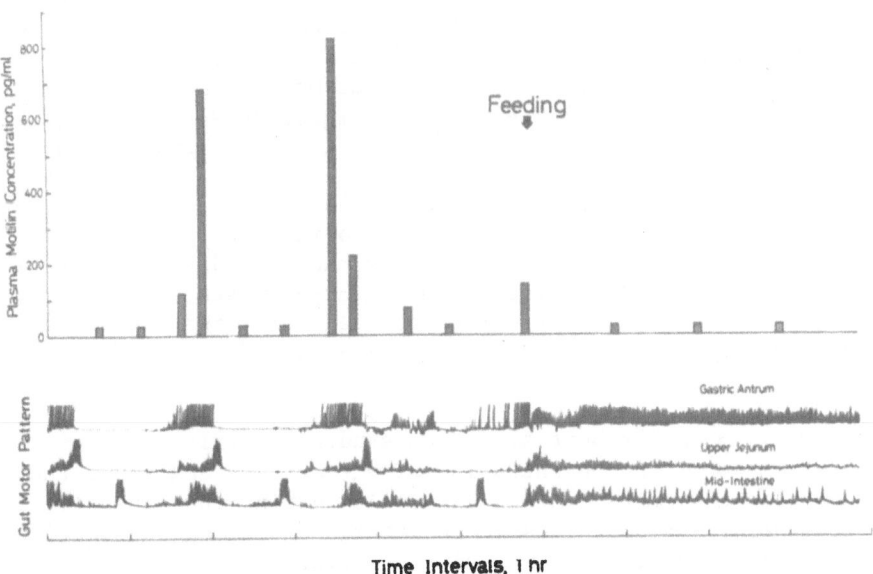

Figure 9. Parallel changes in gastric contractile activity and plasma motilin concentration during the interdigestive state in a conscious dog. However, after feeding, plasma motilin concentration was lowered and gastric contractile pattern changed to the digestive pattern.

DISCUSSION

Brown et al[9] isolated the substance named motilin in 1971. The primary structure of this hormone was also determined by Brown et al[10] and motilin was found to be a linear docosapeptide with the amino acid sequence: H-Phe-Val-Pro-Ile-Phe-Thr-Tyr-Gly-Glu-Leu-Gln-Arg-Met-Glu-Glu-Lys-Glu-Arg-Asn-Lys-Gly-Gln-OH. During the synthesis of this hormone by Wunsch et al[11] it was found in 1973 that the glutamic acid in position 14 was amidated[12]. Later, Yajima et al[1] and Yanaihara et al[21] successfully synthesized the molecule according to the newly revised formula[12]. We have concentrated our effort on the effect of motilin on gastrointestinal motor activity using conscious dogs. The details of these studies were presented at the Erlangen Symposium on Motilin in 1976, which appeared in a recent issue of Scand. J. Gastroent[4]. At that time, we concluded that motilin must be a hormone active only when the animal is in the fasted state and we proposed the name "interdigestive hormone" for this newly identified gut peptide[5]. However, there are still some discrepancies in motilin research. Brown[13] first described a RIA for motilin and demonstrated a concomitant rise in motility of a transplanted gastric pouch and plasma motilin concentration when an alkaline solution was introduced into the duodenum. On the other hand, Mitznegg[14] measured motilin level in man and found that motilin was not elevated in the systemic circulation after duodenal alkalinization. They reported that duodenal acidification induced a rise in plasma motilin concentration in man. This finding tended to argue that motilin exerts its effect during the digestive state. In fact, Ruppin et al[15] found that motilin delayed gastric emptying in man. However, we have demonstrated that motilin has no significant effect upon gut motility during the digestive state in dogs, and Wingate et al[16] have also confirmed that motilin induced the interdigestive myoelectric complex when given to a dog during the interdigestive state. Jennewein et al[17] also observed motor patterns similar to that of interdigestive caudad-moving contractions after i.v. infusion of motilin into dogs during the interdigestive state. These findings were all obtained from dog experiments. Therefore, the discrepancy in motilins's effect between human and dog was considered to be due to a species difference at the Erlangen Symposium in 1976.

However, in the present study, we studied gastric motor activity and plasma motilin concentration simultaneously and found that plasma motilin concentration was always low during the digestive state, but when the stomach was in the period of strong contractions plasma motilin concentration was always elevated. Moreover, even in the interdigestive state when the stomach was in the period of motor quiescence, plasma motilin concentration was low in all measurements. Since the interdigestive contractions migrate along the small bowel in a caudad direction, it is considered that endogenous stimulants for motilin release in the duodenum and upper jejunum are

also completely swept along by the contraction front; then plasma
motilin concentration is lowered. However, basal secretion in the
duodenum and from the pancreas will gradually accumulate in the
duodenum and jejunum and stimulate motilin release again. When
motilin concentration reaches a threshold in plasma, interdigestive
contractions are again generated in the stomach and duodenum simultaneously. Therefore, it is concluded that intervals of the cyclic
recurring episodes of strong contractions in the stomach are determined by the rate of the basal secretion of the stimulants for motilin release in the duodenum and the upper jejunum. Therefore, if
blood samples are taken in the fasted state elevated levels of motilin will not always be obtained. Bloom et al[18] showed that the
mean fasting level of plasma motilin was 72 pmol/1 but individual
values showed a skew distribution from less than 5 to 300 pmol/1.
Furthermore, according to a recent report by Eckard and Grace[19],
mean plasma motilin concentration after a 12 hr fast was elevated; 415 pg/ml in 13 healthy volunteers with variation from 200 to
more than 800 pg/ml.

Most recently, Mori et al studied[20] the changes in plasma motilin concentration during the oral glucose tolerance test in healthy
man and diabetics and reported evidence that ingestion of 100 g of
glucose lowered plasma motilin level in all measurements. We have
also confirmed that plasma motilin concentrations were elevated during the period of strong contractions in the stomach in the interdigestive state in healthy volunteers and were suppressed by ingestion
of a meal[21]. Therefore, we consider that motilin may delay gastric
emptying as reported by Ruppin et al[15]; however, the high postprandial
concentration of motilin is not caused by ingested food itself,
but is a remnant of the elevated concentration during the interdigestive state, or some other unknown stimulus. In fact, since Polak
et al[22] reported that motilin cells (EC$_2$) are located in the duodenum and the upper jejunum, it is not likely that ingested food
still in the stomach stimulates endogenous release of motilin.

The present study together with our previous studies[3,4,5] on
gut motor activity and exogenous motilin clearly indicate that the
interdigestive motor activity of the gut is controlled by the plasma
motilin concentration. These interdigestive contractions, which
originate in the LES and stomach and move in a caudad direction to
the terminal ileum, were first noted by Szurszewski in 1969[23] in the
dog small intestine. The existence of these characteristic changes
in interdigestive motor activity were later confirmed by Code and
Marlett[24], Wingate et al[16], Grivel et al[25], and Itoh et al[3,4,7].
Therefore, since the existence of the interdigestive contractions
and the hormone controlling these contractions has been demonstrated,
it now seems important to attempt to show the functional significant
of these interdigestive movements. It is important to learn
more about these contractions observed during the interdigestive state in humans. Cannon and Washburn[26] and Helzel[27] reported

interdigestive contractions in humans; however, the details of the contractions are not clear enough to compare to our motor pattern obtained in dog experiments. Therefore, we tried to demonstrate interdigestive contractions in the stomach in human subjects and confirmed the existence of similar contractions interrupted by long-lasting quiescence. When duration of contractions and quiescence in the interdigestive state was compared in human subjects and dog, it was found that the duration of each period was surprisingly similar in man and dog, shown in Table 5. Further studies in human subjects are needed.

In conclusion, the current status of motilin may be summarized as follows:

1. The whole molecule is essential for biological activity. However, there exists a more potent analogue of motilin; for example, 15-Gln-motilin was found to be approximately 50% more potent than natural motilin$_{1-22}$.

2. Motilin is released at approximately 100 min intervals during the interdigestive state and feeding stops endogenous release. Inhibition seems to occur 1-2 hr after ingestion of food.

3. The high concentration of motilin observed immediately after feeding is not released by ingested food itself, but is the remnant of the interdigestive elevated concentrations, or some other stimulus.

4. Motilin is slightly inactivated by a single passage through the liver. Major sites for metabolizing motilin other than the liver are suggested.

5. Motilin is active during the interdigestive state and controls the interdigestive contractions of the gastrointestinal tract from the LES to the terminal ileum.

6. The role of motilin in the control of LES function is important in relation to GE reflux.

7. Motilin stimulates pepsin output from the Heidenhain pouch during the interdigestive state.

8. Gastrin inhibits motilin's effect on smooth muscle of the gut but the secretin family of peptides has no significant influence upon motilin-induced contraction.

9. Diseases due to hyper- or hypo-motilinemia are not known.

TABLE 5 - Comparison of interdigestive motor pattern in man and dog

	Period of	
	Contractions	Quiescence
Dog	24.7 ± 0.22	77.0 ± 0.48
Man	29.6 ± 1.92	84.9 ± 11.5

Mean ± SE (min)

ACKNOWLEDGEMENTS

This work was in part supported by a Grant for Cancer Research to Z. Itoh in 1976-77 from the Ministry of Public Welfare and Health. Excellent animal care was provided by Mr. T. Koganezawa, Supervisor of the Animal House of Gunma University School of Medicine. The manuscript was prepared by Miss M. Koike, G. I. Laboratory. Photographic materials were prepared by Mr. F. Ohshima, Chief, Photo Center, Gunma University Hospital. Trasylol was kindly supplied by Mr. S. Hasegawa, Manager of Tokyo Office, Bayer, Japan.

REFERENCES

1. Yajima J, Kai Y, Kawatani H: Synthesis of the docosapeptide corresponding to the entire amino acid sequence of porcine motilin. JCS Chem Comm 1975: 159-160

2. Shimizu F, Imagawa K, Mihara S, Yanaihara N: Synthesis of porcine motilin by fragment condensation using three protected peptide fragments. Bull Chem Soc Jap 49: 3594-3596, 1976

3. Itoh Z, Aizawa I, Takeuchi S, Couch EF: Hunger contractions and motilin. In Proceeding of 5th International Symposium on Gastrointestinal Motility, edited by G Vantrappen. Typoff-Press, Herentals, Belgium, 1975, p. 48-55

4. Itoh Z, Honda R, Hiwatashi K, Takeuchi S, Aizawa I, Takayanagi
    R, Couch EF: Motilin-induced mechanical activity in the
    canine alimentary tract. Scand J Gastroent 11(Suppl 39): 93-
    110, 1976

5. Itoh Z, Takeuchi S, Aizawa I: Effect of synthetic motilin on
    gastric motor activity in conscious dogs. In abstract of
    5th International Congress of Endocrinology, Hamburg, 1976,
    p. 181

6. Itoh Z, Carlton N, Lucien HW, Schally AV: Long-term tubing
    implantation into the external jugular vein for injection
    or infusion in the dog. Surgery 66: 768-770, 1969

7. Itoh Z, Aizawa I, Takeuchi S, Takayanagi R: Diurnal changes in
    gastric motor activity in conscious dogs. Am J Dig Dis 22:
    117-124, 1977

8. Mori K, Seino Y, Taminato A, Goto Y, Inoue Y, Kadowaki S,
    Matsukura S, Imura H: Radioimmunoassay for motilin. Present-
    ed at the 34th Annual Meeting of Japan Endocrinological
    Society at Okayama in May, 1977

9. Brown JC, Mutt V and Dryburgh JR: The further purification of
    motilin, a gastric motor activity stimulating polypeptide
    from the mucosa of the small intestine of dogs. Can J Physic
    Pharmacol 49: 399-405, 1971

10. Brown JC, Cook MA and Dryburgh JR: Motilin, a gastric motor
    activity stimulating polypeptide: The complete amino acid
    sequence. Can J Biochem 51: 533-537, 1973

11. Wünsch E, Brown JC, Deimer KH, Drees F, Jaeger E, Musiol J,
    Scharf R, Stocker H, Thamm P and Wendlberger G: Z Naturforsc
    28C: 235-240, 1973

12. Schubert H and Brown JC: Correction to the amino acid sequence
    of porcine motilin. Can J Biochem 52: 7-8, 1974

13. Dryburgh JR, Brown JC: Radioimmunoassay for motilin. Gastroenter
    logy 68: 1169-1176, 1975

14. Mitznegg P, Bloom SR, Christofides N, Besterman H, Domschke W,
    Domschke S, Wünsch E, Demling L: Release of motilin in man.
    Scand J Gastroent 11(Suppl 39): 53-56, 1976

15. Ruppin H, Domschke S, Domschke W, Wünsch E, Jaeger E, Demling L:
    Effects of 13-nle-motilin in man - Inhibition of gastric
    evacuation and stimulation of pepsin secretion. Scand J
    Gastroent 10: 199-202, 1975

16. Wingate DL, Ruppin H, Green WER, Thompson HH: Motilin-induced
     electrical activity in the canine gastrointestinal tract.
     Scand J Gastroent 11 (Suppl 39): 111-118, 1976

17. Jennewein HM, Bauer R, Hummelt H, Lepsin G, Siewert R: Motilin
     effects on gastrointestinal motility and lower esophageal
     sphincter (LES) pressure in dogs. Scand J Gastroent 11
     (Suppl 39): 63-65, 1976

18. Bloom SR, Mitznegg P, Bryant MG: Measurement of human plasma
     motilin. Scand J Gastroent 11 (Suppl 39): 47-52, 1976

19. Eckardt V, Grace ND: Lower esophageal sphincter pressure and
     serum motilin levels. Am J Dig Dis 21: 1008-1011, 1976

20. Mori K, Seino Y, Taminato Y, Ikeda M, Matsukura S, Imura H,
     Yanaihara N: Secretory pattern of plasma motilin. Present-
     ed at the 43rd Annual Meeting of the Japanese Society of
     Gastroenterology in Tokyo in April, 1977

21. Itoh Z, Takahashi I, and Aizawa I: Measurements of plasma moti-
     lin concentration together with monitoring gastric movements
     in man (in preparation)

22. Polak JM, Heitz P and Pearse AGE: Differential localisation of
     substance P and motilin: Scand J Gastroent 11 (Suppl 39):
     39-42, 1976

23. Szurszewski JH: A migrating electric complex of the canine small
     intestine. Am J Physiol 217: 1757-1763, 1969

24. Code CF and Marlett JA: The interdigestive myo-electric complex
     of the stomach and small bowel of dogs. J Physiol 246: 289-
     309, 1975

25. Grivel ML and Ruckebusch Y: The propagation of segmental con-
     tractions along the small intestine. J Physiol 227: 611-
     625, 1972

26. Cannon WB and Washburn AL: An explanation of hunger. Am J
     Physiol 29: 441-454, 1911-12C

27. Hoelzel F: The relation between the secretory and motor activ-
     ity in the fasting stomach (man). Am J Physiol 73: 463-469,
     1925

GLUCAGON SECRETION INDUCED BY BOMBESIN IN MAN

F. Fallucca*, G. F. Delle Fave**, S. Gambardella*,
C. Mirabella*, L. De Magistris** and R. Carratu'**

*Cattedra di Medicina Costituzionale ed Endocrinologia
**Cattedra di gastroenterologia

Università di Roma, Policlinico Umberto I

INTRODUCTION

It has been shown that bombesin (BBS), an active tetradecapeptide isolated from the skin of frogs[1], displays a number of pharmacological actions, including a hyperglycemic action in the rat
and the dog and an increase in immunoreactive insulin levels in
peripheral blood of the dog have been reported[1].

In man the effect of BBS on carbohydrate metabolism is unknown; moreover no data, in animal or in man, are available on the
effects of BBS on pancreatic alpha cell function and consequently
on glucagon secretion.

The present study was undertaken to elucidate the effects of
BBS on alpha and beta cells of the pancreas in man.

MATERIALS AND METHODS

The design of the study consisted of three thirty-minute periods;
during thefirst and the last periods an intravenous infusion of
saline was given; BBS infusion (10 ng/kg-min) was given during the
second.

Blood samples were taken at 0, 5, 10, 20, 30, and 60 min from
the start of BBS infusion.

Each sample was analyzed for blood glucose, plasma insulin[2] and pancreatic glucagon[3] (IRG), using the 30 K antiserum in the RIA of glucagon.

The study was performed in 8 normal volunteers and in two patients with Zolinger-Ellison syndrome (ZES).

RESULTS

BBS infusion induced negligible changes of blood glucose and insulin, while it increased IRG levels significantly at 10 min and 20 min. Peak IRG levels during BBS were also significantly higher than fasting values (Fig. 1).

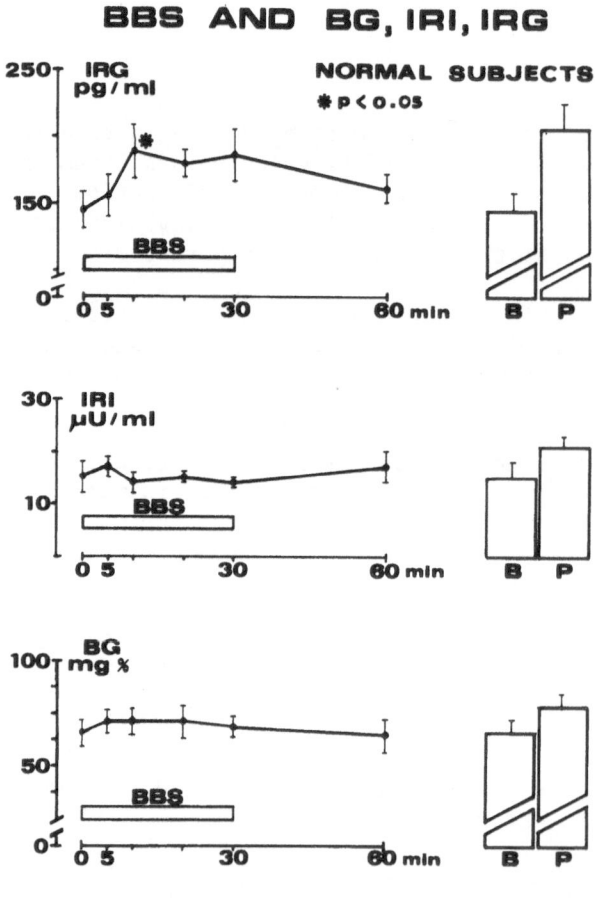

Figure 1

CONCLUSIONS

Bombesin (at the dose employed) stimulates (directly or indirectly) pancreatic glucagon secretion in man. This ability may explain
the hyperglycemic effect of BBS demonstrated in animals (where
greater doses were employed).

REFERENCES

1. Anastasi A, Erspamer V, Bucci M: Isolation and structure of
        bombesin and alytesin, two analogous peptides from the skin
        of the European amphibians bombina and alytes. Experientia
        27, 166, 1971

2. Hales CN, Randle PJ: Immunoassay of insulin with insulin-anti-
        body precipitate. Biochem. J. 88, 137, 1963

3. Aguilar-Parada E, Eisentraut AM, Unger RH: Pancreatic glucagon
        secretion in normal and diabetic subjects. Am. J. Med. Sci.
        257, 415, 1969

POLYPEPTIDES IN BRAIN AND GUT: CHOLECYSTOKININ-LIKE PEPTIDES

G. J. Dockray

Physiological Laboratory, University of Liverpool

Liverpool L69 3BX   England

The recognition that the same active polypeptide may occur in both brain and alimentary tract can be traced to the demonstration by von Euler and Gaddum of substance P in these two tissues[1]. Recent work suggests that there are a number of other peptides which may also be present in both brain and gut[2]. Thus substance P and neurotensin have been isolated from both tissues, somatostatin and enkephalin have been isolated from brain and identified in the gut by radioimmunoassay and immunocytochemistry, and vasoactive intestinal peptide has been isolated from gut and identified in brain by radioimmunoassay. In addition, there is now evidence, reviewed below, that factors resembling COOH-terminal fragments of the gut hormone cholecystokinin can be added to the list of brain-gut peptides[3]. The importance of cholecystokinin in the regulation of pancreatic secretion and gall bladder contraction has long been recognised, and in the present context it is interesting to note that exogenously administered cholecystokinin also has behavioural actions, notably influencing feeding and satiety[4].

The presence of gastrin-like immunoreactivity in brain extracts was first reported by Vanderhaeghen, Signeau and Gepts[5]. Gastrin has been isolated from antral mucosa in the form of peptides of 17 and 34 residues (G17 and G34), and the COOH-terminal pentapeptide of these molecules is identical to that of the two forms of cholecystokinin isolated from intestine (33 and 39 residues, abbreviated to CCK33 and CCK39), and the amphibian skin peptide caerulein[6]. Antisera specific for the COOH-terminus of one of these peptides may cross-react with the others, and the work summarised here was undertaken to clarify the immunochemical relationships between the brain activity and the gut forms of gastrin and CCK.

Crude extracts of cerebral cortex (hog, dog, sheep) or whole brain (rat, guinea pig) were prepared by methods known to extract gastrin; that is to say tissues were briefly boiled in water, homogenised and centrifuged. The supernatant solutions were then fractionated by gel filtration on Sephadex columns which had been calibrated with standard forms of gastrin, CCK and related molecules. Radioimmunoassay using antisera specific for the COOH-terminus of gastrin revealed three peaks of immunoreactive material in the column eluates (Fig. 1). In all species the principal component (II) emerged in a position similar to that of the COOH-terminal octapeptide (CCK8) of cholecystokinin. There was a minor peak (I) in the void volume on Sephadex G25 which again emerged in the void after re-fractionation on Sephadex G50, indicating a component substantially larger than any of the presently characterized forms of gastrin and CCK. A third component (III) emerged after ^{125}I on Sephadex G25, and is likely to be a small COOH-terminal fragment of gastrin or cholecystokinin, possibly the pentapeptide or tetrapeptide.

The immunochemical properties of the main brain component (II) were studied in radioimmunoassays using antisera of different specificity. Antisera known to be specific for the NH_2-terminus of G17[7], or for intact G17[8], did not cross-react significantly with the brain component. In contrast, several antisera specific for the COOH-terminus cross-reacted, although there were quantitative differences between antisera depending on the standard used in the radioimmunoassay. Thus when G17 was the standard there were over 100-fold differences between antisera in estimating the concentration of the brain components (Table 1). In contrast, when CCK8 was the standard there were only minor differences (2-3 fold) in estimates of the concentration of the brain components (Table 1). The conclusion from these results is that the immunochemical properties of the brain components resemble those of CCK8 more closely than G17. Similar results have been obtained with the two minor components (I and III)[3].

The suggestion that the brain factors resemble CCK rather than gastrin is supported by comparison of the immunoreactive components in brain with those in antral mucosa and small intestine. The major immunoreactive component seen when antral mucosal extracts were fractionated by gel filtration corresponded to G17, and a corresponding factor could not be identified in brain extracts. However, in boiling water extracts of hog duodenum or jejunum there was a major component which emerged in a similar position to CCK8 and the main brain peak. There were also minor peaks of immunoreactivity in the intestinal extracts resembling peaks I and III in brain. In addition, boiling water extracts of hog intestine contained a major component which emerged in a similar position to G17. However, this component was distinguishable from G17 on the basis of its pattern of cross-reactivity with different antisera (when it resembled a CCK-like peptide), and because unlike G17 it was degraded by trypsin to a smaller component emerging just before CCK8 on Sephadex G25 or G50.

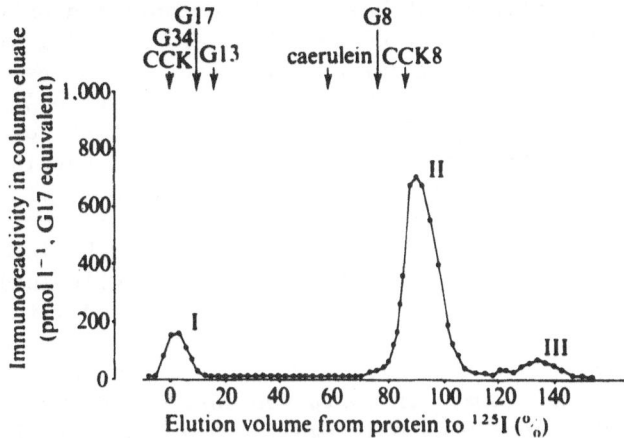

FIGURE 1. Fractionation of an extract of hog cerebral cortex on Sephadex G25 superfine (1 x 100 cm). Fractions of 1.0 ml collected every 10 min. Column eluted with 0.02 M sodium barbital pH 8.4, containing 0.02% sodium azide, at 4°C. The column was previously calibrated with 20% pure porcine CCK, pure natural human G34-I and G17-I, caerulein, synthetic peptides corresponding to the COOH-terminal tridecapeptide (G13) and octapeptide (G8) of G17-I, and CCK8. Elution volumes of the standards are indicated by arrows. In all column runs the void volume was marked by protein estimated by absorption at 280 nm, and the salt region was marked by Na^{125}I. Elution volume is expressed as percentage from void (0%) to ^{125}I (100%). Concentration of immunoreactive material expressed relative to standard G17. Roman numerals identify the peaks of immunoreactivity. (Reproduced from Nature with permission.)

		antiserum		
Standard	C	2716	1296	L2
G17	2,314	177	119	23
CCK8	1,170	520	735	1,355

TABLE 1. Concentration of the main brain component (II) in samples pooled from Sephadex G25 eluates estimated by radioimmunoassay with four antisera cross-reacting at the COOH-terminus of gastrin and CCK and expressed against G17 and CCK8 standards. For further details see ref 3.

Boiling water extracts of hog intestine contained only trace
amounts of activity resembling CCK33. However, material with the
gel filtration and immunochemical properties of CCK33 was obtained
in high yield when small intestine was extracted with boiling 0.5M
acetic acid. Pure CCK33 and CCK39 emerge in a similar position on
Sephadex G50 and so the corresponding peak in the intestinal extracts
is probably a mixture of these two peptides. The poor yield of
CCK33-like activity in boiling water extracts is not likely to be
the result of degradation of the peptide, for when the insoluble
residue of these extracts was treated with 0.5M acetic acid this
component was obtained in good yield (+ 80%, compared with direct
acid extraction). Presumably CCK33 is co-precipitated with denatured
protein at neutral pH in the boiling water extracts, but is readily
soluble in acid conditions. In sharp contrast, acid extracts of
hog cerebral cortex contained only small amounts of material with
the properties of CCK33. Again this observation is not likely to
be due to an artifact of extraction, since pure CCK33 added to acid
extracts could be recovered in good yield. It would appear therefore that in both brain and intestine there is material with the
immunochemical and gel filtration properties of CCK8, and that in
intestine, but not in brain, there are at least two other major CCK-
like components, one of which is probably CCK33.

In radioimmunoassays using ^{125}I labelled CCK8 and an antiserum
raised against CCK8 which cross-reacts with CCK8 and CCK33 (ratio
of molar potencies CCK8:CCK33, 1.0:0.6) the concentration of peak II
material in hog cerebral cortex (boiling water extract) was approximately 100 pmol/g (CCK8 standard) while the concentration of CCK33-
like material was less than 5 pmol/g (CCK33 standard; acid extracts).
By comparison the concentration of CCK8-like material in hog jejunum
was 15-20 pmol/g and that of CCK33-like activity was 25-30 pmol/g.

A full understanding of the structural relationships between the
brain components and the gut hormones requires the purification and
elucidation of structure of these factors. The purification of component II from sheep brain is now well advanced in our laboratory
and has already raised some interesting questions. Thus on both ion-
exchange chromatography and on electrophoresis we have found that the
main brain component (II) can be resolved into at least two, and
possibly three, immunoreactive factors (R.A. Gregory and G.J. Dockray,
unpublished observations). It is tempting to speculate that by
analogy with gastrin two of these components correspond to identical
peptides, one possessing a sulphated tyrosine residue, the other an
unsulphated tyrosine residue. So far as is known CCK occurs in intestine in the sulphated form only, and there has been no suggestion
of the natural occurrence of unsulphated CCK which would in any case
have low activity on gall bladder and pancreas.

The results presented here are consistent with the view that in both brain and gut there is synthesis of a large precursor peptide (e.g. brain component I) which is subsequently converted by proteolytic enzymes, perhaps sequentially, to CCK33 and CCK8. It would seem that in brain there is almost complete conversion of precursor to CCK8-like components, whereas in intestine there is incomplete conversion leading to the accumulation of CCK33-like and CCK8-like factors, and intermediates. If this hypothesis is correct then one would expect to find in brain the NH_2-terminal fragment produced when CCK8 is liberated by cleavage of the precursor. The availability of antisera specific for the mid-portion of CCK33, such as that described by Polak et al.[9], could be used to search for and characterize such a fragment. More detailed studies on the cellular and sub-cellular localization of the CCK-like peptides in brain will provide a foundation for physiological studies on their role in the central nervous system.

REFERENCES

1. Euler, US von, Gaddum JH: An unidentified depressor substance in certain tissue extracts. J. Physiol 72: 74-87, 1931

2. Pearse AGE: Peptides in brain and gut. Nature 262: 92-94, 1976

3. Docray GJ: Immunochemical evidence of cholecystokinin-like peptides in brain. Nature 264: 568-570, 1976

4. Sturdevant RAL, Goetz H: Cholecystokinin both stimulates and inhibits human food intake. Nature 261: 713-715, 1976

5. Vanderhaeghen JJ, Signeau JC, Gepts W: New peptide in the vertebrate CNS reacting with antigastrin antibodies. Nature 257: 604-605, 1975

6. Grossman MI: Gastrointestinal Hormones. In, Peptide hormones. Edited by JA Parsons, London, Macmillan Press, 1976, pp 106-116

7. Dockray GJ, Walsh JH: Amino-terminal gastrin fragment in serum of Zollinger-Ellison syndrome patients. Gastroenterology 68: 222-230, 1975

8. Dockray GJ, Taylor IL: Heptadecapeptide gastrin: measurement in blood by specific radioimmunoassay. Gastroenterology 71: 971-977, 1976

9. Polak JM, Szelke M, Bloom SR, et al: Use of synthetic fragments for specific immunostaining of CCK cells. Gastroenterology 72: A-12/822, 1977.

MOTILIN-, SUBSTANCE P- AND SOMATOSTATIN-LIKE IMMUNOREACTIVITIES IN EXTRACTS FROM DOG, TUPAIA AND MONKEY BRAIN AND GI TRACT

C. Yanaihara, H. Sato, N. Yanaihara, S. Naruse, W.G. Forssmann*, V. Helmstaedter*, T. Fujita**, K. Yamaguchi*** and K. Abe***

Laboratory of Bioorganic Chemistry, Shizuoka College of Pharmacy, Shizuoka, Japan, *Department of Anatomy, University of Heidelberg, Germany, **Third Department of Anatomy, School of Medicine, University of Niigata, Niigata, Japan, and ***Endocrine Division, National Cancer Center Research Institute, Tokyo, Japan

Motilin, a 22-amino acid polypeptide, has been isolated from porcine duodenal and jejunal mucosa as a principle which exerts motor-stimulating effects on gastrointestinal smooth muscle[1]. Dryburgh and Brown[2] have developed a radioimmunoassay and demonstrated the increase of immunoreactive motilin in the circulation of the dog by strong duodenal alkalization. Subsequently, the measurement of motilin immunoreactivity in human plasma has been demonstrated[3]. Using radioimmunoassay technique, the distribution of immunoassayable motilin has been determined in boiling-water extracts of tissues from various parts of the human gastrointestinal tract and other organs. The highest concentration of immunoreactive motilin was found in the duodenum, with little in the colon, liver, pancreas and brain, suggesting that the role of motilin may relate to digestive function[3]. Histochemical studies[4] have also shown that motilin is present in the EC-cells of the mammalian small intestine, mostly in duodenum and jejunum, but not in gastric and lower intestinal EC-cells. Forssmann et al.[5] have demonstrated immunohistochemically using antisera against our synthetic motilin that the motilin cells are present predominantly in duodenum and upper jejunum in Tupaia belangeri and human biopsy materials. The motilin cells have been shown to be differentiated, by simultaneous formaldehyde-induced fluorescence and immunofluorescence treatment, from the serotonin-containing EC-cells which did not react with the anti-synthetic motilin antisera.

Substance P[6,7], which is considered to play a role as neuro-transmitter[8], was shown to be present in the gut intramural nervous plexuses and also in the EC-cells of gut mucosa[9,10]. Recently, Polak et al.[11] revealed the presence of at least two different types of EC-cells by the fact that substance P and motilin were found in different EC-cells.

Somatostatin, which was isolated originally from ovine hypothalamic tissue[12], has been well-known to play a significant role in relation to gastrointestinal physiology[13]. Recent immunohistochemical studies[14-18] and radioimmunoassays[13,19,20] revealed the presence of somatostatin-like immunoreactivity both in the nervous system and in the gastrointestinal tract and pancreas. In fact, it was reported that in addition to substance P and somatostatin, immunoreactive VIP[21-23], gastrin[24] and CCK-PZ[25] were present in both gut and brain, although the physiological significance of the existence of these polypeptides in endocrine and nerve cells remains as yet unclear. Detailed information on the quantitative distributions of these hormones may be of importance in order to understand the physiological functions of these hormones.

We have now compared the concentration of motilin with those of substance P and somatostatin in various parts of the gastrointestinal tract, the pineal body and pituitary. Some brain tissues were also examined by the radioimmunoassay technique. We described previously the synthesis of porcine motilin having immunological and biological properties identical with those of natural porcine motilin[26]. Using the synthetic motilin, production of antisera specific to motilin and development of radioimmunoassay for the hormone were performed. This has then allowed us to compare immunoreactive motilin content with those of substance P and somatostatin in various tissue extracts.

MATERIALS AND METHODS

Synthetic polypeptides
Porcine motilin[26], substance P[27], tyrosyl-substance P[27], somatostatin[28] and tyrosyl-somatostatin[28] were synthesized by the conventional method for peptide synthesis in the Laboratory of Bio-organic Chemistry, Shizuoka College of Pharmacy, and before use the purities were assessed.

Tissue extraction
Tissues from dogs, tupaias and monkeys were minced and extracted with 5-fold their weight of boiling water for 5 min. After acidifying with acetic acid (to 1M concentration) on cooling, the suspension was centrifuged at 3,000 rpm for 20 min at 4°. This extraction was repeated twice and the combined supernatants were concentrated and

lyophilized. The dried material was dissolved in a small amount of
1M acetic acid and the insoluble material was removed by centrifugation. The solution was then submitted to gel-filtration on
Sephadex G-25 (1.5 x 90 cm) using 1M acetic acid as eluent. Fractions containing substances of molecular weight over 1,000 were
pooled, lyophilized and used for radioimmunoassays.

Antisera

Antisera to highly purified synthetic porcine motilin were
raised in guinea pigs and mixed-bred rabbits. Synthetic porcine
motilin (5.0 mg) was dissolved in 0.9% NaCl (0.5 ml) and the solution was mixed with 50% aqueous polyvinylpyrrolidone (PVP, MV 25,000-
30,000, Merck, Darmstadt). The mixture was emulsified with complete
Freund's adjuvant (2.5 ml, Calbiochem, Calif.) and the emulsion was
injected subcutaneously at multiple sites in three rabbits weighing
2.5-2.8 kg and three guinea pigs. A total of 1.0 ml of the emulsion
was used for rabbits and 0.5 ml for guinea pigs. Each of the animals received the 2nd to 8th immunizations in the same manner with
reduced amounts of motilin (0.5-1.0 mg). The series of immunizations were performed at 2 week intervals, and blood was obtained
10 days after the 4th and 5th injections. Finally, the animals
were bled 10 days after the last injection. Among antisera obtained, GP1102 elicited in a guinea pig was used routinely for assays.

The antiserum to substance P (R-400) used in this study was the
one described previously[29]. The antisomatostatin antiserum (AA #101)
was a generous gift from Prof. A. Arimura, Tulane University, New
Orleans.

Radioimmunoassay for motilin

A modification of the method of Morgan and Lazarow[30] was employed. The labelled antigen was prepared by radioiodination of
synthetic porcine motilin by the chloramine T method[31] and purified
by gel-filtration on Sephadex G-10 (1 x 25 cm) using 0.01M phosphate-
saline buffer (pH 7.4) as eluent. The standard diluent was 0.01M
phosphate buffer (pH 7.4) containing 1% BSA, 0.025M EDTA and 0.14M
NaCl. To the standard diluent (0.4 ml) in each tube were added
standard motilin or unknown sample (0.1 ml) and antiserum GP1102
(final dilution 1: 42,000) (0.1 ml). The mixture was kept at 4°C
for 48 hr and then ^{125}I-motilin (0.1 ml) was added. After a further 48 hr incubation at 4°C, goat antiserum to rabbit γ-globulin
(0.1 ml) was added. The mixture was incubated at 4°C for 24 hr
and centrifuged at 3,000 rpm for 30 min at 4°C. The supernatant
was removed by aspiration and the precipitate was counted. The
assay could detect at least 5 pg motilin per tube. The antiserum
GP1102 did not crossreact with synthetic bovine substance P, synthetic ovine somatostatin, synthetic porcine VIP[32] or secretin[33],
synthetic human gastrin[34], natural porcine glucagon or insulin
(Lilly Research Laboratories, Ind.). The C-terminal fragments of

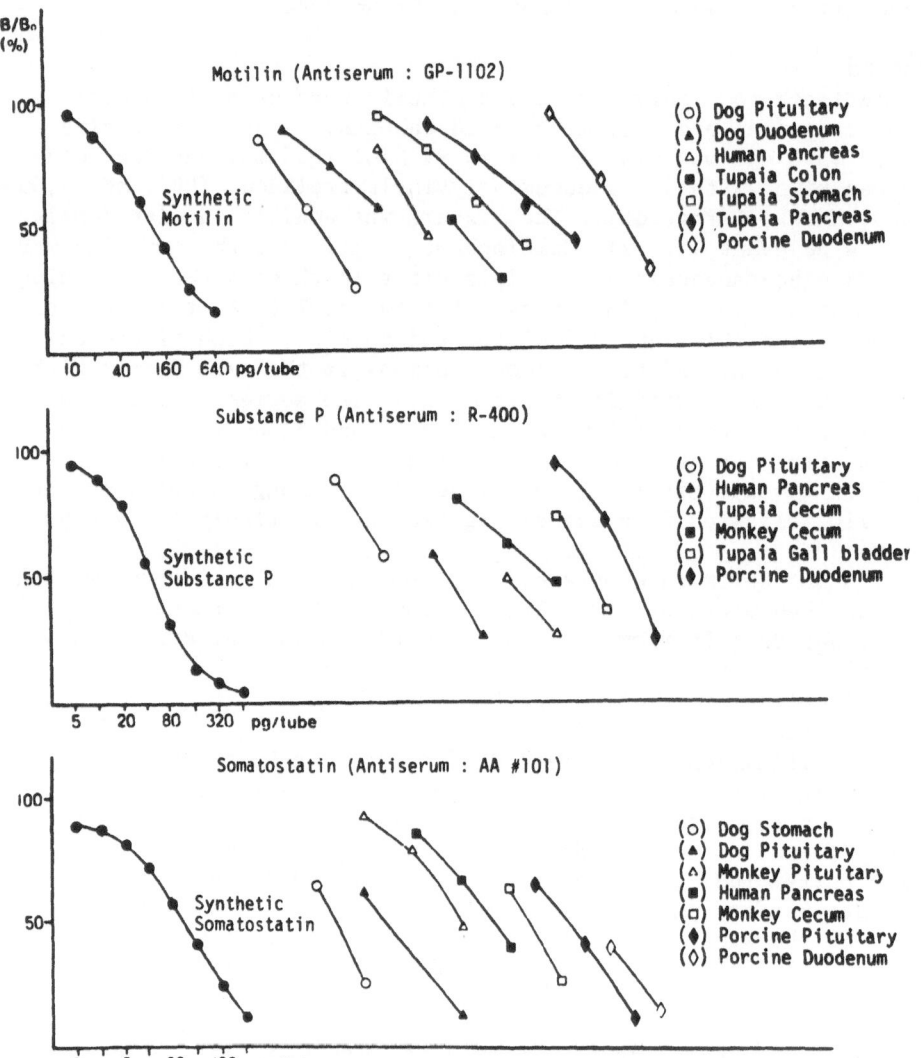

FIGURE 1. Dose-response curves of motilin-, substance P- and somatostatin-like immunoreactivities in various tissue extracts.

motilin, motilin[7-22] and motilin[12-22], exhibited approximately 1%
crossreactivities as compared with synthetic motilin, while the
N-terminal fragment motilin[1-6] did not crossreact in this system.

Radioimmunoassay for substance P

The assay was performed in the system as described previously
using [125] I-tyrosyl-substance P as tracer and highly purified synthetic substance P as standard[29]. The minimum detectable dose was
5 pg per tube. In this assay, there were no crossreactivities with
synthetic ovine somatostatin, synthetic porcine motilin, VIP or
secretin, synthetic human gastrin, natural porcine glucagon or
insulin. The antiserum R-400 had been well characterized immunologically[32].

Radioimmunoassay for somatostatin

The assay was carried out essentially in a manner similar to
that described by Arimura et al.[33]. The labelled antigen was prepared with tyrosyl-somatostatin by the enzymic method with lacto-
peroxidase and [125] I-Na and was purified by ion exchange chromatography on CM-Sephadex C-25 with a linear ammonium acetate gradient
(0.01M-0.5M). The highly purified synthetic somatostatin was used
as standard. The minimum detectable dose of somatostatin in this
system was 2.5-5 pg per tube. No crossreactivities with synthetic
bovine substance P, synthetic porcine motilin, VIP, secretin,
synthetic human gastrin, synthetic LH-RH, natural porcine glucagon, or insulin were observed in the assay. Synthetic somatostatin
fragments H-Cys-Lys-Asn-Phe-Phe-Trp-Lys-OH and H-Lys-Asn-Phe-Phe-
Trp-Lys-Thr-Phe-Thr-Ser-Ser-OH did not crossreact with the antiserum at concentrations up to 50 ng per tube.

RESULTS

Specificity of the assays for some of the extracts were examined in the radioimmunoassay systems specific to the respective
hormones and the results are shown in Figure 1. The dose-response
curves of the extracts examined in 4-fold serial dilution were
approximately parallel to that of each of the hormones in the respective assay systems.

*Motilin-, substance P and somatostatin-like immunoreactivities in
GI tract*

Table 1 shows concentrations of immunoassayable motilin, substance P and somatostatin in extracts from the stomach, duodenum,
jejunum, cecum and colon of dog. The concentration of each of the
immunoreactive hormones is expressed as pg weight equivalent of the
respective hormone per mg wet weight of tissue. Immunoreactivities
of the three hormones were detected throughout the intestine. The
highest concentrations of both substance P and somatostatin were
found in the antrum of the stomach, while a higher concentration of

TABLE 1. Motilin-, substance P- and somatostatin-like immunoreactivities in dog tissue extracts.

(pg/mg wet weight of tissue, mean $\pm$ S.E.)(n=3)

Tissue	Motilin	Substance P	Somatostatin
Stomach 1	1.9 $\pm$ 0.6	<0.1 $\pm$ 0.0	6.4 $\pm$ 2.3
2	2.0 $\pm$ 0.9	1.4 $\pm$ 1.0	7.7 $\pm$ 3.8
3	1.4 $\pm$ 0.8	2.7 $\pm$ 2.0	39.8 $\pm$ 17.9
Duodenum	3.9 $\pm$ 0.8	0.7 $\pm$ 0.2	8.1 $\pm$ 0.8
Jejunum 1	2.3 $\pm$ 0.2	0.2 $\pm$ 0.0	7.6 $\pm$ 0.9
2	2.7 $\pm$ 0.2	0.8 $\pm$ 0.5	10.3 $\pm$ 2.6
3	1.2 $\pm$ 0.3	0.9 $\pm$ 0.5	20.8 $\pm$ 6.6
Ileum	0.7 $\pm$ 0.1	0.9 $\pm$ 0.4	16.8 $\pm$ 4.6
Cecum	0.8 $\pm$ 0.6	0.2 $\pm$ 0.1	3.0 $\pm$ 1.6
Colon	0.5 $\pm$ 0.1	0.6 $\pm$ 0.3	8.5 $\pm$ 4.0

Each of the stomach and jejunum was divided equally into 3 portions and they were numbered consecutively from the upper portion.
Immunoreactivities are expressed by weight equivalents of the respective hormones.

TABLE 2. Motilin-, substance P- and somatostatin-like immunoreactivities in monkey tissue extracts.

(pg/mg wet weight of tissue, mean $\pm$ S.E.)(n=4)

Tissue	Motilin	Substance P	Somatostatin
Duodenum	8.4 $\pm$ 2.6	3.4 $\pm$ 2.7	38.2 $\pm$ 17.1
Jejunum 1	5.1 $\pm$ 1.6	1.7 $\pm$ 0.8	28.7 $\pm$ 9.0
2	3.3 $\pm$ 0.4	1.0 $\pm$ 0.3	17.3 $\pm$ 3.8
3	3.0 $\pm$ 0.4	2.8 $\pm$ 0.4	16.0 $\pm$ 2.6
Ileum	4.5 $\pm$ 1.5	1.7 $\pm$ 0.5	30.9 $\pm$ 11.1
Cecum	1.5 $\pm$ 0.5	3.1 $\pm$ 0.9	5.1 $\pm$ 1.4
Colon	1.3 $\pm$ 0.3	1.5 $\pm$ 0.9	2.6 $\pm$ 0.5
Rectum	1.3 $\pm$ 0.2	1.4 $\pm$ 0.1	14.9 $\pm$ 12.2

The jejunum was equally divided into 3 portions and they were numbered consecutively from the upper portion.
Immunoreactivities are expressed by weight equivalents of the respective hormones.

motilin was detected in the duodenum and the upper portion of the jejunum and a reduced concentration was observed toward the ileum and colon. The distribution of motilin immunoreactivity was in good agreement with immunohistochemical observations[4,5]. A relatively high concentration of somatostatin was also found in the lower part of the jejunum and colon. Although somatostatin has been reported to be non-detectable in the cardiac portion of rat stomach, the present result with the dog cardiac portion revealed the presence of a considerable amount of the hormone in the tissue. In general the concentration of substance P was lower than that of motilin and somatostatin in the dog intestine.

Table 2 shows the concentrations of motilin-, substance P- and somatostatin-like immunoreactivities in various parts of the monkey intestine. The highest concentrations of these three hormones were found in the duodenum. Similar distribution patterns of these hormones were observed in dog and monkey intestine. In monkey, as observed in dog, the lower part of the jejunum also contained relatively high concentrations of immunoreactive somatostatin.

Table 3 summarizes the concentrations of immunoreactivities of the three hormones in various parts of GI tract and some other organs in tupaia. A large amount of immunoreactive motilin was detected in the stomach body, the middle of the duodenum and the upper part of the jejunum, as observed in dog, and the colon was also shown to contain a relatively high concentration of immunoreactivity. Immunoreactive motilin was also found in the lower part of the jejunum, the ileum, pancreas and kidney, but the concentrations were much lower. Of the other organs examined, the gall bladder and adrenals contained rather high levels of the immunoreactivity. The concentrations of immunoreactive substance P in all parts of the stomach and duodenum were much higher than in the jejunum and ileum. Both the colon and cecum also contained a relatively high level of immunoreactivity. Among the other organs, immunoreactive substance P was rich in the adrenals and salivary gland. Interestingly, the concentrations of immunoreactive substance P in tupaia tissues were higher than those found in the dog and monkey. In the tupaia tissues, as found in the dog, the highest concentration of immunoreactive somatostatin was detected in the antrum of the stomach. The upper part of both stomach and duodenum also contained high concentrations of immunoreactivity. It was detected in the lower part of the duodenum, the whole jejunum, ileum, colon and other places, but the concentrations were much lower than in the upper portion of the stomach and the duodenum. In the pancreas, the amount of somatostatin immunoreactivity, which was relatively high, was of the same order as that of substance P.

TABLE 3. Motilin-, substance P- and somatostatin-like immunoreactivities in tupaia tissue extracts.

(pg/mg wet weight of tissue, mean ± S.E.)(n=3)

Tissue	Motilin	Substance P	Somatostatin
Stomach 1	5.8 ± 3.8	63.8 ± 26.8	6.2 ± 4.0
2	18.4 ± 5.0	89.1 ± 51.7	14.4 ± 9.0
3	20.9 ± 17.1	27.2 ± 1.8	5.5 ± 1.6
4	6.6 ± 2.5	36.0 ± 25.3	18.1 ± 8.2
Duodenum 1	5.6 ± 1.2	35.0 ± 16.5	15.3 ± 7.6
2	13.8 ± 6.1	38.8 ± 26.8	5.0 ± 2.1
3	9.3 ± 2.8	30.4 ± 22.0	7.6 ± 6.0
4	8.8 ± 2.5	70.1 ± 50.3	2.6 ± 0.9
Jejunum 1	17.3 ± 14.0	8.1 ± 6.9	2.0 ± 0.5
2	19.3 ± 18.2	5.3 ± 1.9	2.2 ± 0.8
3	1.3 ± 0.3	5.5 ± 1.4	0.9 ± 0.3
Ileum	5.2 ± 1.7	11.0 ± 2.9	2.7 ± 0.7
Cecum	18.8 ± 12.5	102.2 ± 75.2	1.6 ± 0.1
Colon	18.8 ± 12.9	30.9 ± 13.5	6.0 ± 5.2
Pancreas	3.1 ± 1.0	11.1 ± 7.6	11.1 ± 6.0
Adrenal	16.8 ± 0.5	609.6 ± 554	3.5 ± 0.5
Kidney	1.5 ± 0.7	3.1 ± 1.5	<0.1 ± 0.0
Salivary gland	7.2 ± 4.0	25.1 ± 13.8	<0.3 ± 0.0
Gall bladder	49.0 ± 7.6	———	<1.9 ± 1.2

Each of the stomach, duodenum and jejunum was equally divided into 3 or 4 portions, which were numbered consecutively from the upper portions.
Immunoreactivities are expressed by weight equivalents of the respective hormones.

Motilin-, substance P- and somatostatin-like immunoreactivities in dog pituitary, pineal and some brain tissues

Table 4 shows the concentrations of motilin-, substance P- and somatostatin-like immunoreactivities in the extracts from dog pituitary, pineal and some brain tissues. Extremely high level of immunoreactive motilin was found in the pineal, which was also shown to contain immunoreactive somatostatin. The concentration of immunoreactive substance P in the pineal could not be estimated, since the concentration was below the detectable range of the assay system used. The concentration of motilin-like immunoreactivity in the anterior pituitary was higher than that in the posterior lobe, while both immunoreactive substance P and somatostatin were rich in the posterior pituitary. It was remarkable that immunoreactive somatostatin was found predominantly in the posterior pituitary among the tissues examined. All other dog brain tissues tested in the present study contained both immunoreactive substance P and somatostatin in higher concentrations than those observed in the dog GI tract. Motilin-like immunoreactivity was also detected in the brain tissues examined but the concentration was much lower than in the pituitary or pineal. Only the hypothalamus contained some motilin-like immunoreactivity.

Chromatographic behavior of immunoreactive motilin in dog duodenum

Figure 2 shows the chromatographic pattern of the dog duodenum extract on a Sephadex G-50 (medium) column using 1M acetic acid as eluent. Each fraction was examined in terms of optical density at 280 nm and motilin-like immunoreactivity. The duodenum extract gave a main immunoreactive peak which eluted in the same position where synthetic motilin did.

DISCUSSION

A radioimmunoassay specific to porcine motilin was developed using highly purified synthetic porcine motilin. The results obtained in the present study revealed that a large amount of immunoreactive motilin is present outside the gastrointestinal tract. However, the central nervous system was found to contain little immunoreactive motilin. The tissue extract samples used for assays were prepared by gel-filtration of the boiling-water tissue extracts on Sephadex G-25. It should be noted that the gel-filtered samples gave better dose-response curves than extracts which were not gel-filtered, especially in the case of the motilin radioimmunoassay. In order to compare contents of the three hormones in the various tissue extracts in parallel, the same batch of preparations of the extracts was used for the respective hormones.

TABLE 4. Motilin-, substance P- and somatostatin-like immunoreactivities in extracts from dog pituitary, pineal and some brain tissues.

(pg/mg wet weight of tissue)[a]

Tissue	n	Motilin	Substance P	Somatostatin
Anterior pituitary	4	139.9 ± 37.4 [c]	57.3 ± 16.0 [c]	< 20.2 ± 8.4 [c]
Posterior pituitary	4	70.8 ± 18.6	184.4 ± 44.6	581.0 ± 140.4
Pineal body	4[b]	1058.1	< 13.5	21.5
Hypothalamus	4	7.2 ± 2.2	83.3 ± 10.7	53.1 ± 4.7
Corpus striatum	4	0.9 ± 0.3	22.4 ± 12.5	12.1 ± 4.2
Cerebral cortex	4	1.0 ± 0.1	11.8 ± 9.2	27.7 ± 1.6
Medulla oblongata	2	2.2 ± 0.2	35.5 ± 4.7	17.1 ± 2.8

a) Immunoreactivities are expressed by weight equivalents of the respective hormones.
b) Four specimens were combined and extracted together.
c) Mean ± S.E.

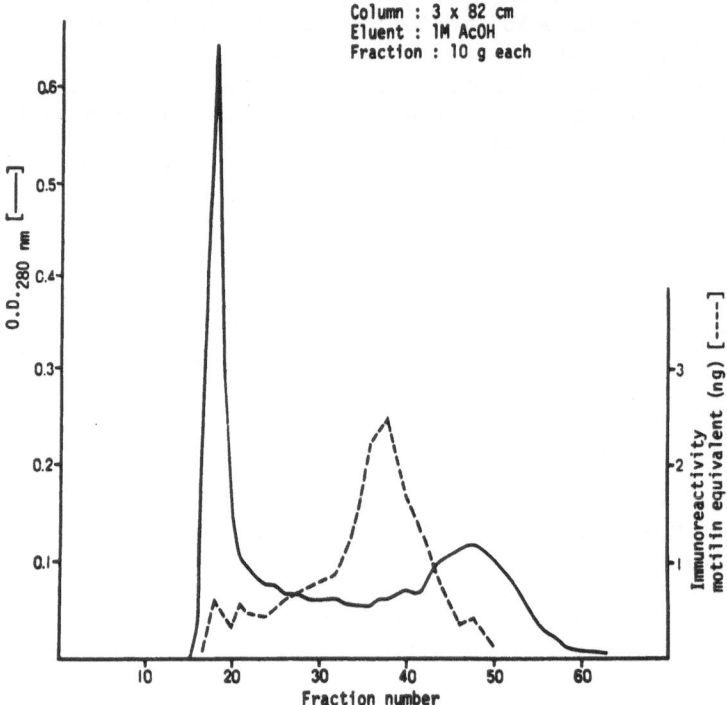

FIGURE 2. *Fractionation of dog duodenum extract by Sephadex G-50.*

The concentrations of immunoreactive motilin and substance P
were found to be much smaller in dog and monkey than in tupaia.
Immunoreactive motilin and substance P in dog duodenum extracts
were only 3.9 $\pm$ 0.8 pg/mg (wet weight of organ) and 0.7 $\pm$ 0.2 pg/mg,
respectively, but 13.8 $\pm$ 6.1 pg/mg and 38.8 $\pm$ 26.8 pg/mg, respectively, in tupaia duodenum. Bloom et al[3] reported a high concentration
of motilin, 65.7 $\pm$ 15.9 pmol/g, in human duodenum extracts. This
value was much higher than the values obtained in the present study
with dog, monkey and tupaia tissues. However, our preliminary results also showed that large quantities of immunoreactive motilin
were found in the lower part of human duodenum, 220 ng/g wet weight,
and in porcine duodenum, 109 ng/g wet weight. These values are
comparable to that in the human duodenum reported by Bloom et al.
As shown in Figure 1, the dose-response curves of the extracts from
various parts of dog, tupaia, human and porcine extracts were approximately parallel to that of synthetic porcine motilin, suggesting
that the motilin in dog, tupaia or human tissues is indistinguishable immunologically from porcine motilin, although the amino acid
sequences of the motilins have not been determined. The reason
for the low quantities of both motilin and substance P in dog and
monkey tissue extracts remains to be clarified. Similarly, substance P and somatostatin in these animals were judged to be immunologically indistinguishable from bovine substance P and ovine somatostatin, respectively, by the fact that the extracts from various
organs examined showed displacement curves nearly parallel to those
of authentic synthetic peptides. In addition, it is remarkable that
tupaia contained high levels of immunoreactive substance P in various
parts of the gastrointestinal tract as compared with those in dog
and monkey.

Somatostatin immunoreactivity was found throughout the GI tracts
in these three species of animals and not much difference was observed in somatostatin content in various tissue extracts among the
animals examined. The kidney, salivary gland and gall bladder in
tupaia did not contain any detectable amount of somatostatin, though
a considerable concentration of immunoreactive substance P was found
in the salivary gland. Substance P has been well-known to stimulate
salivary secretion[36]. Immunoreactive somatostatin in the kidney
and salivary gland has been reported to be undetectable[13]. Nearly
equal levels of substance P and somatostatin immunoreactivities
were found in the tupaia pancreas. The inhibitory effects of somatostatin on glucagon and insulin secretion[13] have been well-known and
substance P has recently been suggested to play a role in glucose
homeostasis[37].

Among the three species of animals examined, the distribution
patterns of motilin, substance P and somatostatin immunoreactivities
were generally similar to each other in their GI tracts. However,
the cecum and colon in one of the three tupaias contained a high
concentration of immunoreactive motilin. On the other hand, the
results with the extracts from dog anterior and posterior pituitaries,

pineal and some brain tissues revealed that motilin, substance P
and somatostatin immunoreactivities are distributed in a manner
characteristic of each of the hormones, suggesting differences in
the cellular origins of these three hormones. Specifically, we have
found large amount of immunoreactive motilin in the anterior pituitary and pineal, while immunoreactive substance P and somatostatin
were mainly located in the posterior pituitary and the nervous
system. Powell et al[38] reported the presence of substance P in
the pineal gland.

Although the gel-filtration of the dog duodenum extract on
Sephadex G-50 gave a main immunoreactive peak which eluted in the
same position where synthetic motilin did,   preliminary results
with a dog hypothalamic extract shows the presence of a big form
of motilin immunoreactivity predominantly.

REFERENCES

1. Brown JC, Mutt V, Dryburgh JR: The further purification of
       motilin, a gastric motor activity stimulating polypeptide
       from the mucosa of the small intestine of hogs. Can J Physiol
       Pharmacol 49: 399-405, 1971

2. Dryburgh JR, Brown JC: Radioimmunoassay of motilin. Gastroentero-
       logy 68: 1169-1176, 1975

3. Bloom SR, Mitznegg P, Bryant MG: Measurement of human plasma
       motilin. Scand J Gastroent 11, suppl 39: 47-52, 1976

4. Pearse AGE, Polak JM, Bloom SR, et al: Enterochromaffin cells
       of the mammalian small intestine as the source of motilin.
       Virchows Arch B Cell Path 16: 111-120, 1974

5. Forssmann WG, Yanaihara N, Helmsteadter V, Grube D: Differential
       demonstration of the motilin-cell and the enterochromaffin-
       cell. Scand J Gastroent 11, suppl 39: 43-45, 1976

6. Von Euler US, Gaddum JH: An unidentified depressor substance in
       certain tissue extracts. J Physiol (Lond) 72: 74-87, 1931

7. Chang MM, Leeman SE: Amino acid sequence of substance P. Nature
       (New Biol) 232: 86-87, 1971

8. Otsuka M, Konishi S: Substance P and excitatory transmitter of
       primary sensory neurons. Cold Spring Harbor Symposia on
       Quantitative Biology XL: 1976, p 135-143

9.   Nilsson G, Larsson LI, Hakanson R, et al: Localization of
     substance P-like immunoreactivity in mouse gut. Histo-
     chemistry 43: 97-99, 1975

10. Pearse AGE, Polak JM: Immunocytochemical localization of sub-
     stance P in mammalian intestine. Histochemistry 41:373-
     375, 1975

11. Polak JM, Philipp H, Pearse AGE: Differential localization of
     substance P and motilin. Scand J Gastroent 11, suppl 39:
     39-42, 1976

12. Brazeau P, Vale W, Burgus R, et al: Hypothalamic polypeptide
     that inhibits the secretion of immunoreactive pituitary
     growth hormone. Science 179: 77-79, 1973

13. Vale W, Brazeau P, Rivier C, et al: Somatostatin. Recent Progr
     Horm Res 31: 365-397, 1975

14. Hokfelt T, Efendic S, Hellerstrom C, et al: Cellular localiza-
     tion of somatostatin in endocrine-like cells and neurons
     of the rat with special references to the A_1-cells of the
     pancreatic islets and to the hypothamus. Acta Endocrinologia
     80, suppl 200: 541, 1975

15. Luft R, Efendic S, Hokfelt T, et al: Immunohistochemical evi-
     dence for the localization of somatostatin-like immunore-
     activity in a cell population of the pancreatic islets.
     Med Bio 52: 428-430, 1975

16. Dubois MP: Immunoreactive somatostatin is present in discrete
     cells of the endocrine pancreas. Proc Natl Acad Sci USA
     72: 1340-1343, 1975

17. Polak JM, Pearse AGE, Grimelius L: Growth-hormone release-
     inhibiting hormone in gastrointestinal and pancreatic D-
     cells. Lancet i: 1220-1222, 1975

18. Orci L, Baetens D, Dubois MP, Rufener C: Evidence for the D-
     cell of the pancreas secreting somatostatin. Horm Metb
     Res 7: 400-402, 1975

19. Arimura A, Sato H, Dupont A, et al: Somatostatin: abundance of
     immunoreactive hormone in rat stomach and pancreas. Science
     189: 1007-1009, 1975

20. Brownstein M, Arimura A, Sato H, et al: The regional distribu-
     tion of somatostatin in the rat brain. Endocrinol 96:
     1456-1461, 1975

21. Brayant MG, Bloom SR, Polak JM, et al: Possible dual role for
     vasoactive intestinal peptide as gastrointestinal hormone
     and neurotransmitter substance. Lancet: 991-993, 1976

22. Larsson LI, Fahrenkrug J, Schaffalitzky De Muckadell O, et al:
     Localization of vasoactive intestinal polypeptide (VIP)
     to central and peripheral neurons. Proc Natl Acad Sci USA
     73: 3197-3200, 1976

23. Said SI, Rosenberg RN: Vasoactive intestinal polypeptide:
     abundant immunoreactivity in neural cell lines and normal
     nervous tissue. Science 192: 907-908, 1976

24. Vanderhaeghen JJ, Signeau JC, Gepts W: New peptide in verte-
     brate CNS reacting with antibodies. Nature 257: 604-605,
     1975

25. Dockray GJ: Immunochemical evidence of cholecystokinin-like
     peptides in brain. Nature 264: 568-570, 1976

26. Shimizu F, Imagawa K, Mihara S, et al: Synthesis of porcine
     motilin by fragment condensation using three protected
     peptide fragments. Bull Chem Soc Japan 49: 3594-3596, 1976

27. Yanaihara N, Yanaihara C, Hirohashi M, et al: Substance P
     analogs: Synthesis, and biological and immunological
     properties. In Substance P. Edited by US von Euler and B
     Pernow. New York, Raven Press, 1977, p 27-33

28. Yanaihara N: in preparation

29. Yanaihara C, Sato H, Hirohashi M, et al: Substance P radio-
     immunoassay using N^{α}-tyrosyl-substance P and demonstration
     of the presence of substance P-like immunoreactivities in
     human blood and porcine tissue extracts. Endocrinol Japon
     23: 457-463, 1976

30. Morgan CR, Lazarow A: Immunoassay of insulin: two antibody
     system. Plasma insulin levels of normal, subdiabetic and
     diabetic rats. Diabetes 12: 115-126, 1963

31. Hunter WM, Greenwood FC: Preparation of iodine-I^{131} labelled
     human growth hormone of high specific activity. Nature
     194: 495-496, 1962

32. Yanaihara N, Sakagami M, Sato H, et al: Immunological aspects
     of secretin, substance P, and VIP. Gastroenterology 72:
     803-810, 1977

33. Yanaihara N, Kubota M, Sakagami M, et al: Synthesis of phenolic group containing analogues of porcine secretin and their immunological properties. J Med Chem 20: 648-655, 1977

34. Yanaihara N: in preparation

35. Arimura A, Sato H, Coy DH, Schally AV: Radioimmunoassay for GH-release inhibiting hormone. Proc Soc Exp Biol Med 148: 784-789. 1975

36. Leeman SE, Mroz EA: Substance P. Life Sciences 15: 2033-2044, 1974

37. Brown M, Vale W: Effects of neurotensin and substance P on plasma insulin, glucagon and glucose levels. Endocrinol 98: 819, 1976

38. Powell D, Skrabanek P, Cannon D: Radioimmunoassay studies in "Substance P". In Substance P. Edited by US von Euler and B Pernow. New York, Raven Press, 1977, in press

PROSTAGLANDINS AND SEROTONIN IN DIARRHEOGENIC SYNDROMES

B. M. Jaffe

Department of Surgery, Washington University School
of Medicine

St. Louis, Missouri U.S.A.

Infusion of prostaglandins E_2 and $F_2\alpha$ (PGE_2 and $PGF_2\alpha$) in man causes diarrhea[1,2]. This effect is due to alterations in both intestinal motility and absorption of water and electrolytes. By virtue of their direct effects on intestinal smooth muscle cells, PGE_2 and $PGF_2\alpha$ contract the longitudinal smooth muscle, but, in addition, $PGF_2\alpha$ also contracts the circular smooth muscle layer[3,4,5]. The increased motility is manifested by abdominal cramps[6] and shortened intestinal transit time[7]. In addition, PGE compounds have been shown to enhance intestinal secretion rather than absorption of water and electrolytes[8-11], an effect which appears to be mediated by cyclic nucleotides[12,13]. This secretory phenomenon contributes significantly to the prostaglandin-induced watery diarrhea. Because of their substantial diarrheogenic properties, we have investigated the role of PGE and PGF compounds in a number of diarrheogenic syndromes. Despite early observations that levels of bioassayable PGF compounds were elevated in a number of endocrine tumors and in the plasma of patients harboring these tumors[14,15], we have consistently found better correlation between diarrhea and circulating levels of PGE.

Normal Controls

Plasma PGE levels were measured in 61 normal controls, 20 men and 41 women, whose ages averaged 34.8 years (range 17-74 years). None of the volunteers had a history of diarrhea. Plasma immunoreactive PGE concentrations[16] averaged 346 $\pm$ 14 pg/ml (mean $\pm$ S.E.M.). Using 2 standard deviations to establish 95% confidence limits, normal PGE levels (using our radioimmunoassay system) range from 252 to 440 pg/ml.

Non-Endocrine Diarrheas

Plasma PGE levels have been measured in 33 patients with chronic watery diarrhea, including 7 with ulcerative colitis, 9 with Crohn's disease, 6 as the result of jejunoileal bypass procedures, and 12 with miscellaneous diagnoses. The age and sex distribution of the patients were similar to that of the control group, ie. 16 males and 17 females and mean age 33.8 years. As shown in Figure 1, PGE levels in this group of patients (mean 351 $\pm$ 26 pg/ml) were similar to those of the normal controls (p > 0.05). None of the specific diarrhea syndromes were associated with elevated PGE levels. Only 3 of the patients (9%) had elevated peripheral PGE levels, and these elevations were marginal.

Zollinger - Ellison Syndrome

Eleven of the 29 patients with the Zollinger-Ellison sydrome had diarrhea, a 38% incidence. This frequency is similar to those previously reported[17,18]. The pathogenesis of the diarrhea in patients with this endocrine syndrome has been clearly ascribed to the effects of gastrin which has both motility[19,20] and secretory effects[21] and to the resultant gastric hypersecretion which causes rapid intestinal transit[22,23], decreased absorption of water and electrolytes[24] and morphologic abnormalities in the jejunum[25]. This observation was substantiated in our study. Overall, serum gastrin levels averaged 6127 pg/ml (normal < 1250 pg/ml), but they were considerably higher in the patients with diarrhea (14245 pg/ml) than in those with normal bowel habits (1278 pg/ml). Thus, in our studies this syndrome functioned as a further control group for patients with endocrine-related diarrheas.

As shown in Figure 2, the patients with the Zollinger-Ellison syndrome (14 males, 15 females) had a normal mean plasma PGE_2 level, 388 $\pm$ 32 pg/ml. Patients with diarrhea and those without this symptom had similar mean levels, 407 $\pm$ 50 and 376 $\pm$ 43 pg/ml, respectively. Only 2 of the 29 patients had significantly elevated peripheral levels of PGE and we have no explanation for the hyperprostaglandinemia in these patients.

Medullary Carcinoma of the Thyroid

(MCT) - Numerous investigators have reported elevated plasma, urine, and tumor tissue concentrations of thyrocalcitonin in patients with medullary carcinomas of the thyroid[26-28]. Although Gray and his coworkers[29] have reported that calcitonin stimulates jejunal secretion of water and electrolytes, there is no evidence that this peptide has any effect on intestinal motility. Since several studies [30,31] have reported that patients with MCT have rapid intestinal transit times, it is likely that at least one other compound is involved in the pathophysiology of the diarrhea. Elevated circulating levels of serotonin have been reported in several MCT patients,[32,33], but this is not a uniform finding, as two of our patients had normal

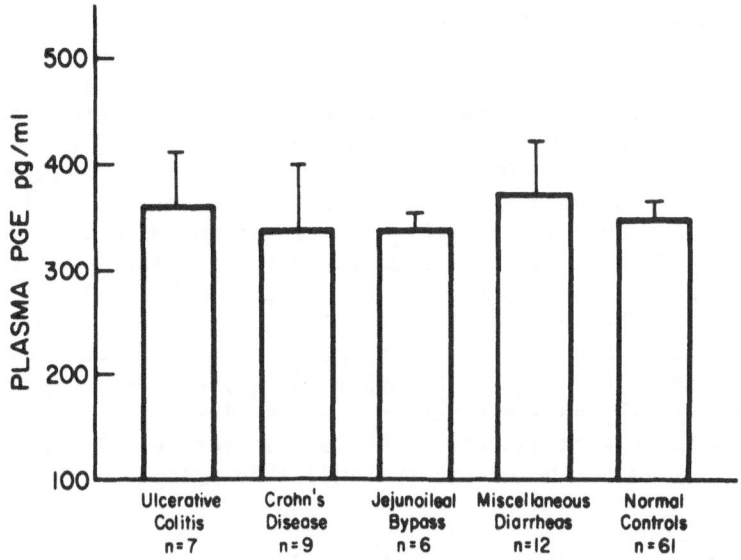

FIGURE 1. Non-endocrine diarrheas

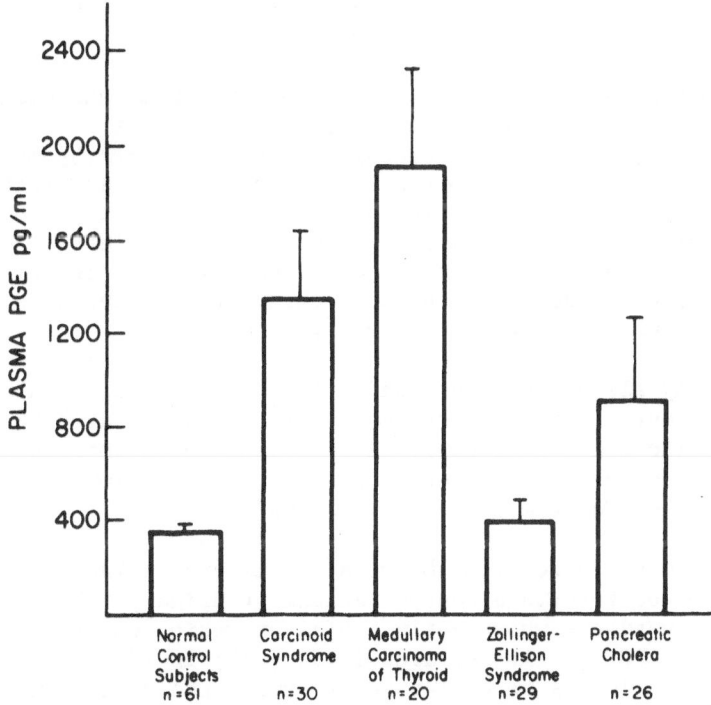

FIGURE 2. Endocrine diarrheogenic syndromes

urinary excretion of 5 HIAA.[34] Prostaglandins are also likely
candidates as mediators of the diarrhea in MCT.

Eighteen of the 20 patients with medullary carcinoma of the
thyroid we have studied (10 males, 10 females) had elevated plasma
levels of PGE. The mean plasma level, 1852 $\pm$ 420 pg/ml, was significantly elevated (p. < 0.001). Circulating thyrocalcitonin concentrations were markedly elevated in all 17 patients in whom the
determination was made (mean 24.5 $\pm$ 6.3 ng/ml; normal < 0.25 ng/ml).
Immunoreactive PGE levels correlated roughly with plasma calcitonin
concentrations. In the 9 patients whose calcitonin levels exceeded
20 ng/ml, PGE levels averaged 2537 pg/ml; in 8 patients with lower
thyrocalcitonin levels (0.6 - 19.9 ng/ml), the mean circulating PGE
concentration was 1403 pg/ml. Diarrhea was a prominant complaint
in 9 patients (45%), an incidence consistent with prior reports[31,35]
Among the patients with diarrhea, PGE levels averaged 2142 pg/ml,
whereas PGE levels in the others averaged 1596 pg/ml.

Carcinoid Syndrome
The carcinoid syndrome is another endocrinopathy which is consistently associated with both diarrhea and elevated plasma PGE
levels. Twentyeight of the 30 patients we have evaluated (12 men,
18 women) complained of diarrhea. Twenty five of the 30 were hyper-
prostaglandinemic; the mean level, 1350 $\pm$ 296 pg/ml, was significant
elevated (p < 0.001). However, among this group, four patients had
diarrhea despite normal plasma concentrations of PGE.

Diarrhea associated with the carcinoid syndrome has been at
least partially ascribed to hyperserotoninemia[36]; it is relieved
by treatment with serotonin antagonists[37] and worsened by monoamine
oxidase inhibitors[38]. Peripheral hyperserotoninemia and increased
daily urinary excretion of 5 HIAA have been consistent features of
this clinical entity[39,40]. Although the predominant effect of serotonin is to increase intestinal tone and motility[41], Kisloff and
Moore have recently demonstrated that this active amine also inhibit
absorption of water and electrolytes[42].

Virtually all our patients had a least one specific indole
parameter measured. Urinary excretion of 5 HIAA averaged 66.8 $\pm$
16.7 mg/day (normal 0-9 mg/day); peripheral blood serotonin levels
(measured either by spectrophotofluorometry or radioimmunoassay[43])
averaged 1655 $\pm$ 604 ng/ml (normal < 325 ng/ml).

There was no correlation between plasma prostaglandin levels
and either blood serotonin concentrations or urinary 5 HIAA excretion. In 14 patients whose daily 5 HIAA excretion was twice normal
(> 20 mg/day), PGE levels averaged 1143 pg/ml; in 10 patients whose
blood serotonin levels were elevated to twice the upper limit of
normal (650 ng/ml), PGE concentrations averaged 1293 pg/ml. Despite
this lack of correlation, there is still reason to consider PGE as

a mediator of carcinoid-related diarrhea. A specific case supports
this suggestion. Prior to resection of a huge ileal carcinoid tumor
involving the regional lymph nodes, Dr. S. had diarrhea and flushing;
his preoperative serotonin and PGE levels were 1170 ng/ml and 568
pg/ml, respectively. Resection of the lesion abolished the flushing
but the diarrhea did not abate; the postoperative serotonin level
was normal (72 ng/ml), but the peripheral venous PGE level was markedly elevated (1264 pg/ml).

Watery Diarrhea, Hypokalemia, Achlorhydria (WDHA) Syndrome
     Although elevated levels of a number of peptides have been
measured in the plasma of patients with the WDHA syndrome, including VIP[44,45] and more recently HPP[46,47], none of these hormones
has been implicated as the diarrheogenic hormone. Although VIP's
spectrum of actions is consistent with the symptoms of the WDHA
syndrome[48], a number of patients with this syndrome have had normal
circulating levels of this peptide. Very little is known about the
biological activity of HPP, and it is therefore difficult to accept
it as playing an etiologic role. Although elevated serotonin levels
and increased rates of urinary excretion of 5 HIAA have been reported
in several patients with the WDHA syndrome[49], it is clear that this
amine is not the mediator of the diarrheagenic syndrome, as it can
not account for the effects noted in these patients. In our experience, serotonin assays have confirmed this lack of involvement.

     Aside from inducing secretory diarrhea, infusions of prostaglan-
dins can mimic the other symptoms of the WDHA syndrome, including
hypochlorhydria[50], flushing[51], hypercalcemia[52], and hyperglycemia[53].
We have previously reported our observations on levels of PGE in 21
patients with the WDHA syndrome[54]. The data described below include
5 additional patients and at least one important new observation.

     We have studied samples from 26 patients, 12 men and 14 women.
The mean age for the population was 56.6 years. All but 6 of the
patients with histologic confirmation of the diagnosis had non-beta
islet cell tumors. Included in this study are four patients with
islet cell hyperplasia and two with adrenal tumors; we have not
assayed plasma samples from patients with pulmonary diarrheagenic
tumors. The mean immunoreactive VIP level was 6978 pg/ml (normal <
100 pg/ml); levels of VIP were elevated in 13/20 patients.

     The mean plasma PGE value was 894 $\pm$ 350 pg/ml (Figure 2) and ten
of the 26 patients had elevated circulating levels of PGE. All seven
patients with markedly elevated plasma concentrations of PGE had
pancreatic tumors. Nineteen of the patients had concurrent measurements of immunoreactive VIP and PGE; 4 patients had simultaneous
elevations of both humoral agents and another four had hyperprostaglandinemia associated with normal VIP levels.

Two patients illustrate the etiologic role of PGE in the WDHA syndrome. The first had a markedly elevated preoperative PGE level, 9939 pg/ml (and a normal plasma VIP concentration). Within a short time after resection of a non-β islet cell tumor of the pancreas, he was asymptomatic and his circulating PGE level had fallen to 464 pg/ml. Eight months later, the patient's diarrhea recurred and his PGE level once again became elevated, 1063 pg/ml. Following resection of a metastatic solitary focus in the left lobe of the liver, he once again became asymptomatic and his plasma PGE returned to normal 246 pg/ml.

The second patient, recently described in detail[55], was unique in that her symptoms disappeared after treatment with indomethacin, an inhibitor of PGE biosynthesis[56]. The patient, a 67 year old female, had a 9-year history of diarrhea. Her daily fecal output ranged from 2-5 liters day and was unresponsive to medication. Her admission K^+ was 2.5 mEq/L, her bicarbonate was 13 mEq/L, and her arterial pH was 7.27. Her plasma level of PGE averaged 977 pg/ml. She needed intravenous fluid supplementation, 2-5 liters, K^+ 140-280 mEq, and bicarbonate, 100-225 mEq per day. Within 24 hours after initiation of indomethacin therapy the patient had normal bowel habits and no longer needed intravenous fluids or electrolytes On indomethacin, her plasma levels of PGE decreased to less than one third of their pretreatment levels. At operation, a non β-islet tumor was removed. The tumor contained 15 times as much immunoreactive PGE[57] (52.5 ng/g) as did the adjacent normal pancreas (3.5 ng/g). Since resection of the lesion, the patient has had normal bowel habits and has required no medication; her postoperative plasma concentration of PGE was 310 pg/ml.

CONCLUSIONS

Although it is not the sole mediator of diarrhea in patients with endocrine diarrheas, PGE seems to play a significant role in the pathogenesis of diarrhea in patients with the carcinoid and WDHA syndromes as well as medullary carcinoma of the thyroid.

REFERENCES

1. Horton EW, Main IHM, Thompson CF, Wright PM: Effect of Orally Administered Prostaglandin E_1 on Gastric Secretion and Gastrointestinal Motility in Man. Gut, 9, 655-658, 1968

2. Karim SMM, Filshie GM: Therapeutic Abortion Using Prostaglandin $F_2\alpha$. Lancet, 1, 157-159, 1970

3. Bennett A, Fleshler B: Prostaglandins and the Gastrointestinal Tract. Gastroenterology, 59, 790-800, 1970

4. Fleshler B, Bennett A: Responses of Human Guinea-Pig and Rat
    Colonic Circular Muscle to Prostaglandins. J. Lab. Clin.
    Med. 75, 872-873, 1969

5. Bennett A, Eley KG, Scholes GB: Effects of Prostaglandins E_1
    and E_2 on Human, Guinea-Pig and Rat Isolated Small Intestine.
    Brit. J. Pharmacol. 34, 630-638, 1968

6. Fried J, Santhanakrishnan, TS, Himizu J, et al: Prostaglandin
    Antagonists: Synthesis and Smooth Muscle Activity. Nature
    (London) 223, 208-210, 1969

7. Misiewicz JJ, Waller SL, Kiley N, Horton EW: Effect of Oral
    Prostaglandin E_1 on Intestinal Transit Time in Man. Lancet
    1, 648-651, 1969

8. Matuchansky C, Bernier JJ: Effect of Prostaglandin E_1 on Glucose,
    Water and Electrolyte Absorption in the Human Jejunum. Gas-
    troent. 64, 1111-1118, 1973

9. Matuchansky C, Mary JY, Bernier JJ: Further Studies on Prosta-
    glandin E_1-Induced Jejunal Secretion of Water and Electro-
    lytes in Man with Special Reference to the Influence of
    Ethacrynic Acid, Furosemide, and Aspirin. Gastroent. 71, 274-
    281, 1976

10. Coupar IM, McColl I: Stimulation of Water and Sodium Secretion
    and Inhibition of Glucose Absorption from the Rat Jejunum
    During Intraarterial Infusions of Prostaglandins. Gut 16,
    759-765, 1975

11. Al-Awqati Q, Greenough WB,III: Prostaglandins Inhibit Intestinal
    Sodium Transport. Nature New Biol., 238, 26-27, 1972

12. Kimberg DV, Field M, Johnson J, Henderson A, Gershon E: Stimu-
    lation of Intestinal Mucosal Adenyl Cyclase by Cholera
    Enterotoxin and Prostaglandins. J. Clin. Invest. 50, 1218-
    1230, 1971

13. Kimberg DV, Field M, Gershon E, Henderson A: Effects of Prosta-
    glandins and Cholera Enterotoxin on Intestinal Mucosal Cyclic
    AMP Accumulation. J. Clin. Invest. 53, 941-949, 1974

14. Sandler M, Karim SMM, Williams ED: Prostaglandins in Amine-Peptide-
    Secreting Tumors. Lancet, 2, 1053-1055, 1968

15. Williams ED, Karim SMM, Sandler M: Prostaglandin Secretion by
    Medullary Carcinoma of the Thyroid. Lancet 1, 22-23, 1968

16.   Jaffe BM, Behrman HR, Parker CW: Radioimmunoassay Measurement
      of Prostaglandins A,E, and F in Human Plasma. J. Clin.
      Invest., 52, 398-405, 1973

17.   Isenberg JI, Walsh JH, and Grossman MI: Zollinger-Ellison
      Syndrome. Gastroenterology 65, 140-165, 1973

18.   Singleton JW, Kern F, Jr., Waddell WR: Diarrhea and Pancreatic
      Islet Cell Tumor. Gastroenterology 49, 197-208, 1965

19.   Logan CJH, Connell AM: The Effect of a Synthetic Gastrin-Like
      Pentapeptide on Intestinal Motility in Man. Lancet 1, 996-
      999, 1966

20.   Smith AN, Hogg D: Effect of Gastrin II on the Motility of the
      Gastrointestinal Tract. Lancet 1, 403-404, 1966

21.   Moshal MG, Broitman SA, Zamcheck N: Gastrin and Absorption.
      A Review. Amer. J. Clin. Nutr. 23, 336-342, 1970

22.   Christoforidis AJ, Nelson SW: Radiological Manifestations of
      Ulcerogenic Tumors of the Pancreas. J. Amer. Med. Assoc.
      198, 511-516, 1966

23.   Shafer WH, McCormack LJ, Hoerr SO: Non-Beta Islet-Cell Carcinoma
      of the Pancreas, with Flushing Attacks and Diarrhea. Report
      of a Case. Cleve. Clin. Q. 32, 13-17, 1965

24.   Adibi SA, Ruiz C, Glaser P, Fogel MR: Effect of Intraluminal
      pH on Absorption Rates of Leucine, Water, and Electrolytes
      in Human Jejunum. Gastroent. 63, 611-618, 1972

25.   Shimoda SS, Rubin CR: The Zollinger-Ellison Syndrome with
      Steatorrhea. I. Anticholinergic Treatment Followed by Total
      Gastrectomy and Colonic Interposition. Gastroent. 55, 695-
      704, 1968

26.   Tashjian AH, Jr., Melvin KEW: Medullary Carcinoma of the Thyroid
      Gland: Studies of Thyrocalcitonin in Plasma and Tumor Ex-
      tracts. N. Engl. J. Med., 279, 279-283, 1968

27.   Tashjian AH, Jr., Howland BG, Melvin KEW, Hill CS, Jr.: Immuno-
      assay of Human Calcitonin: Clinical Measurement, Relation to
      Serum Calcium and Studies in Patients with Medullary Car-
      cinoma. N. Engl. J. Med., 283, 890-895, 1970

28.   Voelkel EF, Tashjian AH, Jr.: Measurement of Thyrocalcitonin-
      like Activity in Urine of Patients with Medullary Carcinoma.
      J. Clin. Endocrinol. Metab., 32, 102-109, 1971

29. Gray TK, Bieberdorf FA, Fordtran JS: Thyrocalcitonin and the Jejunal Absorbtion of Calcium, Water and Electrolytes in Normal Subjects. J. Clin. Invest. 52, 3084-3088, 1973

30. Bernier JJ, Rambaud JC, Cattan D, Prost, A.: Diarrhea Associated with Medullary Carcinoma of the Thyroid. Gut 10, 980-985, 1969

31. Hill CS, Ibanez ML, Samaan N, et al: Medullary (Solid) Carcinoma of the Thyroid Gland: An Analysis of the M.D. Anderson Hospital Experience with Patients with the Tumor, Its Special Features and Its Histogenesis. Medicine, 52, 141-171, 1973

32. Moertel CG, Beahrs OH, Woolner LB, Tyce GM: "Malignant Carcinoid Syndrome" Associated with Non-carcinoid Tumors, N. Engl. J. Med., 273, 244-248, 1965

33. Williams ED: Diarrhoea and Thyroid Carcinoma. Proc. Roy. Soc. Med., 59: 602-603, 1966

34. Kaplan EL, Sizemore GW, Peskin GW, Jaffe BM: Humoral Similarities of Carcinoid Tumors and Medullary Carcinoma of the Thyroid. Surgery, 74, 21-29, 1973

35. Steinfeld CM, Moertel CG, Wollner LB: Diarrhea and Medullary Carcinoma of the Thyroid. Cancer 31, 1237-1239, 1973

36. Misiewicz JJ, Waller SL, Eisner M: Motor Responses of Human Gastrointestinal Tract to 5-Hydroxytryptamine in Vivo and in Vitro. Gut 7, 208-216, 1966

37. Melmon KL, Sjoerdsma A, Oates JA, et al: Treatment of Malabsorbtion and Diarrhea of the Carcinoid Syndrome with Methysergide. Gastroenterology, 48, 18-24, 1965

38. Kabakow B, Weinstein JB, Ross G: Effect of Iproniazid upon Serotonin Metabolism. Fed. Proc. 17, 382, 1958

39. Pernow B, Waldenstrom J: Determination of 5-Hydroxytryptamine, 5-Hydroxyindole Acetic Acid and Histamine in Thirty Three Cases of Carcinoid Tumors (Argentaffinoma). Am. J. Med., 23, 16-25, 1957

40. Page IH, Corcoran AC, Undenfriend S, et al: Argentaffinoma as an Endocrine Tumor. Lancet 1, 198-199, 1955

41. Haverback BJ, Davidson JD: Serotonin and the Gastrointestinal Tract. Gastroenterology 35, 570-578, 1958

42. Kisloff B, Moore EW: Effect of Serotonin on Water and Electro-
    lyte Transport in the In Vivo Rabbit Small Intestine. Gas-
    troenterology 71, 1033-1038, 1976

43. Kellum JM, Jaffe BM: Validation and Application of a Radioimmun
    assay for Serotonin. Gastroenterology, 70, 516, 1976

44. Bloom SR, Polak JM, Pearse AGE: Vasoactive Intestinal Peptide
    and Watery Diarrhea Syndrome. Lancet 2, 14-16, 1973

45. Said SI, Faloona GR: Elevated Plasma and Tissue Levels of Vaso-
    active Intestinal Polypeptide in the Watery Diarrhea Syn-
    drome due to Pancreatic, Bronchogenic and Other Tumors. N.
    Engl. J. Med. 293, 155-160, 1975

46. Polak JM, Adrian TE, Bryant MG, et al: Pancreatic Polypeptide
    in Insulinomas, Gastrinomas, VIPomas and Glucagonomas. Lan-
    cet 1, 328-330, 1976

47. Larsson LI, Schwartz T, Lundquist G, et al: Occurrence of Human
    Pancreatic Polypeptide in Pancreatic Endocrine Tumors. Amer
    J. Pathol. 85, 675-684, 1976

48. Said SI, Makhlouf GM: Vasoactive Intestinal Polypeptide: Spec-
    trum of Biological Activity, In Endocrinology of the Gut,
    WY Chey and FP Brooks, eds., New Jersey, Charles B. Slack,
    1974; pp. 83-87

49. Schmitt MG, Jr., Soergel KH, Hensley GT, Chey WY: Watery Diarrh
    Associated with Pancreatic Islet Cell Carcinoma. Gastro-
    enterology, 69, 206-216, 1975

50. Robert A, Nezamis JE, Phillips JP: Inhibition of Gastric Secre-
    tion by Prostaglandins. Amer. J. Dig. Dis. 12, 1073-1076,
    1967

51. Nakano J, McCurdy JR: Cardiovascular Effects of Prostaglandin
    E_1. J. Pharm. Exp. Therap. 156, 538-547, 1967

52. Franklin RB, Tashjian AH, JR.: Intravenous Infusion of Prosta-
    glandin E_1 Raises Plasma Calcium Concentration in the Rat.
    Endoc. 97, 240-243, 1975

53. Robertson RP, Gavareski DJ, Porte D, Jr., et al: Prostaglandin
    (PG)E_1: Inhibition of Glucose-Stimulated Insulin Secretion
    in the Intact Dog. Clin. Res. 21, 635, 1973

54. Jaffe BM, Condon S: Prostaglandins E and F in Endocrine Diarrhe
    genic Syndromes. Ann. Surg. 184, 516-524, 1976

55. Jaffe BM, Kopen DF, Keskemeti KD, Gingerich RL, Greider M:
         Indomethacin-Sensitive Pancreatic Cholera. N. Engl. J. Med.
         in press

56. Vane JR: Inhibition of Prostaglandin Synthesis as a Mechanism
         of Action for Aspirin-like Drugs. Nat. New Biol. 231, 232-
         235, 1974

57. Jaffe BM, Parker CW, Marshall GR, Needleman P: Renal Concentra-
         tions of Prostaglandin E in Acute and Chronic Renal Ischemia
         Biochem. Biophys. Res. Commun. 49, 799-805, 1972

PROSTAGLANDINS AND GASTROINTESTINAL SECRETION AND MOTILITY

S. J. Konturek

Institute of Physiology

Medical Academy, Kraków, Poland

Prostaglandins (PGs) have been shown to be widely distributed in the gastrointestinal tract and implicated in the regulation of its secretory and motor activities. Despite their high biological activity and presence in the gastrointestinal system, their physiological role and the mechanisms by which they control digestive functions remain unknown. Interpretation of studies on the physiological significance of PGs is hampered by their rapid tissue catabolism which necessitates the administration of very high unphysiological doses of the substances, usually accompanied by various side-effects. Recently, several stable methyl analogs of PGE_2 have been synthetized[1] and proved to be highly effective in their action on various digestive functions, particularly in the inhibition of gastric secretion in animals and in man, in the prevention of experimental peptic ulcer formation in animals[2,3] and in the healing of gastroduodenal ulcers in man[4,5].

The following review summarizes recent information about: 1.) the possible role of endogenous PGs in the inhibition of gastric secretion, 2.) the mechanism by which stable methyl analogs of PGE_2 inhibit gastric secretion induced by a meal, 3.) the effects of PGs on intestinal motility patterns and 4.) the use of PGs in the treatment of gastroduodenal ulcers in man.

The observations that PG-like material is present in the gastric mucosa and secreted into the gastric juice and the finding that PGs of the A and E series may inhibit gastric secretion induced by a variety of stimulants suggest that PGs may play a role as negative feedback inhibitors of gastric secretion. This notion is also supported by two additional findings: 1.) the non-steroid anti-inflammatory drugs, potent inhibitors of PG synthetase, augment gastric

acid response to exogenous stimuli[6] such as pentagastrin whereas
arachidonic acid, a precursor of endogenous PGs, given intravenously is a potent inhibitor of gastric secretion presumably due to its
conversion to PGs[7]. On the other hand, Ramwell[8] reported that
arachidonic acid applied directly to the isolated gastric mucosa may
facilitate acid response to histamine suggesting that it may have
some functions other than that of being the precursor of PGs.

We attempted to determine the influence of arachidonic acid
on gastric secretion and to correlate the changes of gastric secretion with those of immunoreactive PGE content in the gastric juice
stimulated by histamine. A flap of the fundic portion of the dog
stomach pedicled on the splenic vessels was secured between lucite
rings of a partitioned chamber as described previously[9]. Each side
of the chamber was bathed with saline and the effluent of the chamber was collected every 15 min to determine acid and pepsin outputs.
In addition the mucosal blood flow was measured using aminopyrine
clearance adapted to this preparation. Arachidonic acid was given
alone or in combination with indomethacin into the artery supply-
ing the isolated fundic flap. Gastric secretion was stimulated by
intravenous histamine to produce half maximal stimulation of gastric
secretion.

Arachidonic acid given intraarterially caused a dose-dependent
inhibition of histamine-induced gastric acid and pepsin secretion
and this effect was accompanied by a marked rise in the immunoreactive PGE output into the gastric perfusate (Fig. 1). Indomethacin prevented the inhibition of gastric secretion caused by arachidonic acid and abolished the rise in PGE output into the perfusing
fluid (Fig. 2). These results might be interpreted as showing that
arachidonic acid is converted into PGs in the gastric mucosa which
in turn inhibit directly oxyntic gland secretion and reduce gastric
mucosal blood flow. As is well known, arachidonic acid is a precursor not only of PGs but also of various other substances such as
thromboxanes and prostacyclin, which are potent vasoactive substances
and might contribute to the reduction in the gastric mucosal micro-
circulation. Indomethacin prevents all secretory changes induced by
arachidonic acid because it inhibits PG synthetase and prevents the
formation of PGs. These studies suggest that endogenous PGs are
involved in the local regulation of gastric secretion and mucosal
blood flow.

The mechanism of the action of PGs on gastric secretion is
virtually unknown. There may be a variety of routes by which the
inhibition of this secretion by PGs may occur such as: 1.) the
suppression of gastrin release, 2.) an impairment of gastric mucosal
blood flow, 3.) breaking of gastric mucosal barrier, 4.) selective
inhibition of vagal or cholinergic activity and 5.) blockade of
specific receptors in the oxyntic glands. There is, however, one

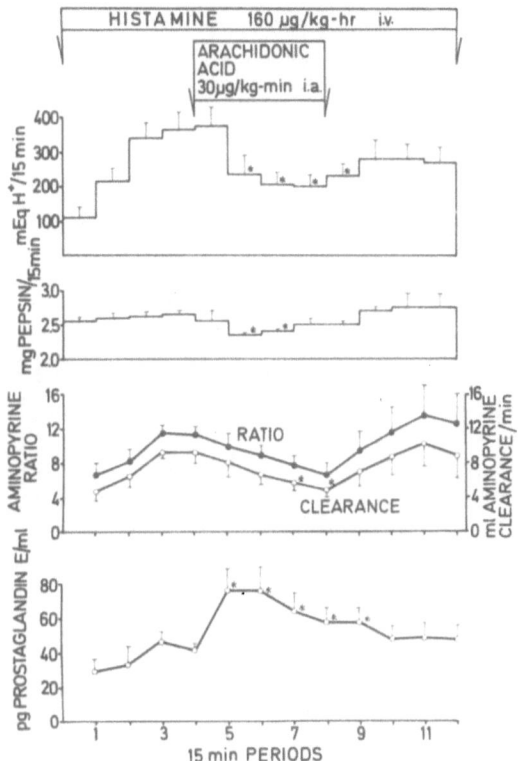

Figure 1. Effect of arachidonic acid infused intraarterially on gastric acid and pepsin responses to histamine, gastric mucosal blood flow and immunoreactive PGE content in the gastric perfusate.

important observation that some stable PGE$_2$ methyl analogs exert their secretory inhibition by local direct contact with the gastric mucosa[2], via direct inhibition of the oxyntic glands and suppression of the antral G-cells[3].

We studied this problem using the whole stomach preparation in dogs with double mucosal bridge at the pylorus to separate the stomach from the duodenum and with a large gastroduodenal cannula to bypass the pylorus between experiments and to disconnect the stomach from the gut during experiments. This preparation was used previously to study separately gastric and intestinal phases of gastric secretion[10]. A meal consisting of 10% liver extract solution introduced into the stomach evoked the gastric phase of gastric acid secretion amounting to about 80% of the maximal response to histamine and was accompanied by a marked rise in serum

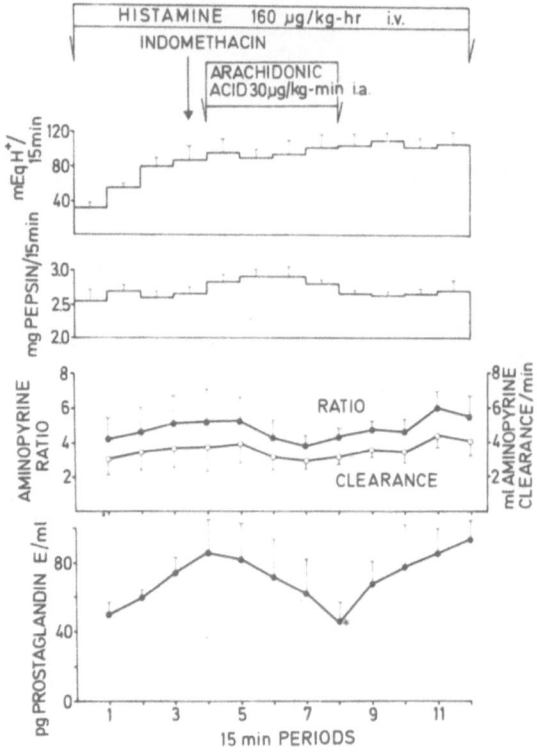

*Figure 2. Pretreatment of the fundic flap with indomethacin pre-
vents the inhibitory effect of arachidonic acid on gastric secretion,
mucosal blood flow and the rise of immunoreactive PGE in the gastric
perfusate.*

gastrin level (Fig. 3). Pretreatment of gastric mucosa with a methyl
analog of PGE_2 (PG-S) almost completely inhibited the acid response
and significantly reduced the serum gastrin response to a gastric
meal. The same dose of PG-S given intraduodenally caused relatively less suppression of gastric acid and serum gastrin responses.
These results seem to indicate that PG-S applied topically is capable of preventing gastric acid and serum gastrin responses to a
gastric meal. This may be simply explained by direct inhibition
of the oxyntic glands and antral G-cells by PG-S.

    This direct action of PG-S also can be demonstrated with re-
gard to intestinal phase stimulation of gastric secretion which can
be conveniently induced by administering a liver extract meal
directly into the gut. With an intestinal meal gastric acid rose
to about 50% of histamine maximum and was accompanied by significant

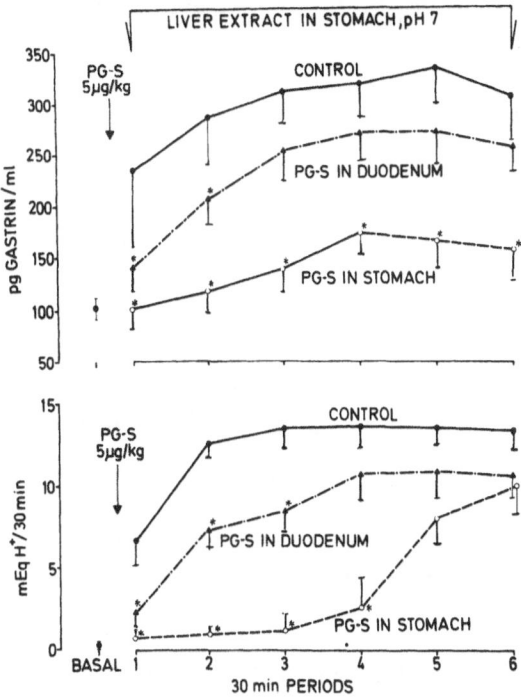

Figure 3. Effect of PG-S given either into the stomach or into the duodenum on gastric acid and serum gastrin responses to liver extract meal in the stomach.   Mean ± SEM of 8 experiments on 4 dogs.

elevation of serum gastrin level. Again, PG-S given directly into the stomach was relatively a more potent inhibitor of gastric secretion and serum gastrin release than when administered intraduodenally. This provides additional evidence that PC-S is a potent local inhibitor of gastric acid and serum gastrin response to a meal.

The finding that PGs act locally should be cautiously interpreted because of the possible disruption of the gastric mucosal barrier by these substances and the possible loss of hydrogen ion from the gastric lumen to the mucosa by acid back-diffusion. This important question has been recently raised by O'Brien and Carter[11], who reported that certain PGE_2 methyl analogs applied in a massive dose to the Heidenhain pouch mucosa may damage the mucosal barrier so that the observed inhibition is more apparent than real. We have repeated the experiment of O'Brien and Carter using PG-S in a similar dose and found that it does not alter the ionic fluxes from control level (Fig. 5) and does not affect the increased mucosal permeability induced by taurocholate (Fig. 6). We also studied this problem in healthy human subjects and found that PG-S does not damage the

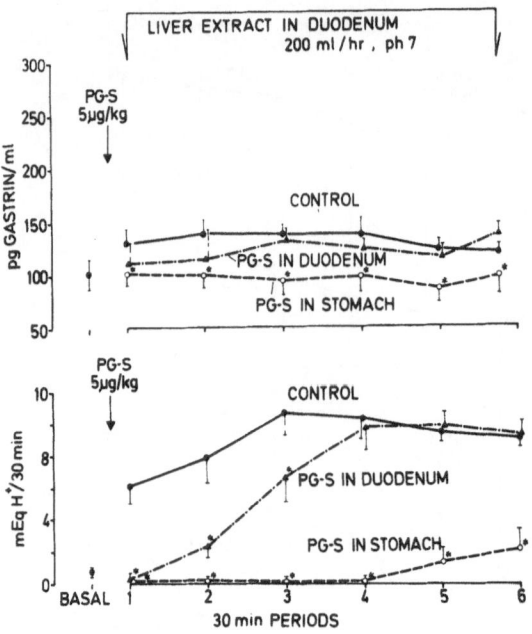

Figure 4. Effect of PG-S given either into the stomach or into the duodenum on gastric acid and serum gastrin responses to liver extract meal in the intestine. Mean ± SEM of 8 experiments on 4 dogs.

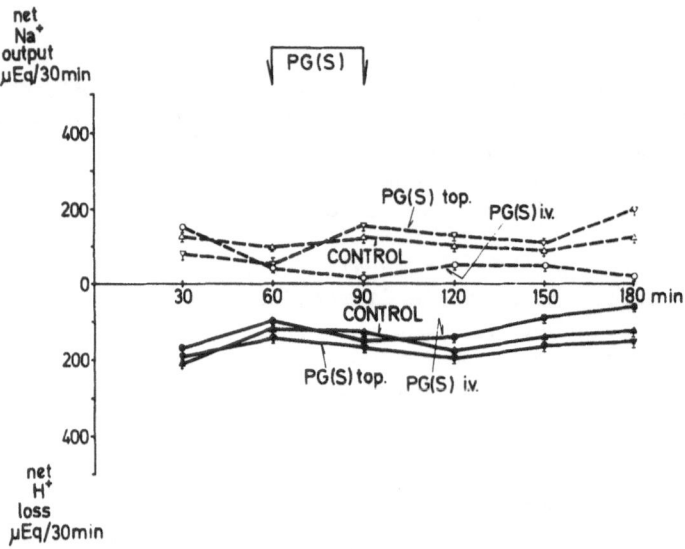

Figure 5. Net Na⁺ output and net H⁺ loss under basal conditions and after topical or intravenous administration of PG-S. Mean ± SEM of 6 experiments on three Heidenhain pouch dogs.

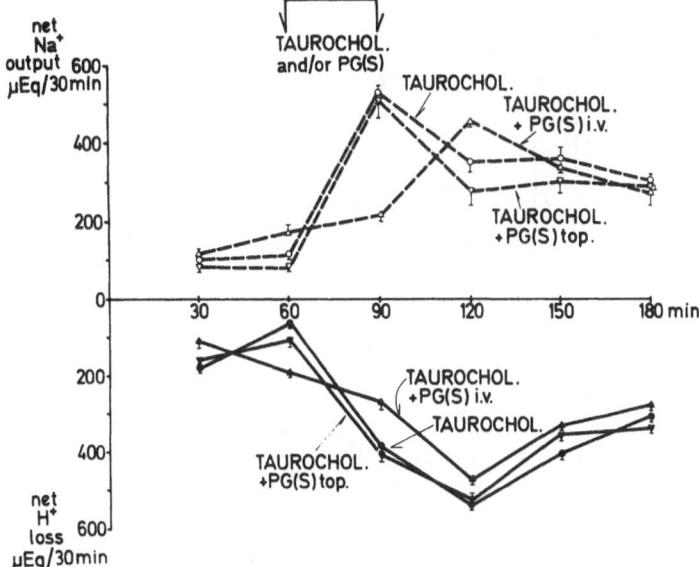

Figure 6. Changes in the net Na+ output and net H+ loss after exposure of the Heidenhain pouch mucosa to taurocholate alone or taurocholate combined with PG-S given topically or intravenously. Mean ± SEM of 6 experiments on three Heidenhain pouch dogs.

mucosal barrier after the administration of a single dose which is effective in gastric acid inhibition.

There is little doubt that in humans PGE_2 methyl analogs are the most potent available inhibitors of gastric secretion induced by a variety of stimuli, particularly by a meal. The inhibition by PGs of postprandial gastric secretion is more effective after oral than intestinal administration and it is accompanied by suppression of serum gastrin response. PGs therefore might be therapeutically useful in the treatment of peptic ulcer and this is supported by two recent double blind clinical trials, one by Fung et al[4] with regard to gastric ulcer and another by Gibinski et al[5] with regard to duodenal ulcer. In both trials, PGE_2 methyl analogs resulted in about 2-3 times higher overall healing rate of endoscopically proven peptic ulcers within about 3 weeks of treatment as compared to placebo. The antiulcer properties of various PGE_2 analogs were best correlated with their antisecretory activity so it may be concluded that these properties are mainly due to the reduction in gastric acid secretion. Immediately after withdrawal of treatment for 3 weeks, the gastric mucosa showed reduced secretion under basal conditions and in response to pentagastrin. It returned to pretreatment responsiveness to secretory stimuli within about one week.

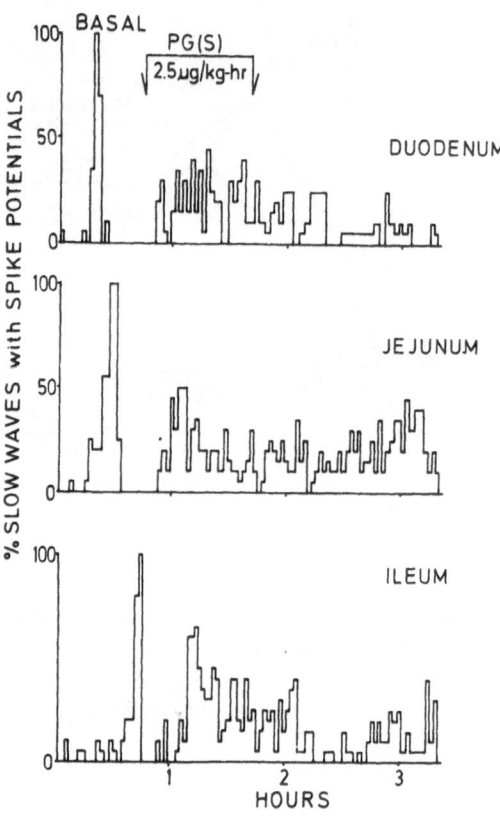

Figure 7. Temporal distribution of slow waves with spike potentials in fasted dog infused with PG-S. Tracings from three electrodes in one dog.

        The difficulties with the use of PG-S as an antiulcer drug are
its side-effects, mainly loose stools and diarrhea. With prolonged
treatment the dose producing 50% inhibition (which was free of side
effects in single doses) resulted in diarrhea, probably due to the
accumulation of the drug in the body. The mechanism of the diarrheogenic property of PGs probably is related to increased intestinal
secretion[12] or "enteropooling" and accelerated intestinal transit
time due at least in part to mechanical effect of the increased
bulk of intestinal content and perhaps to the primary change in
the motility patterns. We tested natural PGs of A, E, and F series
and synthetic methyl analogs of PGE_2 with respect to the absorption
of electrolyte solution from isolated Thiry-Vella loops and found
that natural PGs of A and E series and PGE_2 methyl analogs caused
significant reduction in the net water and sodium absorption.[13]
At higher doses methyl analogs also reversed net absorption of
water and sodium to net secretion, and this is probably the major

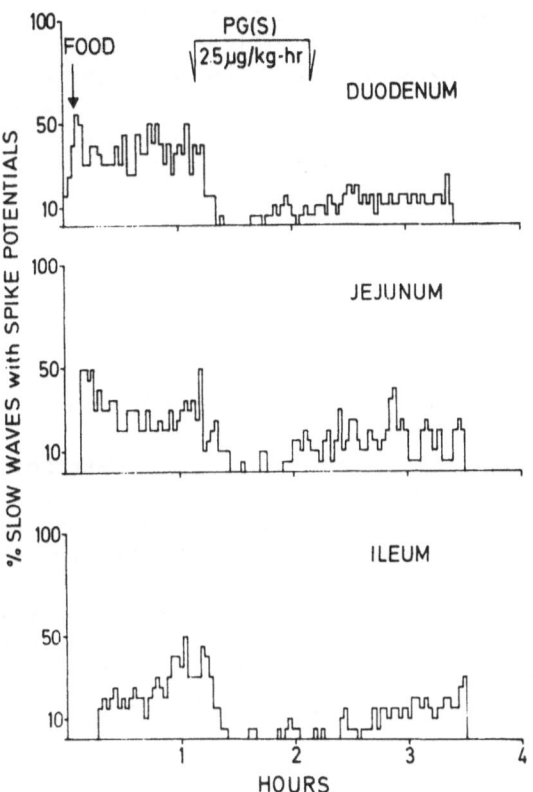

Figure 8. Temporal distribution of slow waves with spike potentials in fed dog infused with PG-S. Tracings from three electrodes in one dog.

mechanism by which PGs cause diarrhea.

We also examined the motility patterns (slow waves and spike potentials) in conscious dogs with implanted electrodes along the gut. In fasted dogs so-called interdigestive myoelectric complexes are present and propagate aborally along the gut. Feeding results in the disappearance of these complexes and in the appearance of more uniform distribution of spike potentials superimposed in a random fashion upon the slow waves. Gastrointestinal hormones, particularly gastrin, CCK and insulin have been implicated in the conversion of fasted to fed patterns of myoelectric activity.[14]

PGs of the A and F series caused little change in motility patterns, but pronounced effects were recorded following PGE$_2$ or PG-S administration. In fasted animals, they increased

the spike activity and blocked the interdigestive myoelectric complexes, indicating an increased motility (Fig. 7). In fed animals,
however, they caused reduction in spike activity and the appearance
of fasted type patterns (Fig. 8). This could be explained by the
predominant inhibitory effect of these PGs on the release of gastroduodenal hormones which are involved in the maintenance of postprandial spike activity of the small bowel.

REFERENCES

1. Yankee EW, Axen W, Bundy GL: Total synthesis of methylprosta-
    glandins. J Am Chem Soc 96: 5865-5876, 1974

2. Robert A, Schultz JR, Nezamis JE, et al: Gastric antisecretory
    and antiulcer properties of PGE_2, 15-methyl PGE_2, and 16,
    16-dimethyl PGE_2. Intravenous, oral and intrajejunal ad-
    ministration. Gastroenterology 70: 359-370, 1976

3. Konturek SJ, Kwiecień N, Swierczek J, et al: Comparison of
    methylated prostaglandin E_2 analogs given orally in the
    inhibition of gastric responses to pentagastrin and peptone
    meal in man. Gastroenterology 70: 683-687, 1976

4. Fung WP, Karim SMM, Tye CY: Effect of 15/R/15 methyl prosta-
    glandin E_2 methyl ester on healing of gastric ulcers.
    Lancet 2: 10-12, 1974

5. Gibiński K, Rybicka J, Mikoś E, et al: Gastroduodenal ulcer
    healing by oral M-prostaglandins. 10th Internat Congress
    Gastroent, Budapest, June 23-29, 1967, p 9

6. Main IH, Whittle BJ: Effect of indomethacin on rat gastric
    acid secretion and gastric mucosal blood flow. Br J Phar-
    macol 47: 666P, 1973

7. Bieck PR, Oates JA, Adkins RB: Inhibition of gastric secretion
    by arachidonic acid in the dog. (Abstract) Clin Res 19:
    387, 1974

8. Ramwell PW, Crane BH: Arachidonic acid and facilitation of
    gastric acid secretion. In: Stimulus-Secretion Coupling
    in the Gastrointestinal Tract. Edited by RM Case and H
    Goebell. Lancaster: MTP Press Ltd 1976, p 107-113

9. Mao CC, Jacobson ED, Shanbour LL: Mucosal cyclic AMP and secre-
    tion in the dog stomach. Am J Physiol 225: 893-896, 1973

10. Konturek SJ, Kaess H, Kwiecień N, et al: Characteristics of
     intestinal phase of gastric secretion. Am J Physiol 230:
     335-340, 1976

11. O Brien PE, Carter DC: Effect of gastric secretory inhibitors
     on the gastric mucosal barrier. Gut 16: 437-442, 1975

12. Konturek SJ, Zawadzka A: Prostaglandins and intestinal secre-
     tion. In: Stimulus-Secretion Coupling in the Gastrointes-
     tinal Tract. Edited by RM Case H Goebell. Lancaster: MTP
     Press Ltd 1976, p 117-120

13. Matuchansky C, Mary JY, Bernier JJ: Further studies on pros-
     taglandin E_1-induced jejunal secretion of water and
     electrolyte in man, with special preference to the in-
     fluence of ethacrynic acid, furosemide, and aspirin.
     Gastroenterology 71: 274-281, 1976

14. Weisbrodt NW, Copeland EW, Thor PJ, et al: The myoelectric
     activity of the small intestine of the dog during total
     parenteral nutrition. Proc Soc Exp Biol Med 153: 121-124,
     1976

RADIOIMMUNOASSAY OF SECRETIN

P. L. Rayford, A. Schafmayer and J. C. Thompson

Department of Surgery, The University of Texas
Medical Branch

Galveston, Texas

Several sensitive radioimmunoassay systems for secretin have been developed and reported[1-6]. The development of these assay systems depended upon the generation of a specific secretin antiserum and the iodination of secretin (natural porcine secretin, synthetic secretin or a tyrosyl derivative of secretin). This manuscript reports the results of studies conducted in our laboratory to validate a secretin radioimmunoassay.

MATERIALS AND METHODS

Radioimmunoassay Technique

Details concerning technical methods and procedures (buffer system, production of antisera, purification of labeled secretin, conduct of the radioimmunoassay and statistical analysis) used to develop the assay have been reported previously[4]. The secretin antiserum used in these studies bound more than 90% of labeled 6-tyrosyl secretin when used in excess and 30% to 35% of labeled secretin when used at a final dilution of 1:100,000, the dilution used in our studies.

RESULTS

Pure natural secretin was used as a reference standard. The amount of secretin added into assay tubes ranged from 0.01 to 1 ng of pure natural secretin. Figure 1 shows two plots for the dose-response curve. For each curve, the percent of labeled secretin bound to

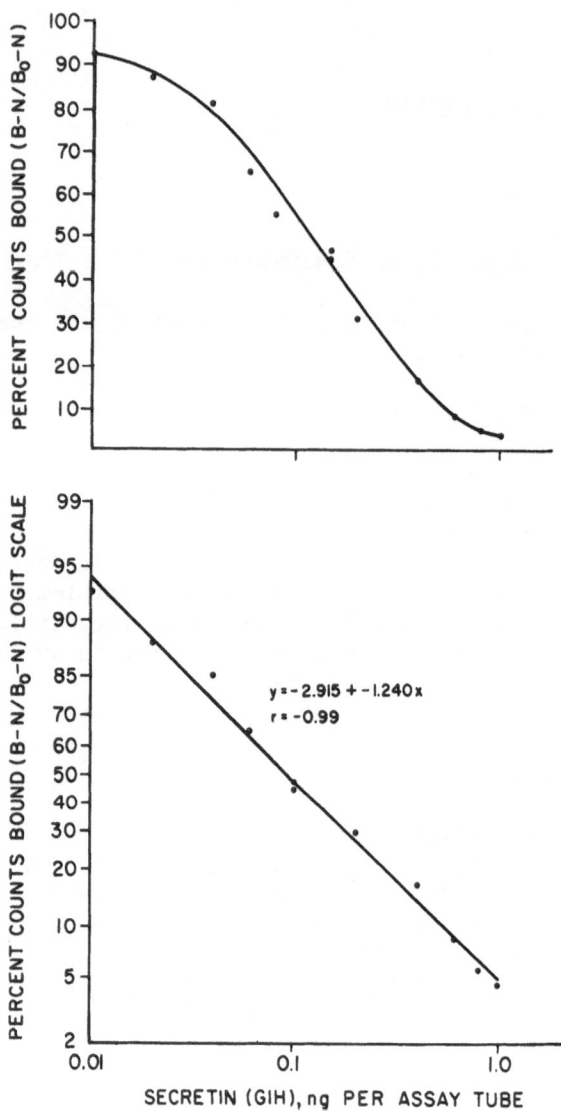

FIGURE 1. Inhibition lines generated by graded amounts of secretin in the secretin radioimmunoassay. The data are plotted as percent bound on an arithmetic (top panel) scale and as percent bound on a logit (bottom panel) scale, each against log dose (0.01 to 1.0 ng) of secretin.

antibody was calculated by the formula:

$$\frac{B-N}{B_0-N},$$

where: B = labeled secretin bound in tubes containing
              graded amounts of unlabeled hormone;
          B_0 = labeled secretin bound to antibody in tubes
              containing no unlabeled hormone;
          N = labeled secretin precipitated in tubes contain-
              ing no antibody.

In the top part of the figure, percent bound is plotted on an arith-
matic scale and the dose of secretin added is plotted on a logarith-
mic scale. When the data are plotted in this manner, a sigmoid curve
is generated. In the bottom part of the figure, percent bound is
plotted on a logit scale against logarithm of dose. The use of
logit mathematical transformation of the response variable results
in the generation of an inhibition line that does not depart significantly from linearity; thus, by regression analysis, the equation
of the line for dose interpolation and correlation coefficient (r =
−0.98) can be calculated.

Specificity of Antiserum

The response of graded amounts of several polypeptides were
measured in our secretin radioimmunoassay (Fig. 2). Glucagon, gastric inhibitory peptide (GIP), synthetic human gastrin I (SHG I),
and pure cholecystokinin (CCK) had no cross-reaction in the assay
system. There was some cross-reaction with vasoactive intestinal
peptide (VIP); however, the potency of the VIP preparation was less
than 0.05% of that of natural secretin.

Measurement of Secretin in Plasma

The ability of graded volumes of dog plasma to displace labeled
secretin from combination with antibody was tested in the assay
system (Fig. 3). The dose-response curve generated by volumes of
dog plasma ranging from 50 to 300 μl was parallel to the dose-response
curve generated for natural secretin. The results indicate that
secretin in volumes of plasma up to 300 μl can be measured accurately in the radioimmunoassay system.

In 12 separate secretin radioimmunoassays conducted over a
period of 6 months, we measured basal circulating levels of secretin in peripheral plasma of man and dogs (Fig. 4). In 96 individuals,
mean basal secretin was 99 ± 40 pg/ml (range 33 to 265 pg/ml) and
basal secretin in 108 dogs was 115 ± 65 pg/ml (range 33 to 301 pg/ml).

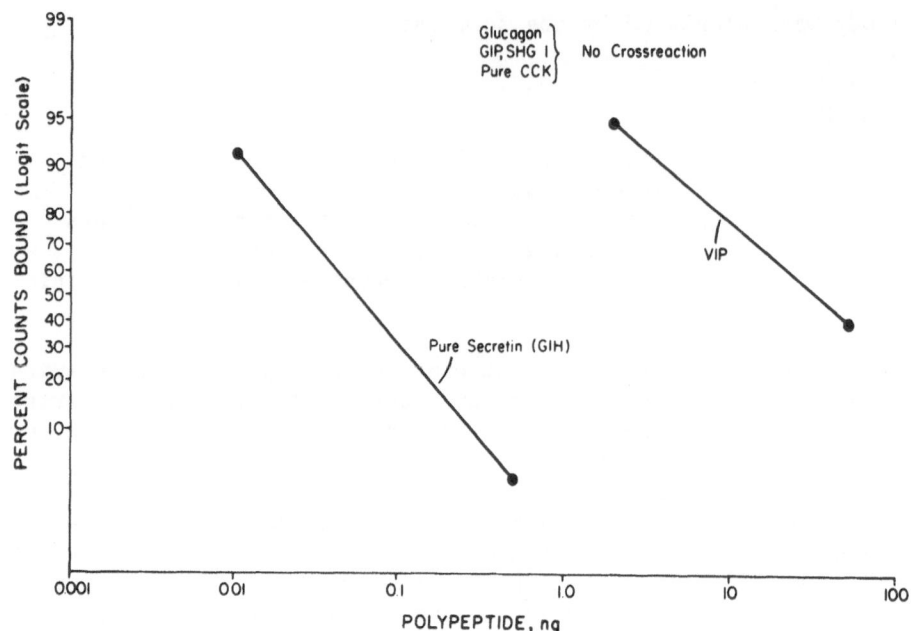

FIGURE 2.   Log-logit plots of inhibition lines generated by graded amounts of secretin and other polypeptides in the secretin radio-immunoassay.

Correlation of Biologic and Immunologic Activity of Secretin

We conducted biologic studies in which we performed simultaneous bioassay and immunoassay of secretin. Five dogs were prepared with chronic pancreatic fistulas and with venous catheters. Pancreatic juice and blood samples were taken during basal and at regular intervals during the infusion of graded amounts of secretin ranging from 0.0625 to 1 U/kg-hr.

Volume, bicarbonate and protein output were measured in pancreatic juice and secretin was measured in plasma samples (Fig. 5). In the top panels, HCO_3 and volume were both increased significantly above basal with 0.125 and higher doses of secretin. Plasma secretin levels were slightly but not significantly increased at 0.125 U secretin/kg-hr; however, doses of secretin from 0.25 to 1 U/kg-hr resulted in significant increases in circulating levels of secretin. Protein was not affected by secretin infusions.

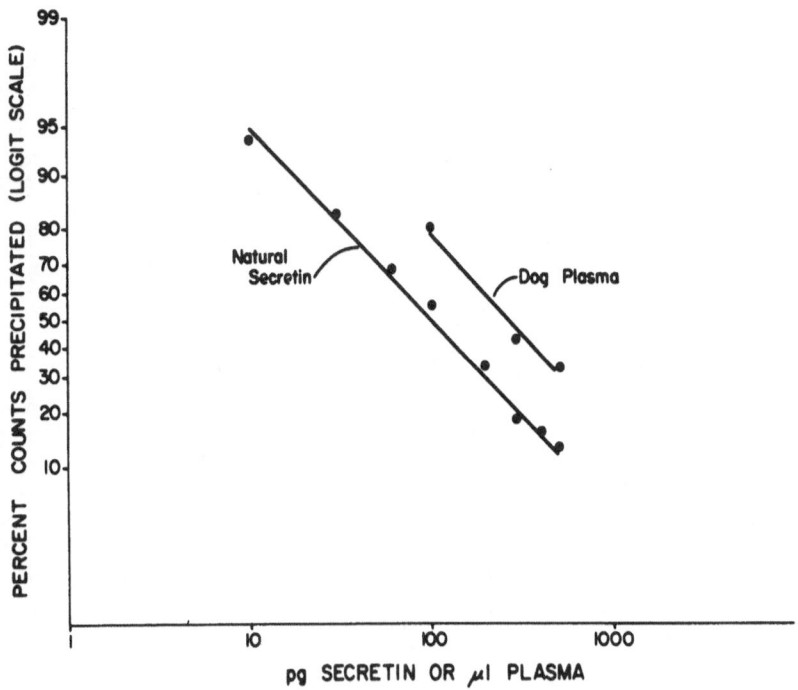

FIGURE 3. *Log-logit plots of inhibition lines generated by graded amounts of secretin or serial dilutions of dog plasma in the secretin radioimmunoassay.*

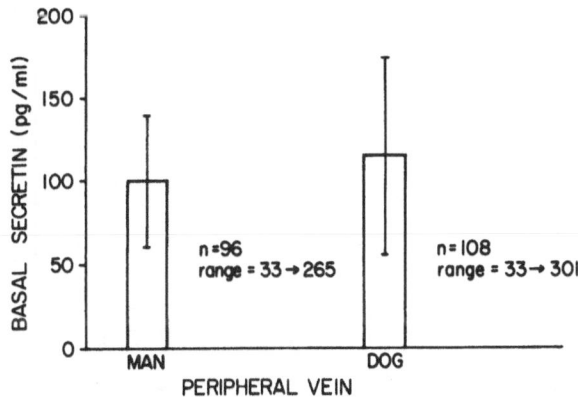

FIGURE 4. *Basal levels of secretin (pg/ml±SE) measured in the peripheral vein of man and dog with 12 different secretin radioimmunoassays.*

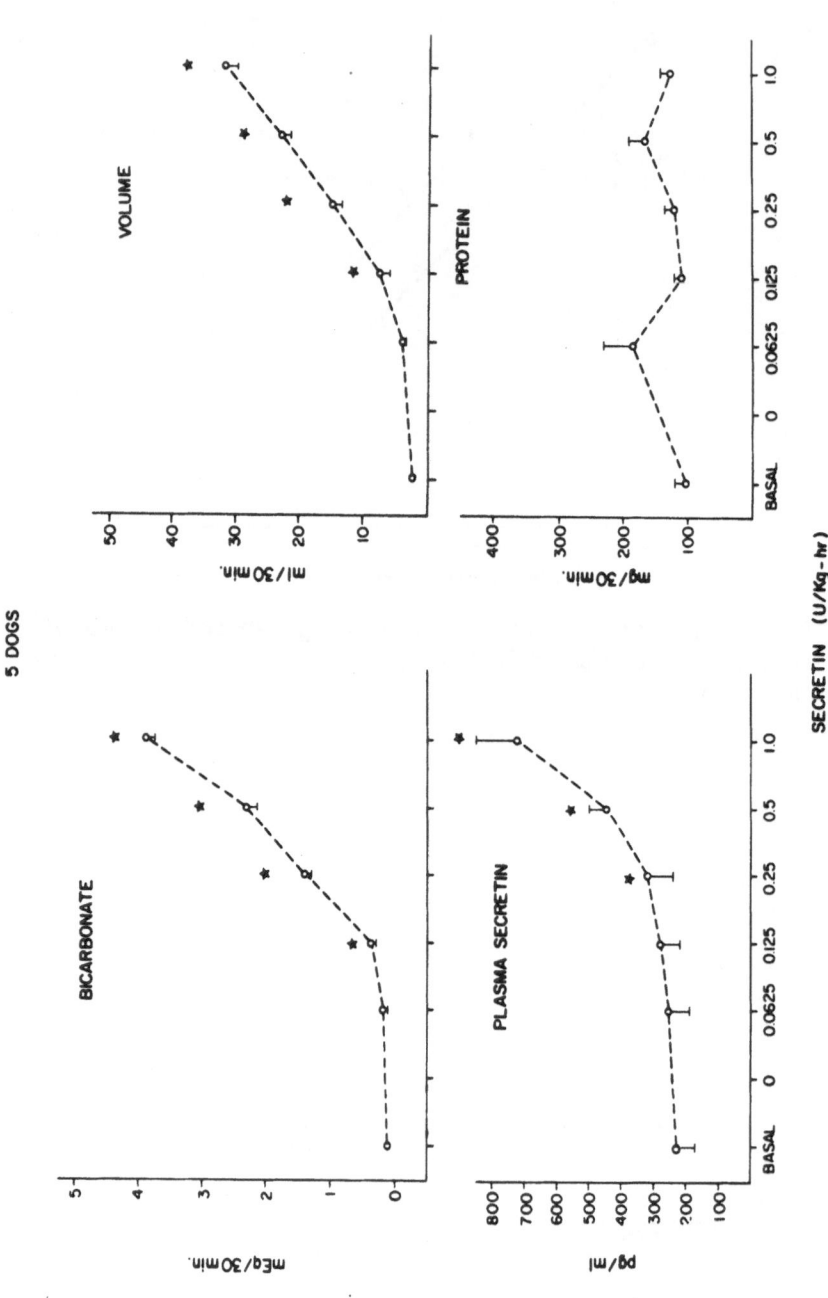

FIGURE 5. Simultaneous measurements of pancreatic secretion of bicarbonate, volume and protein and of plasma secretin in dogs before and after iv infusions with graded amounts of secretin.

DISCUSSION

In the assay reported here, the dose-response curve of pure natural secretin ranged from 0.01 to 1 ng per assay tube. The amount of displacement found with 0.01 ng of secretin (10 pg) was significantly different from that found when no secretin was added to the assay. Since our results suggest that secretin in volumes of plasma up to 300 µl could be measured in the assay, the sensitivity of the assay was 33.3 pg/ml.

For dose interpolation and for comparing slopes of dose-response curves, we used the logit transformation of the response variable and log dose of test substance measured. This mathematical transformation, originally described by Rodbard and colleagues[7], yields a better linear representation of the dose-response data when compared with that obtained with arithmatic-log dose plots. In addition, logit transformation of the response variable allows for calculations by linear regression procedures of slopes, intercepts and correlation coefficients of inhibition lines. Use of a mathematical equation of the dose-response line allows for calculation of amounts of hormone in test samples by desk top calculators or small computer systems. It should be noted, however, that responses (bound) at the extreme ends of the curve may produce significant error in dose interpolation. We generally restrict our analyses to those responses that lie between 10% and 90% bound.

The specificity of the radioimmunoassay was tested by measuring the ability of several peptides to displace labeled secretin from combination with antibody. Of the peptides tested, only VIP showed any cross-reactivity in the assay system. The cross-reactivity observed could be due to VIP contamination with secretin or to VIP possessing antigenic determinants that are similar to secretin. The degree of cross-reactivity, however, was less than 0.05% and should not interfere with the ability of the radioimmunoassay to selectively measure secretin in test samples.

Measurements of basal plasma secretin levels by radioimmunoassay reported by different investigators[4,6,8] are at variance. Basal levels of secretin measured in our radioimmunoassay were about 100 pg/ml in both man (n=96) and dog (n-108). These values represent those obtained from separate assays conducted over an extended period of time. We have found basal secretin to range from below the sensitivity of our radioimmunoassay (<33 pg/ml) to around 300 pg/ml. The mean values reported here represent a good approximation of the median level for basal secretin as measured by our radioimmunoassay.

We have consistently found good correlation between pancreatic output (bicarbonate and volume) and plasma secretin in dogs after the administration of graded doses of secretin or after duodenal acidification with HCl[9]. In this study, significant increases in

pancreatic bicarbonate and volume were found with 0.125 U/kg-hr of
secretin, whereas at this dose level, increases in plasma secretin
measured by radioimmunoassay were not significantly different from
basal.   This slight discrepancy between bioassay and immunoassay
results could be attributed, at least in part, to the variation of
basal secretin levels in individual dogs.   It is important, however,
to note that increases in secretin above basal levels were measured
in 6 of the 8 dogs given 0.125 U/kg-hr.   Pancreatic output of bicarbonate, of volume, and of secretin levels were positively related
to the amount of secretin administered.

REFERENCES

1.   Bloom SR, Ogawa O: Radioimmunoassay of human peripheral plasma
        secretin. J Endocrinol 58: 24, 1973

2.   Boden G, Chey WY: Preparation and specificity of antiserum to
        synthetic secretin and its uses in radioimmunoassay (RIA).
        Endocrinology 92: 1617, 1973

3.   Boehm M, Lee Y, Chey WY: Radioimmunoassay of secretin. I. Pro-
        duction of secretin antibodies and development of the radio-
        immunoassay. In: Endocrinology of the Gut. WY Chey and FP
        Brooks (eds), Charles B. Slack Inc, Thorofare, NJ, 1974, pp
        310-319

4.   Rayford PL, Curtis PJ, Fender HR, Thompson JC: Plasma levels of
        secretin in man and dogs: Validation of a secretin radio-
        immunoassay. Surgery 79: 658-665, 1976

5.   Kolts BE, McGuigan JE: Radioimmunoassay measurement of secretin
        half-life in man. Gastroenterology 72: 55-60, 1977

6.   Straus E, Yalow RS: Fasting and postprandial plasma secretin
        concentrations. Gastroenterology 72: A160/ 1183, 1977

7.   Rodbard D, Rayford PL, Cooper JA, Ross GT: Statistical quality
        control of radioimmunoassays. J Clin Endocrin Metab 28:
        1412-1418, 1968

8.   Chey WY, Hendricks J, Tai HH: Plasma secretin in fasting and
        postprandial states in man. Gastroenterology 72: A133/ 1156,
        1977

9.   Fender HR, Curtis PJ, Rayford PL, Thompson JC: Simultaneous bio-
        assay and radioimmunoassay of secretin. Gastroenterology 72:
        A35/ 1058, 1977

RADIOIMMUNOASSAY OF VASOACTIVE INTESTINAL POLYPEPTIDE

(VIP) IN PLASMA

J. Fahrenkrug and O. B. Schaffalitzky De Muckadell

Department of Clinical Chemistry, Bispebjerg Hospital

Copenhagen, Denmark

A sensitive and specific radioimmunoassay for VIP has been developed which can detect 3.3 pmol liter^{-1} of the peptide in plasma[1]. Antisera to highly purified porcine VIP covalently coupled to bovine serum albumin by the use of carbodiimide were raised in eight rabbits. The immunization dose was 15 nmol (50 µg) VIP per rabbit and injections were given subcutaneously at 8 week[1] intervals. The final dilution, the avidity, and the specificity of each antiserum were determined.

The routine antiserum (No. 5603-6) was used at a final dilution of 1.35×10^6 and reacted with an effective equilibrium constant of 3.5×10^{11} liter mol^{-1}. No crossreactivity was found with porcine gastric inhibitory peptide, porcine pancreatic glucagon, porcine enteroglucagon, human pancreatic polypeptide, synthetic bovine substance P, pure natural porcine secretin, or synthetic ovine somatostatin in concentrations below 10^5 pmol liter^{-1}. ^{125}I-VIP was prepared by a chloramine T method to a specific radioactivity of 900 µCi per nmol peptide. Highly purified porcine VIP was used as standard and antibody-bound and free label were separated by absorption to plasma-coated charcoal. Non-specific interference with the assay system was excluded by extraction of plasma samples with ethanol. The reliability of the assay was investigated by recovery experiments, by serial dilution of plasma samples with high concentration of endogenous VIP, or by immunosorbtion. The within and between assay reproducibility at a concentration of 18.3 pmol liter^{-1} was 1.6 and 2.3 pmol liter^{-1} (1 S.D.), respectively.

Median fasting concentration of VIP in plasma from 74 normal subjects was 7.3 pmol liter^{-1} (range 0-20.0 pmol liter^{-1}). Studies on the release of VIP into the circulation were performed in normal

subjects and anaesthetized pigs. Intraduodenal infusion of hydrochloric acid, fat, or ethanol increased plasma VIP levels significantly, while amino acids, isotonic, and hypertonic glucose or
saline were without effect[2]. After electric stimulation of the
vagal nerves in pigs the concentration of VIP increased significantly in both portal and peripheral plasma[3]. Atropine did not influence this effect of vagal stimulation, while the response was completely abolished by hexamethonium.

REFERENCES

1. Fahrenkrug J, Schaffalitzky de Muckadell OB: Radioimmunoassay
    of vasoactive intestinal polypeptide (VIP) in plasma. J.
    Lab. Clin. Med. 89, 1379-1388, 1977

2. Schaffaltizky de Muckadell OB, Fahrenkrug J, Holst JJ, Lauritsen
    KB: Release of vasoactive intestinal polypeptide (VIP) by
    intraduodenal stimuli. Scand. J. Gastroent., in press

3. Schaffalitzky de Muckadell OB, Fahrenkrug J, Holst JJ: Release
    of vasoactive intestinal polypeptide (VIP) by electric
    stimulation of the vagal nerves. Gastroenterology 72, 373-
    375, 1977

EFFECTS OF BOMBESIN AND CALCIUM ON SERUM GASTRIN LEVELS IN PATIENTS WITH RETAINED OR EXCLUDED ANTRAL MUCOSA

V. Speranza, N. Basso and E. Lezoche

Instituto di III Patologia Chirurgica, Università di Roma

Rome, Italy

SUMMARY

Serum gastrin levels and gastric acid output were determined under basal conditions, following i.v. infusion of bombesin (BBS) (15 ng kg^{-1} min^{-1} for 90 min), and following i.v. infusion of Ca^{++} (4 mg kg^{-1} hr^{-1} for 4 hours) in 28 postoperative peptic ulcer patients. In 5 patients antral mucosa was excluded with the duodenal stump (EAM) and in 8 patients it was retained on the lesser curve of the stomach (RAM). In 15 patients antrectomy was complete (control). Basal gastrin levels were higher in both EAM and RAM patients than in the control group. BBS infusion augmented gastrin levels in all incomplete antrectomy patients. Ca^{++} was less effective. No effect of BBS and Ca^{++} was apparent in the control group. In EAM patients basal and stimulated serum gastrin levels were significantly higher than in RAM patients.

INTRODUCTION

Incomplete antrectomy has been claimed to be related to the recurrence of peptic ulcer in gastrectomized patients.[4-9]

Remnants of antral mucosa may be left on the gastric side (retained antral mucosa, RAM) or on the duodenal stump (excluded antral mucosa, EAM). Although in both instances the pathophysiological mechanisms have been related to hypergastrinemia, no clear data are available. In previous papers it has been shown that bombesin (BBS), a tetradecapeptide (amino acid sequence: Pyr-Gln-Arg-Leu-Gly-Asn-Gln-Trp-Ala-Val-Gly-His-Leu-Met-NH$_2$) isolated by

Erspamer[1], does not stimulate release of extragastric gastrin in patients with a two-thirds gastrectomy[2] but does augment serum gastrin levels in two-thirds gastrectomy patients when remnants of antral mucosa are present.[3]

In this study gastrin levels were measured under basal conditions and following Ca^{++} and BBS infusion in a relatively large group of EAM and RAM patients with recurrent peptic ulcer (stomal ulcer) to evaluate the clinical relevance of RAM and EAM in the pathophysiology of stomal ulcer.

MATERIAL AND METHODS

Twenty-eight stomal ulcer patients with a Billroth II type gastrectomy were studied. The diagnosis was ascertained by x-ray examination by a standard barium meal and/or by endoscopy. During endoscopy biopsies were taken, according to the technique described elsewhere,[7] to ascertain the presence of remnants of antral mucosa. Five patients presented with EAM; 8 patients presented with RAM. In 15 patients antrectomy was complete (control group).

In all patients serum gastrin levels and gastric acid secretion were measured under basal conditions, following an i.v. infusion of BBS (15 ng kg^{-1} min^{-1} for 90 min) and following an i.v. infusion of Ca^{++} (4 mg kg^{-1} hr^{-1} for 4 hours).

All the technical details related to the tests, and to the HCl and gastrin measurement, have been previously described.[2,3]

Statistical significance of observed differences was determined by means of Student's t test. All results were expressed as mean ± standard error of the mean (SEM). The peak gastrin response was defined as the highest gastrin value following the stimulus. The peak acid response was defined as the highest output in four consecutive 15 min periods following the stimulus.

In all patients vagotomy alone (control group) or vagotomy and resection of the gastric (RAM group) or duodenal (EAM group) stump was performed. A BBS infusion test was repeated 10 to 15 days after surgery in all patients who underwent resection.

RESULTS

The mean basal serum gastrin level in the control group was 17 ± 3 picomoles $liter^{-1}$. No significant change occurred following

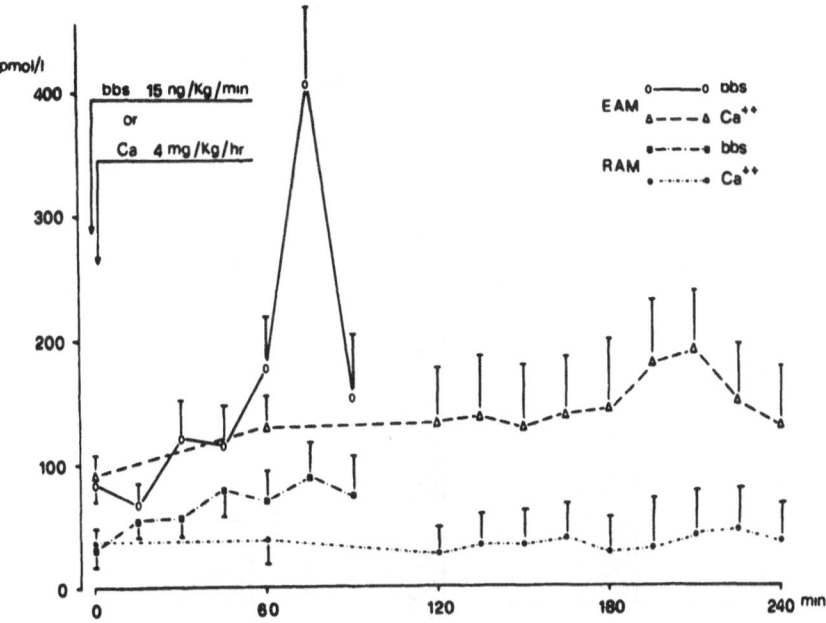

Figure 1. Serum gastrin levels under basal conditions, following BBS infusion, and following Ca⁺⁺ infusion in stomal ulcer patients with RAM and with EAM.

Ca^{++} or BBS infusion. In RAM patients, the mean basal gastrin level was 33 ± 7 picomoles liter^{-1}, significantly higher than in the control group. BBS significantly augmented the serum gastrin levels to a peak response of 86 ± 21 picomoles liter^{-1}. Ca^{++} had little effect (43 ± 10 picomoles liter^{-1}) (Fig. 1).

In EAM patients the mean basal gastrin level was significantly higher (88 ± 12 picomoles liter^{-1}) than in the control group and in the RAM patients. BBS dramatically increased gastrin levels (peak response 409 ± 47 picomoles liter^{-1}). Ca^{++} augmented gastrin levels but to a lesser degree (peak response 188 ± 24 picomoles liter^{-1}) (Fig. 1).

Acid secretion in EAM patients was significantly higher than in the control group and the RAM group. BBS augmented acid secretion both in EAM and RAM patients. No effect was detected in the control group (Fig. 2).

Following resection the BBS infusion did not elicit a gastrin response in any of the patients (Fig. 3).

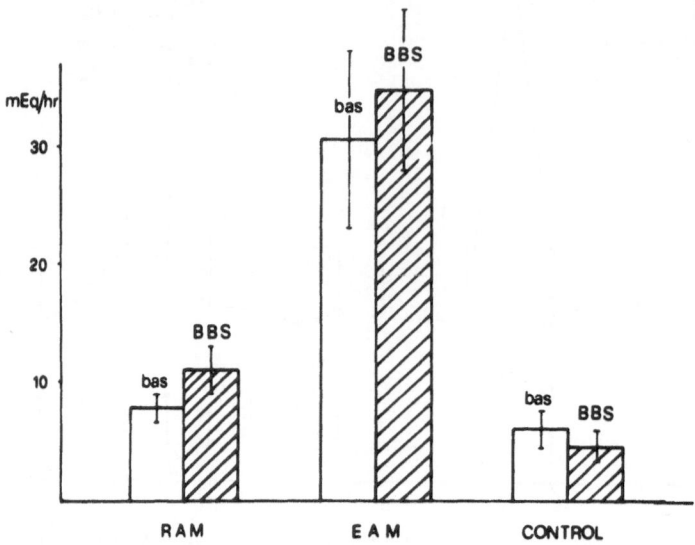

Figure 2. Gastric acid output (peak response) under basal conditions and following BBS infusion in RAM, EAM patients, and in the control group.

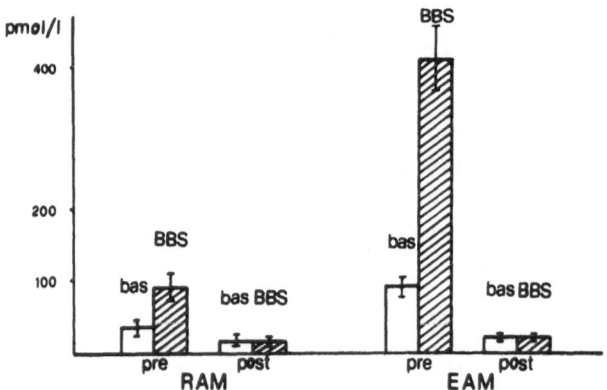

Figure 3. Serum gastrin levels under basal conditions and following BBS infusion (peak response) in patients with RAM and in patients with EAM before and after antral excision.

CONCLUSIONS

In two-thirds gastrectomy patients remnants of antral mucosa, whether "retained" or "excluded", led to basal hypergastrinemia when compared with patients with complete antrectomy. BBS infusion augmented serum gastrin levels in incomplete antrectomy patients. In these patients Ca^{++} was less effective than BBS in stimulating gastrin release. The basal and stimulated gastrin levels of EAM patients were significantly higher than those in RAM patients.

We speculate that the dramatic response to BBS in EAM patients may be explained by: 1) G cell hyperplasia in the excluded antrum; or 2) absent or diminished paracrine inhibitory action of somato-statin.

In our experience incomplete antrectomy has always been associated with recurrent ulcer.

REFERENCES

1. Anastasi A, Erspamer V, Bucci M: Isolation and structure of bombesin and alytesin, two analogous active peptides from the skin of the european amphibians "Bombina" and "Alytes." Specialia 27, 166, 1970

2. Basso N, Lezoche E, Materia A, Giri S, Speranza V: Effect of bombesin on extragastric gastrin in man. Am. J. Dig. Dis. 20, 923-927, 1975

3. Basso N, Lezoche E, Giri S, Percoco M, Speranza V: Acid and gastrin levels after bombesin and calcium infusion in patients with incomplete antrectomy. Am. J. Dig. Dis. 22, 125-128, 1976

4. Harrison RC, Stoller JL: Ulcerogenic potential of the incompletely resected antrum. Am. J. Surg. 122, 198-203, 1971

5. van Heerden JA, Bernatz PE, Rovelstad RA: The retained gastric antrum: clinical consideration. Mayo Clin. Proc. 46, 25-28, 1971

6. Scobie BA, McGill DB, Priestley JT: Excluded gastric antrum simulating the Zollinger-Ellison syndrome. Gastroenterology 47, 184-187, 1964

7. Speranza V, Basso N: Preoperative diagnosis of retained gastric
    antrum by fiber gastroscopic biopsy.   Surg. It. 14, 309-316,
    1971

8. Stoller JL, McDonald TJ, Nunn PJ, Harrison RC: A method of
    measuring the alkaline area of the stomach during operation.
    Surg. Forum 19, 316, 1968

9. Waddel ER: The physiologic significance of retained antral
    tissue after partial gastrectomy.   Ann. Surg. 143, 520-530,
    1956

INDEX